Introduction to Technical Writing: Process and Practice

Introduction to Technical Writing: Process and Practice

Lois Johnson Rew
San Jose State University

St. Martin's Press
NEW YORK

Executive Editor: Susan Anker
Developmental Editor: Michael Weber
Project Management: Editing, Design & Production, Inc.
Cover Design: Darby Downey
Cover Art: J. Michael Lloyd, AIA, CSI

Manufactured in the United States of America.
321
fedc

For information, write: St. Martin's Press, Inc.
175 Fifth Avenue
New York, NY 10010

ISBN: 0-312-00271-8

Acknowledgments

 The text of this book has been printed on recycled paper.

For Bob

To the Instructor

Introduction to Technical Writing: Process and Practice is designed to bridge
the gap between a freshman English course in composition and the world
of work. The students who need a bridge like this are sophomores, juniors,
and seniors in a wide variety of technical majors—from agriculture and
engineering to physics, forestry, nutrition, and computer science. Since a
bridge must be anchored on solid ground at both ends, this book begins
with analogies to problems and situations familiar to students and with
a friendly, supportive tone. Then each chapter discusses a different writing
task, giving examples of both student and professional documents and
discussing how working writers proceed in solving writing problems.

Students in technical and scientific writing classes are by no means
a homogeneous group, and as teachers we need to understand their needs,
their fears, and their strengths. Perhaps the majority of your students will
be problem-solvers from majors like engineering, architecture, food science, and mathematics. They take technical writing because it is required
or because they know they will have to write on the job. They are more
comfortable with numbers, facts, and equations than they are with words
and sentences. Many may fear and dislike writing; students often tell me
they "never were any good in English." Still, these students are involved

in a field that interests them, and they are developing knowledge and skill in a specific discipline. They are also motivated—realizing that this class may be their last stab at learning to write well.

These students respond well to short writing exercises that focus on discrete skills, and *Introduction to Technical Writing* provides many such exercises. Throughout the book you will find concise explanations of process and theory, followed immediately by one or more exercises that allow students to apply what they have learned to a technical writing situation. The book stresses planning and organization, building these broader skills through a progressive series of exercises. A number of these exercises deal with the same topic (for example, bicycles or cafeteria chairs) so that students can practice developing a single writing assignment from idea to draft. In addition to these short exercises, 15 longer, structured writing assignments cover the major forms of technical writing. You can assign as many exercises as your students need, and tailor the number and type of their writing assignments to fit their major fields and their available writing time. If you want your students to work on collaborative projects, you will find in every chapter exercises and writing assignments that call for team effort in writing and editing.

A smaller group of your students may be preparing themselves to be professional writers and editors. These students often are "whole-concept thinkers" from fields like English, psychology, and journalism. They are comfortable in the world of words and sentences, but they lack a technical orientation. They frequently like to write, though their writing may be disorganized and fuzzy because of abstract nouns and weak verbs. The variety of examples in this book will help these students expand their technical vocabulary and understanding, while the exercises in organization, diction, and syntax will sharpen their skills in revision.

Introduction to Technical Writing is divided into three parts. The eight chapters in Part 1 spell out a general process that both students and professionals can use to write technical and scientific documents. Writing on the job is viewed as a process, and the written document—the product—is usually reviewed at several stages for the purpose of improving it. Students, however, still tend to think of a paper as a single assignment to be worked at, turned in, and forgotten as quickly as possible. They need to begin viewing writing as a process, to learn how to give and take criticism during peer review, and to rethink and revise as they work toward the final product.

Part 1 divides the process into eight chapters, covering

- planning
- gathering information
- organizing for the writer
- reviewing the organization

- writing the draft
- revealing the organization to the reader
- reviewing and revising the document
- editing

Even though these steps are discussed separately, students are reminded throughout that writers seldom follow such a straight line of development. The real process is flexible and recursive, often looping back and forth from one step to another. Still, most technical writing students will find security in knowing that a logical, problem-solving sequence is available to them.

Part 2 has five chapters that explore a variety of tools and techniques vital to most kinds of technical writing. The chapters cover

- using a word processor
- defining and comparing
- describing objects and places
- illustrating with graphics
- documenting sources

Because techniques like defining and describing are familiar to students from freshman composition classes, you may wish to begin with them. Others, such as illustrating with graphics, apply more particularly to technical writing situations. The chapter on word processing is included here because, even though the word processor is not a component of technical writing, it is a tool: one that all students need to understand and learn to use. By the time your students begin their professional careers, nearly every occupation will require some familiarity with computers and word processing.

Part 3 covers ten major forms into which most technical writing is cast, with "form" meaning the way a document looks on the page based on organization and layout. This book recognizes that certain forms have evolved because they work successfully to fulfill the purpose of a particular document and to meet its readers' needs. Each chapter defines a form, illustrates it with both student and professional examples, and explains how to adapt the general writing process discussed in Part 1 to this particular form. The forms are

- memorandums and short reports
- business letters
- resumés and job search letters
- instructions and manuals
- procedures
- abstracts and summaries
- proposals
- comparative, feasibility, and recommendation reports
- research reports and other articles for publication
- speeches and oral presentations

The text's three-part structure will give you flexibility in both reading and writing assignments, and the Instructors' Manual suggests possible syllabuses for 10-week and 15-week courses. Major writing assignments are spread throughout the three parts, so you can choose those that apply to your particular students and your academic schedule. If you wish, you can assign an umbrella term project which can include many writing assignments: memos, letters, a proposal, status reports, definition, description, and a long formal report. Alternatively, you can assign only short projects, or a combination of exercises and projects.

Here are the major writing assignments:

1. Choosing a Topic for a Report (Memo Form)   Chapter 1
2. Writing a Proposal for a Term Project Report   Chapter 1, 20
3. Letters that Ask for Information   Chapter 2
4. Writing an Outline for a Term Project   Chapter 3
5. Writing a Status Report   Chapter 3, 14
6. Draft of a Term Project (Comparative or Feasibility) Report   Chapter 5
7. Narrative Report   Chapter 14
8. Interview   Chapter 14
9. Resumé   Chapter 16
10. Job Application Letter   Chapter 16
11. Instructions   Chapter 17
12. Procedure for Generalists   Chapter 18
13. Completing a Formal Report   Chapter 21
14. Popular Article   Chapter 22
15. Article for a Young Audience   Chapter 22

This book would not exist without the help and advice of many people, from students and colleagues at San Jose State University to friends in business and industry across the country. I want to thank them all and acknowledge their contributions.

Students in my technical writing and editing classes have worked with drafts of this manuscript from its inception, and by asking good questions helped me clarify my writing. My thanks to all my students, especially those who contributed examples, who are acknowledged by name in the acknowledgments or at the end of each chapter.

At San Jose State University, Lou Lewandowski, chair of the English Department, mentor, and friend, encouraged me at every turn. My colleagues John Galm, Hans Guth, Allison Heisch, and Patricia Nichols gave me excellent advice. Mark Bussmann, Mina Lunt, and Craig Pollock of the department staff contributed clerical expertise. Lina Melkonian from Career Planning and Placement supplied resumé examples, while Judy Reynolds of Clark Library critiqued Chapter 2. Kenna Mawk, Diane McBurnie, Margee Milks, and Denise Muller tested exercises in their classes. The paper airplane and cafeteria chair exercises were designed by Denise Muller.

For examples and their expert advice I want to thank my friends here in Silicon Valley and all across the country. They include Paula Bell of Silicon Compiler Systems; Sue Birdwell, Xidex Corp.; Dan Brown, Computer Curriculum Corp.; Robert Brown, United Technologies; Bruce Bull, Northern Telecom Business Communication Systems; Bill Burnett, Lockheed Missiles and Space; Michael Calkins, H.M. Goushé; Roger Connally, Fujitsu America; Susan Wertz Constable, Triad Corp.; Jim Fortune, The Lantis Corp.; Elli Gerhardt, an architect; Jim Jahnke, Northrop Services Environmental Sciences; Carl Jones, Lockheed Missiles and Space, for artwork; Richard LeClair, aeronautics professor at SJSU; Kenna Mawk, Intelligenetics; Julienne Morgan, Ford Aerospace; Lori Neumann and Art Walton, IBM; Frank Nichols, NASA Ames Research Center; Thomas Pearsall, University of Minnesota, for audience analysis; Anne Rosenthal, Hoefer Instruments; Kim Shanley, Waterside Associates; Theresa Stover, Qubix Graphic Systems; Shirley Taft, Lawrence Livermore Laboratories; Jacqueline Turner and Bronwyn Wagman, Syva; Jeff Vargas, Hewlett-Packard; and Judy Wilbur, EDS.

My reviewers set high standards and held me to them, and they have made this a better book. I am grateful to Melissa Barth, Appalachian State University; Virginia Book, the University of Nebraska at Lincoln; Stephen C. Brennan, Tulane University; Michael Connaughton, St. Cloud State University; Marc Glasser, Morehead State University; Jack Selzer, Pennsylvania State University; and Thomas Warren, Oklahoma State University.

At St. Martin's Press, Susan Anker and Michael Weber have provided excellent editorial support. I could not have asked for better editors.

Finally, my thanks and my love to my husband Bob. This book is for him.

Overview of the Table of Contents

Contents

To the Student: Answering Your Questions about Technical Writing

If you're like many of the students who appear in my technical writing class at the beginning of a term, you have some questions about technical writing. You may want to ask:

- What is technical writing and what does a technical writer do?
- What is technical editing and what does a technical editor do?

You may especially want to ask:

- Why should I care about my technical writing skill?

and

- How can I use this book?

In this introduction I answer these questions. The answers come not only from my own experience as a teacher, writer, editor, and consultant, but also from statements by managers, writers, and editors in business and industry.

You may be studying technical writing because you are in a technical or scientific major, or because you see technical writing as a possible career choice. In either case, think of this book and your instructor as coaches.

1

Our job is to help you build on the skills you already possess, introduce you to writing processes that will work for you, and give you practice in many types of technical writing. Practice, of course, is hard work— whether you are swimming lengths to build your stamina, singing scales to improve your accuracy, or writing and revising to communicate effectively. As you practice, remember that coaches are not judges but allies— people who want to help you write better and who will cheer you on to victory. We want you to succeed, and we know that you can.

To begin, consider the following questions and answers about technical writing.

Q. WHAT IS TECHNICAL WRITING?

A. Technical writing is a general term that covers writing reports, proposals, manuals, procedures, and instructions. It can even be applied to documents like letters, resumés, memos, and sales brochures. With all this variety, you need a definition that clarifies what technical writing does. Here is one:

> Technical writing is the communication of specific—usually technical—information to an identified reader so that the reader's understanding matches the writer's intention. The writer's responsibility is to make the communication accurate, clear, complete, concise, well organized, and correct.

If you break this definition into its parts, you can get a better idea of what it means.

1. The subject matter is clearly identified—usually from the realm of science or technology.
2. The writing is directed to specific readers who seek information or advice, and who often will make decisions based on what is said.
3. The writing succeeds only if it is understood by the readers; the writer and the readers have, in effect, a contract. Therefore, what you write should be
   - *accurate:* supported and verifiable
   - *clear:* having only one meaning for the reader
   - *complete:* containing all the necessary information
   - *concise:* economical and direct
   - *well organized:* accessible through the logical relationship of ideas, short individual sections, and good headings and layout
   - *correct:* free from errors in form, grammar, usage, and punctuation

If it has these qualities, technical writing can appear in a variety of forms and be used for a variety of purposes within science, business, industry, agriculture, and other technical fields.

Q. WHAT DOES A TECHNICAL WRITER DO?

A. Virtually all people who work in science and technology—like civil engineers, research biochemists, nurses, cattle breeders, and programmers—must write as part of their jobs. In addition, a small group of people are full-time, or professional, technical writers. According to Keith Kreisher of the Society for Technical Communication (quoted in Kleiman 1984, sec. PC,1), their primary job is "to bridge the gap between the technology of the product and the user, to make that technology understandable to the average person."

As computers and other complex technical tools move into the general market, writers are needed who can explain what the tools are and how to use them. Many technical writers have majored in English, liberal arts, or journalism; others come from technical fields but discover that they like to write and choose to do it full time. Technical writers need a feel for language plus excellent writing and editing skills. In addition, they need experience or course work in the appropriate technical field and familiarity with word processing. According to Christine Browning, manager of a software manual writing group at Tandem Computers (quoted in Kleiman 1984), "Technical writing is the best writing job you can have. It's the only one that gives you security. If you can write well and know the science you are writing for, you can get a good job."

What's more, technical writers are well paid; for example, in most parts of the country, entry-level salaries are only slightly lower than those for engineers. The profession is demanding, however. According to technical writer Chris Klemmer (1984), "You should be able to work under pressure and against deadlines. Most projects run late, and the writer is called in at the eleventh hour." In addition, Klemmer says, "you should be able to take criticism. The words you thought were so golden may not seem so to your superior. Be willing and gracious about rewriting."

Q. WHAT IS TECHNICAL EDITING?

A. Technical writing is only one part of the process of producing a technical or scientific document. Technical editing is another; it involves looking critically at the written document for accuracy, completeness, correctness, and effectiveness of presentation. You must always edit what you write, stepping back and reviewing the draft with a critical eye. Editing means correcting grammar and punctuation, but it also involves checking facts and assessing and revising organization. Above all, it means guaranteeing that the reader will understand the writing to mean only what the writer wants it to mean. Editing may even require selecting graphics, compiling a glossary, and laying out each page in preparation for printing.

Like technical writing, then, technical editing includes many tasks.

As a student writer or as a writer in a small company, you usually must be your own technical editor. Be aware, though, that writing and editing tasks are different and that you should not try to do them at the same time. Writing (even technical writing) is creative, so when you write, you need to shut your critical eye for a little while. Don't worry about spelling, punctuation, or getting the exact word. Don't polish each sentence until it gleams like the chrome bumper on an antique car. Let the words flow and get the main ideas down on paper. Then, with a completed draft in your hand, open your critical eye to evaluate what you've written. When you edit, you pick apart and polish, check facts, rearrange sentences, correct spelling and punctuation. One technical editor (quoted in Cavanaugh 1985, sec. PC, 12) put it this way: "The difference between a writer and an editor is that a writer creates something from nothing, while an editor analyzes and works with that creation." This book will teach you both writing and editing skills and tell you at what points you should become an editor.

Q. WHAT DOES A TECHNICAL EDITOR DO?

A. Although you may edit your own material on the job, in many companies you will work with a professional technical editor. Often it's easier for a trained outsider to review a draft than it is for the writer. In addition, editing itself involves many specific skills, and not all scientists and technical specialists have the training or the time to learn them. Many editors have majored in journalism, linguistics, or English and have excellent grammar and punctuation skills, an interest in language, concern for detail, and the ability to work well with people. In many companies editing is an entry-level position that can lead in time to appointment as a writer.

Q. I DON'T WANT TO WORK AS A WRITER OR EDITOR, SO WHY SHOULD I CARE ABOUT TECHNICAL WRITING SKILL?

A. The first reason is that good writing may be essential in landing you a job in your chosen technical field.

You already know that when you apply for a job, you must fill out an application, prepare a resumé, and write a letter of application. All the employer knows about you is what is on those pages, nothing more. That means you will be judged by what you say and how you say it. One executive (McAlister 1984, 47) who is responsible for 100 engineers put it this way: "We consider the ability to communicate to be as important as grade point average when we interview engineering graduates." The district manager for a large chemical company in Illinois said this about an application he rejected (Galm 1977):

> This application is pretty pathetic in terms of language usage. . . .The young man applying for a position with us will probably not be hired as a sales engineer because the ability to communicate is even more important than his technical skills and other fine qualities. This young man obviously has a lot going for him and would be just the kind of person we would like to hire—except that he obviously can't write an adult sentence.

A. Second, advancement in your profession may depend on your ability to write well.

One of the ways you increase your visibility to managers is through your writing. A software manager for a high-tech firm told students (Passen 1985) that because top managers are constantly busy, they must delegate responsibilities, and they "need to be able to trust the writer." In addition, since writing is a permanent record, it is often used at employee evaluation time. A recent survey of a research and development group in an oil company (MIT 1984) revealed that "evaluations in over half the categories in the annual . . . performance reviews are based at least partly on an individual's writing."

A. Third, as a scientific or technical professional, you'll write more than you'd expect.

The oil company survey indicated that all the research and development professionals spent a substantial portion of their time on technical communication, whether writing, reading and editing, or speaking. Managers spent 50 percent of the work week on communications, supervisors spent 40 percent, and staff members spent 35 percent. Here's another way to think about it: a productive worker can easily generate 800 to 1,000 pages of writing in a year, simply by the daily writing of memos, reports, letters, proposals, analyses, work orders, and instructions. That's enough for at least two books!

A. Finally, you are responsible for what you write.

Computers, word processors—and now desktop publishing—make it seem that the writer's job is getting easier. But in fact, computers, laser printers, and graphics software are simply sophisticated tools. By themselves they will not generate text, nor will they ensure that the words they print out (however beautifully) are clear, organized, and correct. That responsibility lies with the writer, and it can be a heavy responsibility. A wrong statement in a set of instructions, for example, could damage equipment or seriously injure the operator, perhaps leading to an expensive lawsuit.

In addition, as the chairman of a large forest-products company said (Hahn, 1985 sec. F, 7):

> If your writing has even a minor error in it, your reader may question how much care and thought really went into it. And be sure to punctuate carefully. Even a single misplaced comma can change the whole meaning of a sentence. Grammar, spelling, and punctuation are the cornerstones of good writing and should be carefully considered at all times.

What's more, you can't depend on the department secretary to type and correct rough drafts, for secretaries have nearly disappeared. The director of technology in one large telecommunications company told me that his department had 100 professionals and only two secretaries (Bull 1985). When you have your own computer terminal, you'll do your own typing, and you will be responsible for organization, accuracy, and correctness.

Q. HOW CAN I USE THIS BOOK?

A. *Introduction to Technical Writing* will help you become a better technical writer and editor. The eight chapters in Part 1 will guide you through processes you can follow in writing anything from a short letter to a long report. Those eight chapters deal with

- planning
- gathering information
- organizing
- reviewing your organization
- writing the draft
- revealing your organization to the reader
- reviewing and revising the document
- editing

For ease of discussion, each chapter isolates one step in the process. In practice, though, you'll probably find yourself dropping back to earlier steps from time to time. When you gather information, for example, you may want to return to your plans to refine them. And when you review your organization, you may have to gather more information to fill some holes. If you think of each chapter as a step on a staircase, you can see how this works. As Figure I-1 shows, the planning step underlies the whole staircase, so you can easily drop back to that step when you need to—and the same is true for each subsequent step in the process. Together, the stair steps will lead you where you want to go with your writing.

Part 2 discusses five specific tools and techniques you can use to develop and clarify your technical writing assignments. You may have learned some of them in a more general way in freshman composition, but now you can apply them to your major field. You will learn how to

- use a word processor
- define technical terms
- describe objects and processes
- illustrate with graphics
- document your sources

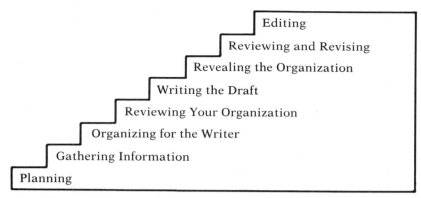

Figure I-1   **Steps in the process of technical writing**

Word processing is a tool rather than a technique, but whatever your field, you need to learn and practice using a word processor. Otherwise, you can use any combination of these techniques in a single writing project.

Finally, each of the 10 chapters in Part 3 takes up one of the common forms in which technical writing appears. You will learn how to write in each of those forms, and you can study both student and professional examples. This section begins with the documents you'll write most often—memos and short reports—and moves to longer and more complex projects like comparative, feasibility, and research reports. Your instructor may ask you to read chapters in any order because the book is designed for flexible reading and writing arrangements. The 10 chapters in Part 3 cover

- memorandums and informal reports
- business letters
- resumés and job search letters
- instructions and manuals
- procedures
- abstracts and summaries
- proposals
- comparative, feasibility, and recommendation reports
- research reports and other articles for publication
- speeches and oral presentations

This book also includes many kinds of exercises to help you practice and improve your skill in technical writing and editing. Some of those exercises are on the same topic—bicycles or cafeteria chairs, for example— so you can learn how to develop a writing assignment from the first idea to a written draft. Every chapter defines the terms that are used, and a glossary near the end of the book pulls all those definitions together. The

Handbook Appendix at the end contains a guide and exercises for punctuation, numbers, capitalization, and spelling. Throughout the book, you'll find many cross-references to other sections of the text. In addition, you can always use the index to find specific topics.

I've talked about the importance of careful editing and of clear technical writing. Now let me be your coach, encouraging you to try your hand at it and do your best. No matter what your previous experience with writing has been, keep an open mind about the tasks you are assigned. As you read this book and learn more about technical writing and editing, and as you apply your new skills to your major field, I predict that you will learn to respect and even enjoy the technical writing process. Whatever your technical specialty may be, *you can become an effective technical writer*. With practice and determination, you can do it! Begin now by reading about *planning*, the first step in the writing process.

REFERENCES

Bull, Bruce. 1985. Northern Telecom Business Communication Systems. Lecture at San Jose State University. 2 May.

Cavanaugh, Wanda. 1985. Untwisting their words. *San Jose Mercury*, 3 March, Section PC p. 12.

Galm, John. 1977. Personal letter. 11 July.

Hahn, T. Marshall, Jr. 1985. Don't let bad writing damage your career. *San Jose Mercury*, 8 May, Section F, p. 7.

Kleiman, Carol. 1984. Computer boom helps give lift to technical writers. *San Jose Mercury*, 24 June, Section PC p. 1.

Klemmer, Chris. 1984. Silicon Valley *Connection*, (STC Newsletter) March.

McAlister, James. 1984. Why engineers fail. *Machine Design*, 23 February, p. 47.

MIT Industrial Liaison Program. 1984. *Communication skills: A top priority for engineers and scientists*. Report No. 32929.

Passen, Gary. 1985. Telenova. Lecture at San Jose State University. 15 March.

PART ONE

THE PROCESS OF TECHNICAL WRITING

How does a big multistory structure like an office building get built? How do builders get from a vacant lot and a mental picture of a building to a solid and useful structure? You don't have to be an architect or contractor to know that much work and many people are involved in the process of constructing any large building. The key to that construction, in fact, is the word *process*; the project will be successful if it follows a methodical, step-by-step approach, from architect's drawing to excavation and ironwork to interior finishing. In addition, inspectors need to review the work at various stages in the process to make sure that all is going well.

When you write a technical document—whether it is a one-page letter or a 25-page report—you can also follow a process, and the step-by-step approach will make your writing task easier and more successful. The eight chapters in this section explain the general process, with each chapter explaining one step. The process is, however, ongoing and interlocking. Thus, even though you will be learning about the process one step at a time, you should understand that when you write, you will often move back and forth from one step to another. As you study the various chapters in Part 1, you may also want to read about some of the specific forms in which technical writing appears. You will find each form discussed in detail in Part 3.

CHAPTER 1

Planning

If you have ever stopped to watch the construction of a big building, you've probably seen two or three construction supervisors gathered around an unrolled set of blueprints—studying the plans, pointing to various sections, and then looking out at the construction site. That set of blueprints is very complex, containing detailed drawings for grading and foundation work, for structural support, for plumbing, for heating and air-conditioning, and for electrical service. When architects plan, they must consider the building's purpose and the client's needs and wants, but they must also study the site and such legal matters as zoning, building codes, and fire regulations. The plans that emerge will usually be a compromise among many factors. When they are finished, though, those plans will guide the construction crew as it turns the once abstract *idea* for a building into a solid *structure* of steel, glass, and stone. Detailed planning is the key to a successful building project.

Completing a technical writing project requires that same kind of careful planning; thus, you can write successfully if you first spend some time thinking about your document's purpose and your reader's needs, and then balancing those matters with length and coverage, tone, form, and available time for writing. You too need some blueprints, and you can begin to create them using the activities listed on the preceding page. Whether you are writing a half-page memo or a 50-page report, you will want to carry out each of these activities before you begin writing. This chapter will help you.

1.1 DETERMINING THE PURPOSE

One question to ask in planning a technical writing project is "What is my goal or purpose in this piece of writing?" As a student, it may be hard to determine a writing purpose because until now your primary goal has been simply to pass your writing class with a decent grade. You have probably tried to please your instructor with whatever you wrote, but the purpose of the writing itself was not important to you. As a working professional, though, writing will be your way of communicating what you have learned to other people. If you do not understand your purpose before you write, your reader will not understand what you are saying.

For example, suppose your manager says: "Contact Laura and Bill over in the design lab and find out how that new control box they're working on is going to affect our power consumption. Then write it up for me. I need some solid information for my meeting on Monday."

If you analyze your manager's request, you can see that your assignment is to *determine* and then *explain in writing* the impact of the new control box design on power consumption, an explanation that may involve your *evaluation, judgment,* or *recommendation*. If you write a description of the control box, if you tell how the control box operates, or if you compare the new control box to an existing one, you will have failed in your writing assignment because you did not clarify your purpose. What's more, your manager will not have the solid information needed for that Monday meeting.

How might that purpose statement look? Perhaps like this:

> This memo explains the electrical power requirements of the newly designed A26 control and the consequent need for changes in our cabling.

A Definition of Purpose

The *purpose* in writing is the goal, the intended or desired result: in other words, the reason for doing the writing.

Knowing your purpose early in the planning step will help you focus on what you are trying to accomplish. How do you discover that purpose from the writing situation? Your first task is to study the situation carefully, looking for key words. In the example above, the manager gives you some key words: *new control box, affect,* and *power consumption.* Of course you need more specifics than these, but you're also told where to look for the information: *the design lab* and *Laura and Bill.* When you talk to Laura and Bill, you can refine your key words: control box becomes *A26 controller,* and power consumption becomes *electrical power requirements.* The key word *affect* tells you that your purpose will be to explain. By analyzing the key words and asking "What is the result I want from this?" you can write a purpose statement based on a writing situation.

Often you can explicitly state your purpose at the beginning of your document, which will help your readers know what to look for when they read. Sometimes you will have more than one purpose. For example, your stated purpose in a brochure might be to describe how an automatic livestock watering system functions. But behind that purpose is another: to convince the reader that your company's watering system is more efficient and easier to install than competing systems.

You need to remember that in technical writing, *you* are the expert, the person who knows something. Your readers seek information, knowledge—even recommendations—from you, and they depend on your accuracy and clear view of what you are doing. That means you can't assume reader knowledge about your topic or reader understanding of your purpose. A good way to ensure that readers will understand your purpose is to state it at the beginning of the paper or section.

Possible Purposes for Technical Writing

Listed below are some major purposes for writing a technical document, each followed by an example of a purpose statement. Notice from the examples how many ways a purpose statement can be written.

To define what something is or means

> Within the next two years you will be asked to choose a long-distance telephone company, and in the process you will be faced with many new terms. To help you understand those terms, we have prepared a glossary of telephone terminology. Learning "phonespeak" should help you make better decisions as the time draws near to choose a long-distance company.

To describe an object or a process (what something looks like or how something works)

Until very recently, most wood-burning stoves had overall energy efficiency ratings of 45 to 60 percent. Today a new generation of wood-burning stoves has overall energy efficiencies as high as 75 percent. The new technology that makes these high efficiencies possible is the catalytic combustor. The purpose of this paper is to describe how a catalytic combustor works in a wood-burning stove (Davis 1985).

To instruct (tell how to do something)

This Tutorial is designed to give you, the user, an opportunity to learn how the basic elements of the Tri-dent Practice Management System work. A small data base is included and can be restored to its original form whenever you wish. In this way we hope that new staff members can quickly and easily learn how to operate the system through the Tutorial. Also, this sample data base will allow you to experiment with the system without fear of corrupting your actual, real-life data (Triad 1984).

To analyze as a basis for a conclusion

This paper discusses the sources of drinking water in the Lorimer Valley and the ways in which those water resources are managed to provide an adequate supply of safe drinking water to the Valley's 400,000 residents. It discusses contamination problems that have affected the valley's drinking water or may do so in the future and the activities of various governmental agencies to respond to those problems.

To report status or progress

To: L. Rew
From: LaVada Clayton
Date: November 8, 1985
Subject: Status Report on the Technical Evaluation of COM Recorders
Reporting Period: October 28, 1985, to November 8, 1985

To explain

There is some confusion and inconsistency about the preparation of work package schedules and the various reviews, milestones, and activities to be included. The Management Manual spells out the review requirements well enough, but does not discuss the schedule requirements. This memo is intended to outline the various schedule requirements and to suggest the structure to be used in developing a work package schedule (Nichols 1984).

To interpret, evaluate, judge, or recommend

This report evaluates objective information on a range of players as a guide for recommending the purchase of a compact disc player. The sales pitch from a commissioned salesperson can be persuasive, but as a basis for making a choice, it is the wrong medium. The salesperson will suggest buying the brand in stock, or will say his or hers is "one of the best" on the market. That type of meaningless phrase can only mislead or confuse the buyer (Vargas 1984).

To propose, sell, or persuade

In response to your Request for Proposal, this memo proposes a study of office telephone systems to determine the best system for Consolidated Life Insurance Company (Edwards 1985).

To request information

I am interested in obtaining information about 6-inch to 10-inch reflecting telescopes and accessories produced by your company (Lankford 1983).

As you plan, remember that (1) you must understand your purpose before you begin to write, and (2) you should plan to let the reader know the purpose early in the document.

CHECKLIST FOR PLANNING PURPOSE

In determining the purpose of a document, ask these questions:

1. Why am I asked to do this writing? What is my intended result?
2. What are the key words that will help me?
3. Which of the following verbs describes my purpose? Is it to

_____ define?   _____ report status or progress?
_____ describe an object?   _____ explain?
_____ describe a process?   _____ interpret, evaluate, judge,
_____ instruct?   or recommend?
_____ analyze?   _____ propose, sell, persuade?
                                     _____ request information?
                                     _____ other? (name it)

EXERCISE 1-1 Writing Meaningful Purpose Statements

Directions: Following are five situations that require a technical writing response. Read each situation carefully and then choose a verb from the checklist that clearly describes the purpose of the writing. Use that verb to construct a good purpose statement. It's possible to have more than one purpose for a given situation.

Situation 1. TTT Plastics Co. has designed a locking ski rack that will fit on the roof of most cars. The ski rack has passed all its functional and design tests, and the first racks will be built and ready to sell in two months. As one of the designers, you are asked to write up the brochure that will be packed with each ski rack, telling customers the correct installation procedures.

Situation 2.   Because of widespread concern about environmental pollution, the Rock Springs Water Co. has decided to send all its customers a booklet about the composition of the water that it supplies. Your job is to write that booklet, and you have access to the studies that have been done on water from the three wells that Rock Springs Water Co. operates.

Situation 3.   Ralston Engineering has the contract to build the new turbine on the Silver Bend Dam on the Green River. The contract specifies that reports must be submitted at monthly intervals to the Green River Watershed Advisory Committee. Those reports must include work done, problems, and schedules. You are to write the first report.

Situation 4.   To ensure that mechanics understand potential problems with the AM-FM radios in its automobiles, Duvall Automobile Company wants a general information section in the front of the manual. This section will include information about how AM and FM reception works and will also tell what the terms *range, flutter,* and *capture effect* mean. You are to write this section of the manual.

Situation 5.   Diamond Bicycle Manufacturing Co. has successfully tested its new alloys for bicycle frames and wheels. The alloys will reduce the cost of its lightweight 10-speed bicycles and thus open the market to casual riders. To attract and instruct potential buyers, Diamond Manufacturing wants a well-written consumer booklet that will tell inexperienced bicyclists all about the parts of a bicycle and how it works. Your job is to write the brochure.

1.2 FOCUSING ON THE READER

Another important planning question is "Who will read this?" Here you consider what is called "reader" or "audience," and your answer to the question will influence how you write and what you include. As you think about potential readers, remember what you have decided about your purpose, because reader and purpose often shape each other. You can evaluate by reader type, needs, and number.

Types of Readers

You can communicate a clear technical message if you have a good idea of the type of reader you are addressing, and if you *write to that reader*.
   Have you ever tried to read one of the legal notices published in your

local newspaper? Those notices sometimes concern class action suits and are published to alert citizens who might wish to join the suit. However, with their legal jargon and complicated sentences, these notices are almost incomprehensible, and probably only a few people understand them well enough to take action. Legal notices may be written *for* intended readers, but they are not written *to* those readers.

Readers can be classified by *background* (what they know) and *function* (what they do).

Background Suppose you were asked to write a discussion of photosynthesis—the forming of carbohydrates by plants from carbon dioxide and water. Because of their differing education and background, your discussion would need to be different for each of the following readers:

- a high school junior in a biology class
- a college junior in a botany class
- a college-graduate laboratory technician at a company that produces herbicides
- a scientist with a Ph.D. doing research on the properties of chloroplast membranes

Understanding your reader's background will help you determine matters like vocabulary (how many technical terms you can use), motivation (why the reader is reading this), and potential bias (what the reader already believes). Above all, though, you need to know reader background to determine *content*, because what is necessary information for some readers is unnecessary information for others.

Function On and off the job, readers fall into five major types: *generalists, managers, operators, technicians,* and *specialists.* As Figure 1-1 shows, you can arrange these five types along a continuum from the least technical to the most technical.

Because their needs differ, each type of reader is described in the following paragraphs to help you determine how to write to that reader.

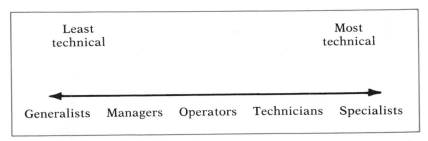

Figure 1-1 Types of technical readers

Generalists. Generalists are consumers, general readers, or laypersons. These readers have no special knowledge of the topic and are reading out of interest and for information and knowledge. All of us are generalists when we are reading outside our own specialty. An oil geologist reading about heart transplants is a generalist, as is a heart surgeon reading about coastal oil exploration. Much technical writing is aimed at generalists— instructions, descriptions, procedures, definitions, and reports are frequently aimed at first-time users or are intended for information purposes. In addition, many magazines are aimed at generalists, ranging all the way from the simplified *Popular Science* to the scholarly *Scientific American*.

When you write for generalists, you should make your writing "user friendly"; that is,

- free of specialized language (jargon)
- easy to understand, with clear organization and examples and comparisons to what is familiar, if possible, yet avoiding equations and formulas
- interesting, with anecdotes and background material
- accurate, by double-checking facts and verifying any conclusions

Here is Albert Einstein writing for generalists about a Euclidean continuum (quoted in O'Hayre 1966, 84):

The surface of a marble table is spread out before me. I can get from any one point on this table to any other point by passing continuously from one point to a neighboring one and repeating the process a large number of times, or, in other words, by going from point to point without executing jumps. We express this property on the surface by describing the latter as a continuum.

Note Einstein's comparison to a marble table—a familiar image that is easy to visualize. Notice also the simple language and the definition—a way of adding to the generalist's knowledge.

Suppose you were writing to generalists about airplane controls and their effect on takeoffs. Your explanation might begin like this:

What makes an airplane climb? You may have watched the pilot in a flying movie pull back on the stick to lift the plane over a sudden obstacle like a bridge or a tall tree. In reality, though, it's not the stick that makes an airplane climb—it's power. It's true that pulling back on the stick controls the elevators and will move the nose up momentarily, but without increased power, the speed will soon decrease and the airplane will actually sink! Power from the throttle makes an airplane climb and pulling back on the stick only controls the speed.

Without "talking down" to such a reader, the explanation begins with something the reader knows, then progresses to simple explanations of needed terms.

Managers. Managers are only one step removed from generalists: they also read technical material for information, but they want that information to help them make decisions. Concerned with the day-to-day activities of a business or industry, managers must balance the varying demands of time, resources, people, and costs—but ultimately, they must produce a product or service. Managers are also called executives and administrators.

Some managers are technically trained in fields like engineering, agriculture, or science; others are generalists from fields like business, sociology, or psychology. Even those who are technically trained, though, often become generalists before very long. Management itself takes so much time that few managers can stay on the "cutting edge" in their technical fields.

If your reader is a manager, you should remember these things:

- Avoid jargon and long explanations of theory.
- Organize the document to emphasize your conclusions and recommendations (maybe by putting them at the beginning).
- Check your facts and verify your conclusions. Because you are the authority on the subject, a manager relies on you for your technical expertise.
- Supply background information if necessary, but remember that your reader is busy and therefore unwilling to wade through unnecessary material.
- Be specific, providing the details that will help the manager make decisions.

If you had studied airplanes suitable for short-field takeoffs and landings in order to make a recommendation, you might write to a manager like this:

> As a result of our study, we strongly recommend that Mountain Mining acquire a twin-engine STOL (short takeoff and landing) airplane for use by survey parties in Alaska and the Yukon Territory. STOL characteristics such as full-span flaps and drooped ailerons will allow our crews to take off and land on short runways (less than 1000 feet) in the rugged mountain areas, while giving a payload capacity of 2 tons—enough for a crew of 10 plus equipment. Details and specific recommendations appear in Section 2 of this report.

Notice the emphasis on the recommendation, with only minimal technical detail. Notice too the specifics that are given—runway accessibility and payload.

Operators. Like generalists and managers, operators read for information, but they read a technical document because they need to know how to perform some operation. When you write for operators, make your material clear and easy to understand—usually in the form of instructions with a clear 1-2-3 method of organization. Be careful, though, not to talk down to operators; they may be well educated and very knowledgeable in their fields (for example, the operator of a nuclear energy plant or the operator of a flight simulator).

In writing for operators:

- Use reasonably short sentences and standard English word order, not because operators are uneducated, but because they may be reading and performing an operation at the same time, or they may need to remember the steps as they carry out the operation.
- Avoid equations, but include tables and graphs to make any math easily accessible.
- Ensure that all the information operators need is at their fingertips in one document, so they will not have to go to a second source.
- Check the accuracy of the instructions, making sure that all steps are included and in the proper sequence.

When flying an airplane, a pilot is an operator, and a standard lesson for new pilots is how to make a takeoff from a short field over a 50-foot obstacle. If you were writing to pilots on this subject, you might write like this:

To make a short-field takeoff over a 50-foot obstacle:
1. Start the takeoff at the extreme end of the runway. Set 15 degrees of flaps. Hold the stick back and smoothly open the throttle.
2. As the engine develops full power and the plane is accelerating smoothly, apply right rudder to correct for torque. The plane will lift off when the minimum flying speed is reached.
3. Attain and maintain the airspeed for the recommended *maximum angle of climb*. Keep the throttle wide open.
4. At about 100 feet above the ground, assume a normal climb and use normal climb power. Ease the flaps off.

These instructions assume knowledge of some standard terms and activities (this is not the first flying lesson), but the sequence of actions is clear and simply stated.

Technicians. Technicians as readers share many characteristics with operators, but they also share characteristics with the last type of audience—specialists. Technicians are a bridge between operators and specialists: they usually have considerable practical, hands-on experience, and they are interested both in what to do and how things work. Technicians frequently build and maintain complex equipment, translating the theory of specialists into machines and processes. Thus, they are interested in the practical application of theory, not theory itself.

As with any of the five types of reader, the educational level of technicians varies. Some will have college degrees, while others will have reached their status through years of practical experience. Like operators, technicians often need to perform operations, but the operations are more likely to be varied and individual rather than repetitive. For this reason, technicians need

- background information and some theory, especially if it will help them understand *why* a particular application or function is necessary
- explanations or definitions of technical jargon
- tables, graphs, and other illustrations, but a minimum of complex mathematics
- comparisons and analogies to introduce new information
- accurate presentation of data

A technician might modify or maintain an existing airplane for short-field takeoff and landing capability; he or she might also build a prototype of a new STOL model. To do that, the technician needs detailed technical information. You might explain the function of different types of airplane flaps to a technician like this:

> Plain trailing edge flaps operate by increasing wing camber, resulting in an increased maximum lift coefficient, decreased stall angle, and no change in lift curve slope, as illustrated in Figure 7.3. Fowler flaps provide an increase in area by aft translation, as well as a camber increase. They also make use of slots, or gaps, between the main wing and flap, and between the various flap elements, to provide natural boundary layer control to delay separation. Figure 7.4 shows the lift characteristics of a single-slotted 30 percent chord Fowler flap (Kohlman 1981, 124-5).

Even without the figures, which are not included here, you can see how much more technical this information is than the paragraph written for generalists, yet how direct and specific the writing is.

Specialists. Specialists, or experts, usually have an advanced degree, years of experience in their fields, or both, and are, therefore, the most technically oriented of all reader types. When you write on a technical subject, readers usually assume that you are knowledgeable in that area; in other words, that you have done the research or design and are reporting your findings. Thus, when you write for specialists, you are writing to your peers—but you need to remember that they may not share all the areas of your own expertise. For specialists:

- You can assume a large shared technical vocabulary, but you still may need to define any terms with unusual or restricted meanings.
- You should be careful to follow accepted scientific methods in the work and in the writing.
- You can provide theory, equations, and detailed supporting data.
- You should refer readers to additional studies or sources.

As a student, you write to specialists when you submit papers to your instructors, but in this case, the specialists probably know more about the subject than you do. As you gain expertise in your field, though, specialists who are your peers will become important readers.

A specialist in aeronautical engineering who is designing a wing would already know the equation for lift ($L = C_L S \rho / 2 V^2$), which contains the term C_L for the coefficient of lift. You might write for such a specialist in this way:

> The maximum value of C_L, which occurs just prior to the stall, is denoted by $C_{L,max}$ . . . In comparison with the standard NACA airfoils having the same thicknesses, these new Ls(1)-04xx airfoils all have
>
> 1. Approximately 30 percent higher $C_{L,max}$
> 2. Approximately a 50 percent increase in the ratio of lift to drag (L/D) at a lift coefficient of 1.0. This value of $C_L = 1.0$ is typical of the climb lift coefficient for general aviation aircraft, and a high value of L/D greatly improves the climb performance (Anderson 1984, 223).

Notice that the explanation relies on equations and very technical terms, which the writer assumes the specialist reader understands.

Needs of Readers

You will also want to determine your readers' needs. In other words, *why* do they want to read what you have written? Readers' needs are heavily influenced by the *area* in which they work. In an industrial environment, for example, a reader might work in any of the following seven areas.

Research   The emphasis in research is on finding and refining new materials and methods in a given technology. Researchers constantly strive to extend the outer limits of the current technology; frequently, instead of working on a specific product, they study things that seem to have no immediate practical value. Most readers in a research environment are *specialists* or *technicians*, and they will be looking for data that can be synthesized and applied to new situations. Researchers ask questions like these: What is it? Where does it come from? How does it change? Why? What else works like this? What if . . . ?

Preliminary Design   In the design area, readers will be the planners who set standards for new products, determining materials, spelling out requirements, and establishing operating goals. Again, these readers tend to be *specialists*. Based on their previous experience with products, they plan months and years in advance. Designers ask questions like these: What should it be made of? What must it do? How fast should it perform? How reliable will it be?

Development   People in development labs translate the criteria and standards of the designers into detailed specifications and then into working models. While they may be *specialists* or *technicians* in their field, they are less interested in theory than in practice because they must convert design theory into performance. Developers ask questions like these: How can we meet that standard? What changes are necessary in the design? How can we adapt or increase the current technology? How does what you are telling me mesh with what I already know? How can we stay within the cost projections?

Production   Readers in production or manufacturing must take the processes developed in the laboratory on one or two prototypes and make those processes work efficiently in mass production. They can be *specialists, technicians, managers,* and/or *operators*. Their concerns are intensely practical, centering on these questions: How can we speed up that process? How can the reliability be tested? How can we cut costs? How can we ensure accuracy?

Marketing   Readers in marketing are concerned with selling the product. They must focus on customer needs and from those needs forecast sales. They are concerned with costs, service, and customer relations. They may be *specialists, generalists,* or any type in between, but again, their concerns are practical. Their questions might be the following: How can we meet that specific market need? What will give us a price advantage? What will be the primary selling points?

Installation and Maintenance   *Operators* and *technicians* are the readers in this area. They want "how to do it" instructions and descriptions. They ask questions like these: What do I do when X happens? How do I fix Y when it fails?

Administration   Administrators must coordinate the activities of all the other groups to ensure that products or services are delivered on time, with low production costs, and with reliability. Administrators must worry about people as well as products: *managers* and *generalists* make up this group. Their questions include these: Where is the most logical place to produce X? How can we increase production at that plant? Who is the best manager for that project?

While the needs of readers in other environments may be different from those in industry, they too will require answers to a wide variety of questions. In agriculture, for example, you might find researchers, planners, developers, farm owners, field workers, and administrators.

Number of Readers

You cannot assume that only one person is going to read what you've written, so you should never dash off even a memo or a letter without careful planning. That memo or letter might go to a reader you hadn't anticipated, and it could have repercussions for both you and your employer. Therefore, a third question to ask is "How many people will read this?" You'll want to think in terms of the primary reader, the secondary reader, and multiple readers.

The Primary Reader   The primary reader is the person who makes decisions or acts on the information you provide in a piece of technical writing. Often the primary reader is the person who assigned or requested the document. Sometimes, though, you'll be asked to write something by your supervisor, who will then transmit the information to the primary reader. You should characterize your primary reader by type and needs, and you should also, if possible, know the person's name. If you are writing instructions to consumers, you won't know a specific name, but it will help to visualize a real person you know and write to that person.

The Secondary Reader   Frequently, what you write will also be read by a secondary reader like a specialist or technician, who may be reading for information or advice. This reader is usually one step removed from the decision or action, but wants or needs your input. If you can, you should also classify secondary readers by type and purpose, and you should know the person's name.

Multiple Readers   What if you discover that your primary and secondary readers are of different types or have different needs? Then you should (1) design your document for the primary reader, and (2) modify it for any secondary readers.

For example, suppose you are writing a report, "Improving the Efficiency of Greenfield Generating Station No. 4." Your primary reader is a manager who needs to make decisions based on your conclusions. You will focus on monetary implications, keep technical terms to a minimum, and eliminate theory and equations. You know, however, that technicians and specialists in fossil fuels will also read this report. For them, you might add an appendix to cover the theory and include the equations and other information they need and want.

As a student, suppose that you are writing a report, "The Feasibility of Providing 500 Student Bike Lockers near Campus Buildings." The campus facilities manager might be the primary reader, but your writing instructor (who assigned the report) would be a secondary reader. The primary reader has different needs—cost, specifications, demand for the project, alternatives—than does the instructor, who wants to see how well you can research, organize information, and write.

As you can see, you must consider many factors to understand your potential reader. All these possibilities may seem overwhelming, but if you work with a Planning Sheet like the one in section 1.6, you can quickly characterize your readers. At the planning stage, it pays to take the time to understand your potential reader or readers. Then when you organize your material and begin to write, you can be assured that you will meet your readers' needs.

CHECKLIST FOR DETERMINING THE READER

1. Who is my primary reader?

  _____ generalist   Name: _____
  _____ manager   Job title: _____
  _____ operator   Background: _____
  _____ technician   Needs: _____
  _____ specialist

2. Who is my secondary reader?

  _____ generalist   Name: _____
  _____ manager   Job title: _____
  _____ operator   Background: _____
  _____ technician   Needs: _____
  _____ specialist

3. How will I meet the needs of multiple readers?

4. Which content factors do I need to stress for my particular readers?

  _____ avoiding technical jargon
  _____ using examples and comparisons
  _____ providing interest through examples and anecdotes
  _____ giving facts and statistics
  _____ presenting theory
  _____ emphasizing conclusions and recommendations
  _____ providing background
  _____ organizing in a 1-2-3 sequence
  _____ using short sentences
  _____ avoiding equations and substituting tables and graphs
  _____ having all the information in one document
  _____ explaining technical jargon
  _____ assuming understanding of jargon
  _____ providing equations and detailed supporting data
  _____ referring readers to additional studies or data

EXERCISE 1-2   Determining Potential Readers

Directions: In a small group of four to six students, discuss one of the situations below and choose from the checklist as many categories of readers as seem to apply. Elect a spokesperson from your group to defend your choices to the rest of the class. Be prepared to discuss why you have selected the specific potential readers and how those readers would influence the content of the writing.

1. Henry Chan will graduate from General University in June with a BS in chemistry. He has a B average, and a B+ average in his major. He hopes to work in the laboratory of a pharmaceutical firm; his primary interest is in chemical synthesis and purification. He has already written his resumé and now must write a letter to accompany that resumé as he applies to Optimum Drug Co. Who are his readers?

2. Paula Swink works for a paper manufacturing company that produces cartons of various types for containers. The company has designed a carton that can be used as a single file drawer; it is a double-sided box with a separate cover. Both box and cover are precut and scored for easy folding but are to be shipped flat and unassembled. Paula's current assignment is to write the instructions telling how to assemble the carton. Who are her readers?

3. Asa Tower works at Precision Mfg. Co. Yesterday Precision's Plant #2 suffered damage from a flash fire. The fire started when sparks from a carbide cutting tool used by workers repairing the roof ignited a bin of aluminum shavings. Asa's assignment is to describe the damage. Who might read his report?

4. Carol Rojanski works for Quality Metals Corp., a small tool manufacturing company. The company wants to purchase a computer-aided design (CAD) system to increase its tool design productivity. This is a major purchase, with $200,000 allocated in the budget. Six different CAD systems are currently on the market that seem to meet some of the criteria of Quality Metals engineers. Carol's job is to evaluate the six systems based on five criteria that have been established and, in a formal report, to recommend one system for purchase. Who might read her report?

1.3 LIMITING THE SCOPE OF THE TOPIC

Still another question you'll want to ask as you plan is "What is the scope of my topic?"

A Definition of Topic and Scope

The *topic* is what is to be treated in a document or part of a document. In other words, the topic is the *content*, while the purpose is what you want to accomplish with that content. At the beginning, you need to plan both what will and what will not be treated and, by focusing on your topic, to set limits beyond which you will not go, thus setting the *scope* of the discussion.

Professional Topics and Student Topics

The task of choosing a topic and limiting its scope may differ for professionals and students. A professional is usually given the topic, which is rather narrowly defined. For example, on the job you would probably not be asked to write about something as broad as "geothermal activity." If you worked in that field, your assigned topic might be as specific as "Potential Power Sources from Geothermal Activity in the Northwest Section of Wyoming." The scope of your discussion is already defined. As a technical writing student, however, you are often writing in simulated conditions, and sometimes you will be able to choose your own topic. When you choose, you must also decide how much you are going to include—or the scope.

Even as a writing professional, you will need to set your scope, simply because you have a limited amount of time to write, and your readers have a limited amount of time to read. If you are asked for a "short" report on the causes of a switching failure, be aware that the request implies "done quickly" and "to be read quickly." On the other hand, if your assignment is to produce a "thorough" evaluation of methane gas migration from filled city-dump properties, you can expect more time for writing (maybe!) and a potentially longer document.

The key word in setting scope is *limit*. You can decide on the limits of your topic by considering:

- *Time.* How much time do you have to write? How much time will the primary reader allot to reading?
- *Length.* How long is your written piece expected to be? For example, readers only spend 15 to 20 seconds reading a resumé for the first time. Those readers expect a resumé to be limited to one or two pages.
- *Useful life.* How long will this document be used? Will it be read and used only for a week? Or will it be filed as a permanent or five-year reference?

The primary considerations, though, are fulfilling your purpose and meeting your readers' needs. What you include in your document should help advance those goals.

CHECKLIST FOR CHOOSING AND LIMITING A TOPIC

1. What is my purpose?
2. Who is my intended reader?
3. What is my topic (the content of my writing project)?
4. How much time do I have to write?
5. How long should this document be?
6. How long will this information be used (one week, six months, five years?)
7. How much shall I include?

WRITING ASSIGNMENT #1 CHOOSING A TOPIC FOR A REPORT

Directions: Follow your instructor's directions to choose a topic for your term project. Possibilities include:

- a research project that reports the current thinking (within the last four years only) on a topic appropriate to your major
- an interview (supported by previous research) with someone working in the field you are preparing to enter
- a comparative study that researches a consumer product, evaluates competing brands of that product, and recommends one brand for purchase
- a collaborative feasibility study of a change or addition needed at your college campus in general or in a particular department. Campus possibilities might include a coffeehouse, ice cream parlor, pub, record store, or travel agency—or even a parking garage or automated bank teller stations. Department possibilities might include physical objects (like adding or changing a laboratory or a major piece of equipment) or programs (like adding or changing a major or minor). This will be a group project by four to six students, which takes early and detailed planning.
- others, suggested by your instructor

When you have chosen your project, write your instructor a *memo* (see Chapter 14 for information about memos and for proper memo form). Include the following information:

1. the specific topic and scope of your project
2. reasons for your choice
3. purpose of the project
4. anticipated reader or readers
5. request for permission to pursue the project

1.4 PLANNING THE PROPER TONE

When you were a teenager, did an angry parent ever tell you, "I don't like your tone. I don't like your tone at all!" What did you say to provoke that outburst? What did your angry parent mean by *tone*? More than likely, the anger was provoked not so much by what you said (the content of your message) as by how you said it (the manner of your presentation). Sometimes your angry parent may even have misinterpreted your attitude, influenced in gauging your tone by his or her own emotions.

If you transfer an experience like this to the realm of technical writing, you can understand how complex the idea of tone is. Just as the tone of your voice can alienate a listener, so the tone of your writing can alienate a reader. Knowing that, it makes sense to carefully plan the tone of anything you write.

A Definition of Tone

Tone is the attitude you project to your readers based on the manner in which you write. Readers react to what you say, and their reaction is based on how you "come across" to them. Tone thus involves both writer and reader, and it is projected through content (what you say) and writing style (how you say it).

Varieties of Tone

You can divide adjectives describing tone into three categories. Look at the three lists below and determine which list has positive words, which negative, and which neutral.

assertive	sarcastic	professional
informal	condescending	straightforward
kind	stuffy	firm
courteous	conceited	no-nonsense
cheerful	begging	cool
tactful	pompous	indifferent
friendly	hostile	objective
polite	rude	detached
confident	brusque	impersonal
knowledgeable	blunt	serious
lively	angry	formal

You probably labeled the first list positive, the second negative, and the third neutral. After reading the lists, you might think that to "sound technical" when you write, your tone should come from the neutral list. But neutrality won't always work.

Suppose you are writing a job application letter. Do you really want the reader to think you are "cool" and "detached"? When you write a proposal (selling your services to undertake a project), do you want to be thought "indifferent" and "objective"? On the other hand, if you write instructions for installing brake shoes on an automobile, do you want the tone to be "pompous" and "condescending"? As you can see from these examples, you need to *choose* your intended tone; you can't just let it happen.

Factors That Influence Tone

When you are planning, you can choose a tone based on three factors:

- your position as writer in relation to the reader (also called your "stance")
- the content of the document
- the form the document will take

First, you must consider your *stance;* where are you positioned relative to your reader? In any technical writing situation, you can be writing *up* to superiors, *horizontally* to peers, or *down* to subordinates. You can also be writing inside or outside the company. Figure 1-2 shows how stance might look.

Your relationship to readers also depends on their position. Before you decide on tone, you need to put yourself in your readers' shoes: where do they see themselves in relation to you? Their perception of the relationship will also influence tone.

Second, think about your topic or content. Is the content of your message positive (good news), negative (bad news), or neutral (informational)? When you write a status report to your manager giving the solution to a vexing problem, the content of the message is positive. When you write a letter to a job applicant you're not going to hire, the content is negative. And when a report discusses the operation of the air scrubber for the laboratory cleanroom, the content is neutral.

If the content is positive, planning tone is not too difficult because the reader's primary response will be to the good news. But even here you could spoil good news with a tone that's grudging or condescending. If the content is neutral, you should be careful not to intrude on that neutrality. And if the content is bad news, you must plan tone very carefully. If you don't, your document may end up crumpled in the wastebasket— and only partly read! Perhaps even worse, you may provoke an angry response.

Finally, determine the form the document will take. The two forms most likely to be tone sensitive are letters and memos, because both are written personally from you and directly to a specific reader. A third tone-sensitive form is the proposal, in which you want to persuade the reader to accept your plans and ideas. However, even seemingly objective forms like instructions, descriptions, and reports can be tone sensitive if you have not thought about your intended reader.

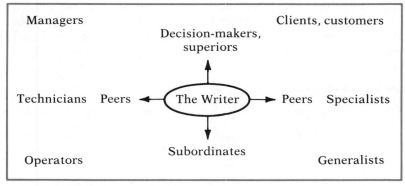

Figure 1-2   Writer's stance in planning tone

Ways to Control Tone

Whether the reader finds your tone pushy or cordial, rude or polite, depends on

- how you use words
- how much "personalness" you include
- whether you stress the reader (*you*) or the writer (*I, we*)

Be aware that when you accuse ("you failed to," "you should have"), the reader will react emotionally. When you talk down ("you should have known," "everybody knows"), the reader will respond negatively. The words you choose directly influence the tone you project.

If you totally remove yourself and your personality from your writing, you will often produce a heap of lifeless words. Great scientific writers like Thomas Huxley, Charles Darwin, and Albert Einstein were not afraid to be present as persons in what they wrote. Later writers (those of 30-40 years ago) tried to be "only scientists," and they wrote with third-person pronouns, in the passive voice and past tense, trying to reveal almost nothing about themselves in their writings. But such total self-obliteration is not necessary most of the time. In letters and memos, for example, readers will be more accepting and interested if you write to them with the same tone you would use in talking to them. That means using their name and their company name and using pronouns like *I* and *you*. Reports tend to be more neutral and objective, and often use third-person pronouns (*it, he, she, they*), especially in sections of description and explanation. In conclusion and recommendation sections, though, contemporary technical writers frequently use the first person (*I, we*) to convey the sense of one real person writing to another real person.

Remember, however, that the reader should be at the center of the communication, not the writer. Thus,

NOT: We have reviewed your complaint about the missing parts in order #2116, and we will make the following adjustment. (Writer centered)

BUT: You are correct; order #2116 should have included four sets of bearings, and we're sorry for the inconvenience. Your new invoice reflects the price adjustment and extends your discount period an additional 30 days. (Reader centered)

CHECKLIST FOR PLANNING TONE

1. In relation to the reader am I writing

_____ up?   _____ down?   _____ horizontally?

2. Is the content of my document

_____ good news?   _____ bad news?   _____ neutral?

3. Which adjectives describe the tone I want to project?

_____ positive?   _____ negative?   _____ neutral?

EXERCISE 1-3   Evaluating Tone

Directions: The following excerpts come from a variety of technical writing situations. After reading each, choose two to four adjectives from the lists in section 1.4 that describe the dominant tone. Then write a paragraph telling how you would react as a reader to what is written and why you would react that way. Submit your explanations to your instructor and be prepared to discuss them in class.

1. Memo within a company:
To: Supervisors on All Shifts
From: Manager of Production
We have several problems in our coating area in regard to substrate rejects that each of you must address. Define the cause and initiate corrective action on a sustained basis in regard to your responsible shift. . . . Effective immediately, the following will be done on all shifts without fail. . . . If you have a problem in the implementation of this directive in any way, please see me, as compliance is mandatory, effective immediately.

2. From a package, directed to the consumer:
This is the all-time indispensable kitchen tool—the vegetable peeler. You'll use it, use it, use it! It'll never let you down until the blade dulls, and eventually, the blade will dull. We think you should take a good look at your peelers. Chances are they're several years old and don't do the job they used to do. Modern technology hasn't worked out a way to keep the blade sharp forever. We have managed to keep this tool inexpensive and easily affordable—for a kitchen workhorse, that is.

3. From a newsletter to fellow employees:
I and my colleagues on the Part-Time Employee Council are appalled at the apathy and general laziness of the part-time employees regarding issues of concern to us. If you want to improve the conditions under which you work, you must work for it. You must work for it because no one, not even those of us who are committed to improving our situation, can do it alone or for you.

4. From a brochure for customers:
Have you ever wondered what's in a glass of water? Most of us have. That's why we are sending you this first report on water quality. At Springview Water Company, we are committed to providing a safe and reliable supply of high-quality drinking water. The water delivered to our customers meets all existing safe drinking water standards. We hope you will take this opportunity to learn more about your water supply and the standards that measure its safety, now and in the future.

5. From an investment letter to stockholders:
By far the most impressive thing Fifield Co. has going for it is its management. There are thousands of public companies out there, and this is the

only legitimate one I have been able to find which is being successfully run by beautiful women! I know of few public companies which have a woman president. Three of the four board members are businesswomen with extensive business experience. . . . The most impressive thing to me about Fifield Co.'s management (besides the beautiful women) is their philosophy about entrepreneuring. No salaries, low salaries, hard work, keep expenses down, and come up with fresh ideas. Not a bad formula for success. That's management! That's a formula I have never seen fail when applied. The problem is that it is not applied often enough. As you can see, I highly recommend this company.

EXERCISE 1-4   Revising for Tone

Directions: Revise each paragraph in the previous exercise so the tone is appropriate for the indicated readers. Be prepared to discuss the specific changes you have made and why you made them.

1.5 CHOOSING A SUITABLE FORM AND STYLE

An important part of planning involves choosing the form and style in which your writing will appear. In Part 3 of this book, "The Forms of Technical Writing," you will find examples of each of the 10 major forms in which technical writing usually appears. When you begin any writing project, you can refer to these forms for models and examples. Now, at the planning stage, you simply need to consider which of these possible forms to choose for a particular assignment.

Definitions of Form and Format

The *form* is the way the written material looks based on organization and structure (for example, sections and paragraphs), while the *format* is the way the material looks based on type face, type size, and page design. These two terms are often used interchangeably, and from the definitions, you can easily see why. Think of it this way: *format* is a way of showing the reader the *form* you have chosen as a writer.

Factors That Determine Form

You have already learned about the major factors that determine form: purpose, reader, and topic. Often, on the job, the form of presentation will be dictated by your assignment: "Karen, write a letter to the transit authority detailing our objections to the projected overpass." "Bob, prepare a memo for my signature spelling out the new procedures for testing the modules before they are packaged." "Stinson, you'll be working on our proposal for the automatic ticket dispenser starting the first of the month." In each of these cases, someone else has determined the purpose, reader, and topic and has chosen the form. When you choose the form yourself, you should also be guided by purpose, reader, and topic. Technical writing forms have evolved out of practical experience, and each form meets a special need.

Definitions of Style

Whether a piece of writing should be formal or informal in style also depends on purpose, reader, and topic. While you are planning, think about the degree of formality required for the occasion. *Formal style* is writing that is highly structured, impersonal, and deliberate. *Informal style* is writing that is casual and conversational, like the ordinary, everyday language of speakers.

The key words that clarify the differences are *conversational* and *speakers*. If you write the way you speak, your writing will be conversational and informal. You will include contractions like *can't, won't,* and *I'll.* You will refer to people by their first names, and you'll talk directly to the reader, using the pronoun *you.* You'll use short sentences and the active voice. Because it projects a friendly tone, informal writing can be effective in one-to-one situations like memos and letters, especially if you already know the person to whom you are writing.

However, much technical writing is directed to readers personally unknown to the writer, who may be affronted by an informal approach. For example, many people dislike being called by their first names by a stranger. Also, if your document will go to more than one reader or be used over a long period of time, you may need to write more formally because those readers are less likely to know you. That means using third-person pronouns (*it, he, she, they*) instead of *I* and *you,* and eliminating contractions. It may mean longer and more complex sentences and increased use of the passive voice. Formal writing distances the reader from the writer.

A third possibility is to write in a *semiformal style.* Currently called "user friendly," semiformal style uses the pronouns *I* and *you* extensively and may even use contractions occasionally. In this style, you do not address the reader by name, but you do write short and uncomplicated sentences in the active voice. Even so, semiformal style is more carefully structured than informal style. This book is written in semiformal style. As you read, notice how often I directly address the reader as *you,* and how many sentences use the active voice. If you count a sample of sentences, you'll find they are fairly short, averaging 18 words. Still, this book is designed as a written document, not a conversation, and I deliberately follow some conventions of formal writing.

The last sentence three paragraphs before this one, illustrates what I mean:

Because it projects a friendly tone, informal writing can be effective in one-to-one situations like memos and letters, especially if you already know the person to whom you are writing.

In informal, conversational style that sentence might end, "know the person you are writing to." In formal style, that sentence might end, ". . . especially if the writer already knows the person to whom he or she is writing." I am more comfortable with the semiformal style that preserves the *you* pronoun of informality but follows the formal style of not ending the sentence with the preposition *to*. Table 1-1 summarizes the characteristics of each style.

Table 1-1   CHARACTERISTICS OF THE THREE WRITING STYLES

Type	Use of contractions	Use of first names	Person	Sentence length	Voice
informal	yes	yes	first and second	short	active
semiformal	sometimes	no	first, second, third	fairly short	primarily active
formal	no	no	third	medium	active or passive

Style is directly related to tone; the style you choose will influence the tone you project. But specific forms of technical writing are typically written either in formal, semiformal, or informal style. Thus, in planning, you need to understand the forms that are available to you as a technical writer.

Ten Major Forms of Technical Writing

The 10 major forms in which technical writing appears are previewed for you in Table 1-2. Study them to see what your choices are. For detailed descriptions and examples of each form, see the appropriate chapter in Part 3.

CHECKLIST FOR CHOOSING FORM

1. What form is most appropriate to fulfill my purpose and meet my reader's needs?

    _____ memo   _____ abstract or summary
    _____ informal report   _____ proposal

Table 1-2 TEN MAJOR FORMS OF TECHNICAL WRITING

Form	Purpose	Reader	Topic	Style	Explained in chapter
1. memos	to communicate *within* your organization, informing, requesting, proposing, confirming	any reader within an organization	no restrictions	semiformal or informal	14
2. business letters	to communicate *outside* your organization, sometimes internally, informing, explaining, confirming, proposing, requesting	any type	no restrictions	formal to informal	15
3. resumés and job search letters	to obtain an interview for a job	generalist, manager, specialist	your qualifications for a specific position	formal	16
4. instructions and manuals	to tell how to do some task	operator	no restriction	semiformal or informal	17
5. procedures	to explain how a natural or mechanical process works	generalist, manager, technician, specialist	no restriction	semiformal or formal	18
6. abstracts and summaries	to give an overview or condensation of a longer document	specialists and technicians all types	no restriction, based on longer document	based on style of longer document	19
7. proposals	to suggest an idea or offer your services in performing a task	manager, specialist	the project that needs to be done	formal, semiformal, informal	20
8. formal and informal reports	to inform, explain, evaluate, recommend	specialist, technician, manager, generalist	varies	semiformal, formal	14 21
9. research reports and articles for publication	to inform, explain, evaluate	specialist, technician, manager, generalist	no restriction	varies by type of publication	22
10. speeches and oral presentations	to inform, explain, recommend, orally rather than in writing	all types	no restriction	formal to informal	23

_____ business letter
_____ resumé, job search
       letter
_____ instructions
_____ procedures

_____ comparative, feasibility,
       or recommendation re-
       port
_____ research report, article
       for publication
_____ speech, oral presentation

2. What style do I want to project?

_____ informal   _____ semiformal   _____ formal

EXERCISE 1-5   Choosing a Suitable Form and Style

Directions: For each of the following writing situations, you are given the purpose, reader, and topic. Refer to Table 1-2 and choose a suitable form and style. If more than one form or style is appropriate, note that. Be prepared to discuss your decisions in class.

1. purpose: to explain how to recharge it
    reader: operator
    topic: the air-conditioning system in an automobile
    form:
    style:

2. purpose: to present the main points, including the conclusion and
                recommendation of a long report
    reader: your boss, a specialist in airport construction
    topic: Midland City's environmental impact report on extend-
                ing the length of two runways at Midland Airport
    form:
    style:

3. purpose: to confirm in writing what was agreed to in a phone call
    reader: Susan Jasper, a mechanical engineer
    topic: the harmonic vibrations induced in the frame that holds
                the X-160 motor
    form:
    style:

4. purpose: to obtain information
    reader: Charles Hammel, Vice-President of Sales at SASA Con-
                tainer Company
    topic: the most efficient way to ship bulk chemicals overseas
    form:
    style:

5. purpose: to report what you did during the month of January
    reader: your immediate supervisor

topic: accomplishments, problems, plans to solve the problems
form:
style:

1.6 SCHEDULING TIME AND RESOURCES

The final step in planning is to schedule your time and resources for the writing task. A simple writing assignment can be shoe-horned into available minutes in your work day, but a complex assignment needs a detailed work plan.

A Definition of Schedule

A *schedule* is a detailed work plan for a writing project that extends from the beginning date to the completion date and shows intermediate deadlines.

For writing a short letter or memo, the schedule may be simply a mental note like this: "I have a status report due tomorrow afternoon. It'll take me about an hour to gather the material, organize it, and write a draft. Then I need somebody to check it over—Cheryl might do it—and that could take another hour. I need to allow 15 to 30 minutes to correct the draft and get it printed out. Let's see, that's 2½ hours. I have a meeting most of tomorrow morning. Guess I'd better start on it right after lunch to be sure I get it done."

For instructions, procedures, proposals, or reports of any length from 3 to 500 pages, your schedule should be more elaborate. It should also be written down, so you can refer to it during the writing process to keep yourself on track. Long documents can take from two weeks to two years to write; obviously, the bigger and longer the project, the more carefully you must schedule the parts of the process. Long writing projects often include graphics such as photographs, drawings, tables, and graphs. In scheduling, you must allow time for these to be completed, to be integrated into the text, and to be approved.

In addition, writing projects (especially big ones) are often team written: once the plan is in place, a manager divides responsibilities among several writers. Each writer then contributes his or her assigned portion, and a master writer or the manager combines the portions. Because differing writing styles and approaches must be integrated, team writing can take from 20 to 30 percent more time, and additional reviews may be needed to resolve inconsistencies.

Finally, long documents usually require several levels of review: for technical accuracy, for organization, for clarity at the sentence and paragraph level, and for details of punctuation and spelling. Since these reviewers also have other work to do, you need to schedule their participation and notify them well in advance of the time when you'll need their help.

Even though a long writing project is more complicated than a short letter or memo, in *all* your writing—short or long—you should schedule time for each of these eight basic steps:

1. planning
2. gathering information
3. organizing to prepare to write
4. reviewing your organization
5. writing the draft
6. revealing your organization to the reader
7. reviewing and revising the draft
8. editing

The eight steps form the first eight chapters of this book, and you will learn the details of each step as you continue to read. Remember that the steps often overlap, and you can drop back to a previous step whenever it is necessary.

The Writing Cycle in Industry

Schedulers in industry and business often plan time and resources with task breakdown charts, which show how various tasks can overlap during the writing process. These charts are usually shown as a series of time lines. Figure 1-3 shows an example of the task breakdown chart for a student's term project (Anderberg 1986).

Task breakdown charts are also included in proposals because they show the careful planning that will ensure a project's completion by the deadline.

Industry planners schedule about 6.5 hours' total time for each page of a document, including time for research, writing, editing, completing illustrations, and producing the master page. Even so, writers seldom have enough time to do all the research they would like. When you write, you should allow 2 to 4 hours to write each page, depending on the complexity of the task. Most writers spend 60 to 70 percent of that time on the first draft.

The Writing Cycle in the Classroom

The scheduling problems you face as a student are similar to those you will face as a professional. For example:

- You have to prepare assignments for other classes at the same time. Students sometimes say to me, "You don't understand. I have work to do for other classes too!" Of course I know that, but I must assume

Schedule

Task Description	Month / Week Ending Date	September 12	19	26	October 3	10	17	24	31	November 7	14	21	28	December 5	12	19
1. Gather preliminary information		▲														
2. Proposal Draft[a] -- Review/Edit		▲														
Final -- Review/Edit		▲														
3. Gather information		─────────						▲								
4. Letter of inquiry		────────		▲												
6. Organize information from companies					────────			▲								
7. Contact word processor users					────────────				▽							
8. Combine company/user information								────────				▽				
9. Status report					────	▲										
10. Procedures Draft -- Review/Edit						────	▲									
Final -- Review/Edit						────	▲									
11. Status report								▲								
12. Outline Draft -- Review/Edit							────────		▽							
Final -- Review/Edit							────────		▽							
13. Art/tables							────────────					▽				
14. Status report											────	▽				
15. Formal report Type											────	▽				
Review/Edit											────	▽				
Retype												────	▽			
Send out for editing												────	▽			
Final art/tables												────	▽			
Rewrite/corrections												────	▽			
Due													▽			
16. Oral presentation													────────		▽	

[a]Typing time has been included in both the draft and final schedules

▲ -- Completed
▽ -- Scheduled

Figure 1-3 Task breakdown chart for a writing project

that students have allotted enough time in their schedules to do the work for each class. Likewise, on the job you will seldom have the luxury of doing only one project at a time. Phone calls, meetings, classes, and other assignments will all intervene, and you will be pressed for time.

- Someone else sets the deadlines. In the classroom, your instructor sets the basic schedule by giving due dates for your major assignments. Those dates are based on the college calendar and the instructor's experience. On the job, writing project deadlines are also set by some outside force—perhaps the shipping date of a product or the due date of a proposal. Rarely will you set your own deadline.
- It's hard to estimate how long something will take if you've never done it before. This problem can only be cured by experience, and the only way to get experience is to begin scheduling. Whether in the classroom or on the job, you can begin by

asking questions of those who have done it before
looking at schedules of similar projects
considering your own work habits

and then,

allowing extra time for emergencies.

While you're a student, it's important to learn how to schedule a writing project and then how to work at it a little bit every day. Unfortunately, many students are like firefighters who rush from one blaze to the next. That is, they only do those assignments each day that are due the next day, and they never are able to work ahead. Crisis scheduling like that won't work for a long-term writing project.

How to Schedule a Writing Project

You can successfully schedule any writing project by following these steps:

1. Determine the due date.
2. Complete a detailed planning sheet to clarify purpose, reader, topic, tone, style, and form. A sample planning sheet follows.
3. Start with the due date and work backward to the present, setting interim dates for the following:

In the classroom	In industry
due date	due date
proofing	printing
assembling	approval by management
final typing and illustrations	final page proofs
revising	keyboarding and illustrations
review	text rewriting; illustrations revision
typing	

Continued on p. 44

PLANNING SHEET FOR _____ PROJECT
                        Due Date:_____

The PURPOSE of this project is to:
        _____ define   _____ report status or
        _____ describe an object   progress
        _____ describe a process   _____ explain
        _____ instruct   _____ interpret, evaluate,
        _____ analyze by   judge, or recommend
                _____ classifying   _____ propose, sell,
                _____ dividing   persuade
                _____ comparing   _____ request information
                                        _____ other (name it here)

READER/S
Name of primary reader _____   Job title _____
   type: _____ generalist
        _____ manager   Background _____
        _____ operator   Needs _____
        _____ technician
        _____ specialist

Name of secondary reader _____   Job title _____
   type: _____ generalist
        _____ manager   Background _____
        _____ operator   Needs _____
        _____ technician
        _____ specialist

How to meet needs of multiple readers:

TOPIC
Scope _____

Time allotted for project: by the calendar _____
                            in work–hours time _____

Length of project _____

Useful life _____

TONE
My position relative to the primary reader. I am writing
   _____ up
   _____ down
   _____ horizontally

The content is _____ good news
               _____ bad news
               _____ neutral
                                        (Continued)

I want the tone to be (choose 2–3 adjectives)

_____

FORM

    _____ memo

    _____ informal report

    _____ business letter

    _____ resumé, job search letter

    _____ instructions, manuals

    _____ procedures

    _____ abstract or summary

    _____ proposal

    _____ comparative, feasibility, or recommendation report

    _____ research report, article for publication

    _____ speech or oral presentation

STYLE

    _____ informal

    _____ semiformal

    _____ formal

SCHEDULE

Due date _____

Number of hours projected _____

BREAKDOWN OF SCHEDULE

Phase	Completed By	Total Time Allowed
Planning		
Gathering Information		
Organizing (writing outline)		
Reviewing Organization		
Writing the Draft and Preliminary Illustrations		
Typing		
Reviewing the Document		
Revising		
Final Typing and Illustrating		
Assembling		
Proofing		

first draft and preliminary illustrations
outline revision
review of outline
writing outline
gathering information
planning
assignment

approval by management
first draft
approval by management
research and input
estimates
assignment

4. Ask experienced writers or your instructor to help estimate the time needed for various phases, or estimate these yourself based on your work habits and your other assignments.
5. Revise the schedule to fit the total time you have. As you work backward from the due date setting interim dates, you may arrive at today and be only partway through the list of tasks. Then you have to trim and slice—a few hours from this segment, a day or two from that. But be sure to allow some extra time for emergencies like a broken typewriter or computer.

Then, as you follow your schedule, note places where you had to adjust the time. The next time you schedule a writing project, you'll have a better idea how long things take.

At this point such elaborate scheduling may seem like more trouble than it's worth—especially for a short writing assignment. But remember that you are learning a new skill. If you spend five minutes now carefully planning a memo or letter, you are learning how to apply a useful set of questions to any writing task. When your writing tasks grow more complex, you'll remember that you need to consider purpose, reader, scope, tone, style, and schedule, and you'll build these consideratons in as you go. In the long run, careful planning will save you time and make you a better writer.

EXERCISE 1-6　Planning a Proposal

Directions: Writing Assignment #2 will be to write a proposal to your instructor. The subject will be your choice for a term project. To do this, you will need to read Chapter 20 on proposals. Following your instructor's guidelines about the project, plan your proposal, filling in as many of the blanks as appropriate on a copy of the Planning Sheet. (Make several copies of the Planning Sheet now so you can use one for each writing project.) Submit the Planning Sheet to your instructor if requested.

WRITING ASSIGNMENT #2　Writing a Proposal for a Term Project Report

Purpose: to persuade the reader that your proposal should be accepted

Reader: your instructor or your employer (or both)

secondary reader: your fellow students

Directions: When you have received informal approval of the topic for your term project, your next major writing task is a formal proposal setting forth detailed plans for your project. Even though you may be at the beginning of your technical writing course, now is a good time to begin practicing with more formal kinds of technical writing. Proposal writing will (1) teach you a new skill and (2) force you to plan for your long-term writing project. The requirements for a research proposal, a feasibility proposal, or an interview proposal will be given to you by your instructor. Following the Request for Proposal (RFP) requirements, write the proposal in memo form.

REFERENCES

Anderberg, Nadine. 1986. *The comparison of four word processing systems.* Formal report, San Jose State University.

Anderson, John D., Jr. 1984. *Fundamentals of aerodynamics.* New York: McGraw-Hill, p. 223.

Clayton, LaVada. 1985. *Technical evaluation of COM recorders.* Formal report, San Jose State University.

Davis, Mary M. 1985. *The evaluation and comparison of four wood-burning stoves.* Formal report, San Jose State University.

Edwards, Kellie. 1985. *Office telephone systems: a comparative report.* Formal report, San Jose State University.

Kohlman, David L. 1981. *Introduction to V/STOL airplanes.* Ames: Iowa State UP, pp. 124-125.

Lankford, Terri. 1983. *A comparative study of 8-inch Schmidt-Cassegrain telescopes.* Formal report, San Jose State University.

Nichols, Frank. 1984. Interoffice memo. 6 February.

O'Hayre, John. 1966. *Gobbledygook has gotta go.* U.S. Department of the Interior, Bureau of Land Management. Washington: Government Printing Office, p. 84.

Triad Systems Corp. 1984. *Tri-Dent practice management system operator's guide.* Livermore, CA: Triad Systems Corp.

Vargas, Jeffrey. 1984. *A comparative study of nine compact digital disc players.* Formal report, San Jose State University.

CHAPTER 2

Gathering Information

When you decide to attend a movie on the other side of town, what are your thought processes before you leave for the theater? If you're going to drive, you may check the address of the theater; then if that part of town is unfamiliar, you'll either ask a friend how to get there or consult a city map. If you're taking a bus or subway, you may consult transit maps and schedules. For a jaunt across town, this process will take only five minutes, but those minutes of preparation make your trip easier and more successful.

What if the trip is more extensive? Suppose you are lucky enough to anticipate a three-week visit to Japan. Now your preparation will probably take more than five minutes. You'll want to think about what you already know about Japan, remember places friends have told you to be sure to visit, investigate airlines and airfare, and determine the current rate of exchange. You'll ask about typical weather during your projected visit so you'll know what kind of clothes to pack, and you may read travel books and consult people who have been there before. In this case, your preparation will be extensive, but it will result in a more enjoyable vacation.

In many ways information gathering is similar to trip planning. If you are writing a short memo or letter, you may already know everything you need to say. In that case, information gathering simply means jotting down those ideas for later organizing and writing. For a short report or proposal, you may only need to call a fellow worker or two for their information and expertise. But a long report, proposal, or article for publication may require the use of many kinds of sources; for that you'll need careful preparation and a methodical search procedure. A thorough search procedure will (1) reassure you that you've checked all the reasonable sources, so you can't be criticized for not checking x, y, or z, and (2) save you time in the long run. It's easier to locate a theater with a map than without one, and it's easier to gather information with a plan than without one. This chapter covers the four information-gathering activities listed on the preceding page.

2.1 PREPARING FOR THE SEARCH PROCESS

While you worked on planning, you probably thought in a general way about what you wanted to write, but now that you're about to start gathering information, you need to think more specifically about what you already know and what you need to know.

Determine What You Already Know by Brainstorming

Brainstorming is a good technique to determine what you already know (and don't know) about a subject. Begin with your purpose statement and your topic, and quickly jot down all the things that come to mind related to purpose and topic. Ask yourself the key questions: who, what, why, how, where, and when. Write down ideas as fast as you can without worrying about order or relationship. Use short phrases, abbreviations, key words, sentences—whatever will remind you later of the points you want to include. Write down everything, even those things you're not sure you want to include.

Here are five different ways to brainstorm. Read through the list and choose a way you feel comfortable with. On another project you may want to try a different method until you find the one that works best for you.

1. Use a big sheet of paper and write ideas down as you think of them, leaving some white space around each idea. Later you can

cut this paper into sections for sorting, leaving one idea to each section.

2. Use index cards and write each idea on a separate card. Don't try to save money by combining ideas on one card; if you need to save money, use small, uniform pieces of paper instead of cards.

3. Use a big sheet of paper and begin in the center with what you think is the key word. Circle it. Then jot your ideas down around it, working out from the center. In this "clustering" technique, as you generate ideas, you circle the terms and join them with lines to group related ideas.

4. List your ideas—one after another—on a sheet of paper. Then study that list and group the ideas on another sheet of paper.

5. Talk your ideas into a tape recorder. When you're finished, listen to the tape and group related ideas on paper.

Caution #1: At this point don't try to write polished, complete sentences or paragraphs. You are generating ideas, and it's important now to capture those ideas before they float away like helium-filled balloons. Don't get bogged down in details either; you can always go back later to work out the details.

Caution #2: Your ideas don't need to be correct or proven; being critical at this point can kill spontaneity, destroying potentially good ideas.

Occasionally you'll find that you already know everything you want to say. When that is true, follow the suggestions in Chapter 3 to organize the items on your brainstormed list. Most of the time, though, you will need to do some more thinking. If possible, leave your brainstormed ideas for a while, turning your conscious mind to another task to let your subconscious mind continue working.

EXERCISE 2-1   Practice in Brainstorming: Bicycles

Suppose that you are going to write a short report on 10-speed bicycles suitable for commuting and recreational riding. As part of that report, you must describe a typical 10-speed bicycle. Information from your Planning Sheet appears below.

Purpose: To describe the whole and the parts of a 10-speed bicycle.
Readers: Fellow college students. (How do you describe these readers? Educated? Interested? Nontechnical? Why would they read your report? As potential consumers?)
Topic: Limited to general description of whole and parts. How the parts work and how they are built are not considered.
Tone: Objective, descriptive.

Directions: Choose one of the five ways of brainstorming and generate a list that will describe the parts of a 10-speed bicycle. Since you have not gathered information for this topic, work with what you already know about bicycles. Don't be concerned if you are unfamiliar with the names of some of the parts. Be prepared to discuss your lists in class.

EXERCISE 2-2   Practice in Brainstorming: Describing an Object in Your Major Field

Directions: Think of an object in your major field with which you are familiar. Set up a typical purpose, reader, topic, and tone for the description, following the guildelines in Chapter 1. Then use one of the ways of brainstorming to generate a list that will describe the parts of that object. (For example: an oscilloscope, a microscope, a lathe, a word processor.)

EXERCISE 2-3   Practice in Brainstorming: Term Project

Directions: Once your formal proposal has been approved (see Writing Assignment #2, in Chapter 1), you are ready to begin gathering information. Before you launch that information search, though, you should determine what you already know. Choose one of the five ways to brainstorm and generate a series of key words and ideas to guide you in your information search. If you are working on a collaborative feasibility study, this brainstorming should be done by the whole group. You will find that together you can generate many keywords, which you can then group for each individual member's research.

Get an Overview

Often when you write something technical on the job, you will be writing out of your own experience. You will have inspected, planned, or designed something and will be reporting your results. Because you already have a basic idea of what you're doing, any information you gather will be to fill in details. As a student, though (and often as a professional), you may have to write about a subject that is new to you. Then it's helpful to get an overview of the subject before you begin researching details or interviewing experts. You can find such overviews in several places:

- *Encyclopedias.* Most of us think first of the general encyclopedias like *World Book* and the *Encyclopedia Britannica.* For technical subjects, though, you may want to consult more specialized encyclopedias in scientific and engineering fields. These range from general scientific encyclopedias like the *McGraw-Hill Encyclopedia of Science and Technology* to those in specific fields like the *Encyclopedia of Chemical Technology* and the *U.S. Pharmacopoeia and National Formulary.* Check the reference section of your library and talk to a reference librarian to find the most appropriate general source for your particular project. Because of the long preparation time for encyclopedias, the information is usually not current, but these references will provide useful background.

- *Dictionaries.* Many good technical dictionaries are available to help you define terms. Ask a reference librarian for help.
- *Handbooks, almanacs, yearbooks, and atlases.* Again, each scientific and technical field has many general sources in this category. As with encyclopedias, the information here will not be current, but it will help you understand the subject.
- *A good general book on the topic.* For general books, consult the library catalog under the subject heading. Look for titles with words like *introduction to, basic,* or *general.* Look also for books with the latest publication date. If the subject is totally new to you, you may want to look at a book like a high school text, which can give you a simplified overview.
- *Company records.* If you're working, you can often find earlier reports that have dealt with the same subject in the company files. At this stage you are looking for design documents or those with background or overview statements. Ask colleagues, your company librarian, or the document controller.

While you read overview material, be on the lookout for the key terms writers use in talking about this subject. You'll need up to 10 key terms to help you in further research, so jot them down as you notice them. For example, some key terms for a study of home exercise bicycles might be *ergometer, flywheel, pedaling resistance device, drivetrain, single action, double action, triple action.*

Choose the Most Likely Information Sources

Whether you are a college student or a writing professional, your potential information sources can be divided into three areas: recorded information, knowledgeable people, and personal experience. Recorded information includes books, articles in journals and magazines, company documents and reports, government reports and studies, sound recordings, laser and video disks, and computer data files. Knowledgeable people include fellow workers, consultants, customers, college and university professors, salespeople, and maintenance and repair people. Personal experience is what you have been involved in yourself; it could include results of experiments, investigations, trips, calculations, or previous projects. Figure 2-1 shows the likely information sources in each area.

Plan for Efficient Searching, Note-taking, and Documentation

The planning you need to do for efficient note-taking and documentation is a housekeeping chore—unglamorous but necessary. To understand the importance of this kind of planning, follow Joe P. Smith as he begins gathering information. Joe is a student about to write a recommendation

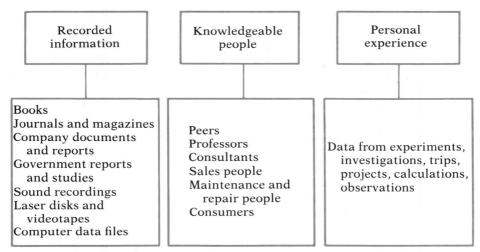

Figure 2-1 **Potential sources of information**

report. Since he's also working, he plans to compare copy machines to determine which one to recommend to his boss, a commercial real estate broker. He has determined his purpose, reader, and potential topic. He's brainstormed to see what he already knows, and now he's ready to begin gathering information—he thinks.

Joe begins his research at the campus library during a free hour between classes. Since he's most familiar with the library catalog, he spends 20 minutes searching in the subject catalog under *machines* (too many choices), *copy* (nothing relevant), and *copiers* (two books on the history of copiers—one eight years old and one five years old). In disgust, Joe next goes to the periodical index he knows best, the *Reader's Guide*, and he pulls out the latest volume. On a sheet of notebook paper, Joe writes down eight potential articles. Then it's time to go to class, so he stuffs the paper in his book bag.

That afternoon Joe calls two office-machine distributors to set up appointments with salespeople. The next day when he arrives for the first interview, the salesperson demonstrates two machines and talks enthusiastically about the characteristics of each. Joe leaves that interview and goes immediately to his second appointment. There he learns about the advantages of still another copy machine. Next, he's introduced to the head of the repair shop, who talks about repair records and implies that a competitor's product is prone to breakdown.

When Joe gets home, his head is spinning with so much information and he's tired, so he watches TV. Two days later he tries to reconstruct what he learned in his library research and his interviews, but all the information has jumbled in his mind like a tossed salad. He looks for his list of article titles, and he can't find it. He didn't take notes at his interviews, and the only written material he has is a sales brochure. He re-members that the repair record is important, and he'd like to use that information. He can't quote the repair person, though, because he forgot to write down his name; now Joe is too embarrassed to call the distributor and ask for that information again. Joe actually found potential leads and some solid information for his report, but because he didn't plan, he didn't finish his library search and couldn't use most of what he heard in his interviews.

To avoid Joe's plight, here are some helpful hints. Before gathering information, find out what kind of documentation (literature citation) you will need to provide in your final paper and what system of documentation you should use. Ask your writing instructor or those working in your major field for advice, and see Chapter 13 for information on documentation systems. The more you know at the beginning about your future needs, the easier you'll find your search and your subsequent writing. Specifically,

- **in searching written sources:**
1. When you gather information about likely sources, use 3 × 5 index cards and note the complete identification of each source on a separate card as soon as you find it. At the top of the card write the author's name (last name first), the title of the article or chapter, and—for books, the title of the book, the publication date, publisher, and library call number; for articles, the title of the periodical, volume number, and date. Later you can use these cards to arrange your List of References section, and you won't have to go back to find missing information. On the back of each card write down how useful this source is and what information you find in it. When you write your List of References, you can use this information to annotate the references. See Chapter 13.
2. Before you start examining the sources themselves, supply yourself with plenty of change for a copy machine, or buy a copycard that can be used instead of change. Rather than copying lengthy material onto index cards, photocopy relevant pages. *Be sure* that you write the author's name or a cross-reference to your identification card on the copied article, and be sure you have the page number.
3. If you do copy material onto index cards, use larger cards (4 × 6) to keep the two kinds of information separate. Write down direct quotations, enclosing them in quotation marks. Double-check them for accuracy. Later, if you wish, you can paraphrase or summarize, but you'll know what the author really said when you are ready to write, and you won't be tempted to plagiarize.

- **in contacting other people for information:**
1. Clearly identify each source on a separate 3 × 5 index card—name, title or position, place of business, phone number, and the date of the interview.

2. Prepare questions in advance (see more on this under Interviews in section 2.3).

3. Consider bringing a tape recorder to any interviews, but ask for permission first. Whether you tape the conversation or not, write down all you can remember *immediately* after the interview, while the material is fresh in your mind.

- **in developing your own information:**

  Keep good records. Carefully document all experiments or observations in a laboratory notebook; keep track of calculations and the methods used to obtain them; make notes of personal inspections. Otherwise, you'll find yourself wasting time backtracking.

Determine the Time Frame of Your Project

The *time frame* of a project refers to how much time you want to cover and how far back you want to go. Knowing the time frame of your project will help determine the limits of your search. For example, if you want only the latest and most current data, you will not look in books because book publishing usually takes at least two years, with the information probably three to four years old. But if you want a broader view, with the history or background of developments that have led up to current technology, books will be useful. For more current information, you should consult the most recent indexes to periodicals in your field, perhaps one regular index and one that includes abstracts. You might also look at government reports.

For the very latest information (newly announced products and technological advances), you may have to talk to people. Even as a student, you can often interview experts in the field if you are willing to call companies and tap into the network of those in the know. One way to begin is to ask professors in fields related to your subject. They may be experts themselves or serve as sources to recommend potential experts. You can also consult the Library of Congress Science *Tracer Bullet*, a short bibliographical guide to current "hot topics" in science and technology. *Current Contents*, a weekly service that reproduces tables of contents from journals in specific fields, is another likely source for very current information.

Find the Key Words Used in Discussing Your Subject

Whether you are searching written sources, writing away for information, or interviewing experts, much of your success will depend on your ability to ask the right kinds of questions. That means you must narrow your topic to a few *key words* and phrases. You can find these in the following ways:

- Review your purpose and topic for the key words in those statements.

- Go back to your page or cards of brainstormed ideas, and from them cull up to 10 key words that describe your subject.
- Study the notes you took while reading overview or background material. Again, pick out the key words used by writers on that subject.
- Consult the *Library of Congress Subject Headings* book. Additional terms listed for an entry (such as those listed for the sample entry in Figure 2-2) will give you more key search words.

Caution: Sometimes the Library of Congress subject headings are outdated. Periodical indexes may include more current information.

Key words have always been important in finding information, but they become critical as more material becomes available for computer searching. Because all indexes do not use the same vocabulary, you also need to think of synonyms for your key words. Unfortunately, indexes often lag several years behind industry in redefining or using new terms, and phrases may appear in the literature long before they are used as index terms.

Once you have a consolidated list of about 10 words or phrases, you are ready to begin searching for information. Whether you search in writ-

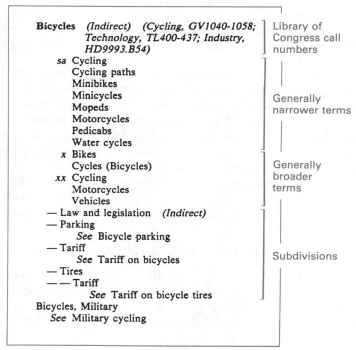

Figure 2-2   Entries in the *Library of Congress Subject Headings* book

ten sources, by interviews, or by personal involvement, if you proceed in a methodical way and write down information *as you find it,* your search will be successful.

CHECKLIST OF ACTIVITIES TO PREPARE FOR THE SEARCH PROCESS

1. What do I already know about the subject? What words and ideas come to me when I brainstorm?
2. What is a likely source to read for an overview? What words and ideas come from an overview?
3. What are my most likely sources of information:
   _____ recorded information
   _____ knowledgeable people
   _____ personal experience
4. Have I planned for the search process with
   _____ index cards?
   _____ change for the copy machine?
   _____ preparation for interviews?
5. How will I set up and maintain good records of my information?
6. What is the time frame of the project?
7. What are the key words used in talking about the subject?

EXERCISE 2-4: Setting Up a Search Process Plan

Directions: Consult your Planning Sheet (see Chapter 1) to refresh your memory about your purpose, reader, topic, tone, form, style, and schedule. Study the time allotments listed under topic and schedule. Then fill out a Search Process Plan like the one here. If requested, turn it in to your instructor.

Search Process Plan

Time available for information gathering: _____
Finish by (date): _____
Preliminary activities completed on (give dates below):
      brainstorming: _____
      securing materials: source cards, note cards, money for copy machine,
          tape recorder: _____
      overview reading: _____

Best sources of information will probably be (number them in the order you expect to use them):
      written: _____
      other people: _____
      personal experience: _____
My time frame is: _____

Purpose statement: _____
Topic: _____
Key search words are (put down at least three or four; more is better):

_____   _____   _____

_____   _____   _____

_____   _____   _____

_____   _____   _____

2.2 SEARCHING WRITTEN SOURCES OF INFORMATION

Gathering written material involves continual decision-making, frustrating dead ends, and occasional pleasant surprises. This section covers strategies for both manual and computer searches; you can read the whole thing to help you decide how to proceed, or you can read those parts that will be most useful to you. *Caution:* In some libraries the computer search is not yet effective or useful. Ask your instructor for advice.

Manual Search

How you approach a manual information search will depend on (1) how many years you want to cover, (2) how much time you can spend looking, and (3) what the most promising sources are. Assuming you have answered these questions, read an overview, and picked out key words or phrases, you are ready to begin looking.

Using library sources   Library sources can be divided into four major categories: books, reference material, periodicals, and government documents. Each category has its own sources, as Figure 2-3 shows.

You may decide to consult all four or only one, but don't be afraid to ask a librarian for help. Each category will allow you access to a different kind of information.

Books. The library catalog is your main guide to finding books. In most libraries, the catalog is arranged by author, title, and subject. Use the author/title catalog only if you *know* the name of the title or author you want; otherwise, use the subject catalog and start looking under your key words.

Bibliographies will also direct you to books. Bibliographies are lists of books on a particular subject with author, title, publisher, and publication date information. You can find bibliographies at the end of general articles or books on your subject, or you can look in the library's *Bibliographic Index* to find specific bibliographies on a subject. For current books,

Figure 2-3   Four major library sources

you can consult *Books in Print*, which lists books by subject, author, and title. You can also sometimes find bibliographies in the subject catalog under Your subject: Bibliographies (for example, Pollution: Bibliographies).

When you find a likely source, copy all the bibliographic information on one of your 3 × 5 index cards (author, title, publisher, city, date, and library call number, if there is one). Continue searching with your key words until you've found a number of sources that sound promising. Now, if you can search your library's computerized checkout system, find out which works should be on the shelves. If after you search the shelves, you cannot find a book, ask for help. If you find a reference to a book or article your library doesn't have, use the Interlibrary Loan Service to get it from another library. However, plan ahead, because it takes at least two weeks to get material through Interlibrary Loan. You may also have to pay for copying costs.

Reference material. Reference books include encyclopedias of all kinds, dictionaries, atlases, handbooks, and yearbooks. They are useful for background information and general factual data. Using your key words, consult three or four if you can. Examples of reference books are the *Chambers Dictionary of Science and Technology*, the *Encyclopedia of How It Works*, and the *Weather Almanac*.

Reference books will also be listed in the library catalog, but searching there may not be the most efficient way to find them. One good source is Sheehy's *Guide to Reference Books*, which lists and annotates some 14,000 reference books by subject, author, and title. Figure 2-4 shows sample entries. Ask a reference librarian where to find it in your library.

Periodicals. Periodicals include newspapers, commercial magazines, and professional journals of all kinds. Periodicals are especially useful for

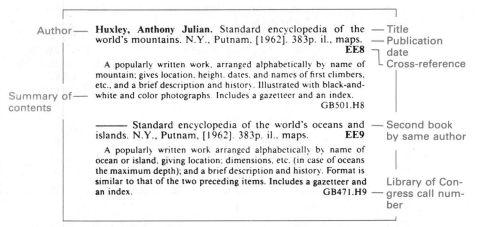

Author——   **Huxley, Anthony Julian.** Standard encyclopedia of the ——Title
world's mountains. N.Y., Putnam, [1962]. 383p. il., maps. ——Publication
EE8 ⌐ date
 ⌐
A popularly written work, arranged alphabetically by name of ⌐ Cross-reference
mountain; gives location, height, dates, and names of first climbers,
etc., and a brief description and history. Illustrated with black-and-
Summary of——   white and color photographs. Includes a gazetteer and an index.
contents   GB501.H8

—————— Standard encyclopedia of the world's oceans and —— Second book
islands. N.Y., Putnam, [1962]. 383p. il., maps.   **EE9**   by same author

A popularly written work arranged alphabetically by name of
ocean or island, giving location; dimensions, etc. (in case of oceans
the maximum depth); and a brief description and history. Format is
similar to that of the two preceding items. Includes a gazetteer and   Library of Con-
an index.   GB471.H9 —— gress call num-
ber

Figure 2-4   Typical entires in Sheehy's *Guide To Reference Books*

locating current information and for providing material in depth rather than breadth. A library at a large university may subscribe to anywhere from 6,000 to 25,000 periodicals and may have 350 or more books that index and abstract periodicals. Indexes provide easy access to the information you seek if you (1) find the appropriate index, and (2) use your key words to search for relevant data.

Many students are familiar with the *Reader's Guide to Periodical Literature*, which indexes articles in popular magazines. For most scientific and technical research, however, the *Reader's Guide* is of limited value because its material is aimed at general readers and is too broad. But many other indexes have the same format, so knowing how to use the *Reader's Guide* will help you use indexes like the *Applied Science and Technology Index*, the *Business Periodicals Index*, and the *General Science Index*. Figure 2-5 shows typical entries from the *Applied Science and Technology Index*.

Most index citations include title, author, title of periodical, date of publication, volume number, and pages. You'll find what you need in these indexes if your subject is indexed and if the article titles are complete enough to highlight any of your key words and phrases. Copy all the key information on your index cards. A few indexes also include abstracts (short summaries of the article contents). You may find abstracts more useful than simple title listings because you'll have a better idea of the article's contents before you search for it. In Figure 2-6 are typical entries from an index called *Geographical Abstracts*.

Don't neglect indexes to newspaper articles. The *New York Times Index* is available in many libraries and provides access to more than 136 years of *Times* articles. Many other newspapers are also indexed, including the *Christian Science Monitor, the Wall Street Journal,* and the *Los Angeles Times.* Use your key words to help you find current articles. If you find an

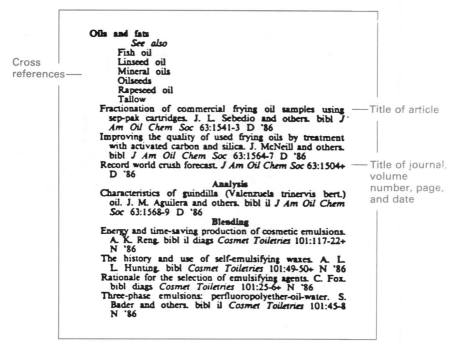

Cross references—

Oils and fats
        See also
    Fish oil
    Linseed oil
    Mineral oils
    Oilseeds
    Rapeseed oil
    Tallow
Fractionation of commercial frying oil samples using —Title of article
    sep-pak cartridges. J. L. Sebedio and others. bibl *J*
    Am Oil Chem Soc 63:1541-3 D '86
Improving the quality of used frying oils by treatment
    with activated carbon and silica. J. McNeill and others.
    bibl *J Am Oil Chem Soc* 63:1564-7 D '86
Record world crush forecast. *J Am Oil Chem Soc* 63:1504+ —Title of journal,
    D '86   volume
                                                               number, page,
                    Analysis   and date
Characteristics of guindilla (Valenzuela trinervis bert.)
    oil. J. M. Aguilera and others. bibl il *J Am Oil Chem*
    Soc 63:1568-9 D '86
                    Blending
Energy and time-saving production of cosmetic emulsions.
    A. K. Reng. bibl il diags *Cosmet Toiletries* 101:117-22+
    N '86
The history and use of self-emulsifying waxes. A. L.
    L. Hunting. bibl *Cosmet Toiletries* 101:49-50+ N '86
Rationale for the selection of emulsifying agents. C. Fox.
    bibl diags *Cosmet Toiletries* 101:25-6+ N '86
Three-phase emulsions: perfluoropolyether-oil-water. S.
    Bader and others. bibl il *Cosmet Toiletries* 101:45-8
    N '86

Figure 2-5   Entries from the *Applied Science and Technology Index*

Article— title

86G/0993   *Tropical cyclone intensity analysis using satellite data.* V. F. DVORAK, (National —Article
Oceanic & Atmospheric Administration, Washington, DC, Satellite Applications Laboratory; NOAA-   identification
TR-NESDIS-11), 1984, 53 pp.
    New and improved techniques for determining tropical cyclone intensity from satellite data have been developed.
This paper contains descriptions of methods designed to be used with visible, enhanced infrared and digital infrared
data. The analysis techniques all use cloud feature measurements and rules based on a model of tropical cyclone
development to arrive at the current and future intensity of a tropical cyclone.   -from *STAR*, 22(23), 1984

86G/0994   The development of a convective rainfall model in Jamaica using radar data. L. A. EYRE
& M. GRIFFITH, in: *Remote sensing: data acquisition, management and applications. Proc. poster
sessions RSS/CERMA conference, London, 1985*, (Remote Sensing Society, University of Reading),
1985, pp 15-24.
    Using data from a RC-32B meteorological radar system, manually digitised cloud/rainfall echoes were plotted on a
matrix of 435 squares covering the island of Jamaica and immediately adjacent seas. During four two-month periods in
1983 and 1984, data from 0900 hours to 1700 hours local time on days experiencing the diurnal convective heating-   —Abstract of
precipitation cycle were analysed. Days with synoptic scale disturbances modifying the diurnal cycle were excluded. In   article contents
this way the pattern of daily convective rainfall over the island was extracted from the radar data, representing a model
for this process apparently consistent and normal for an island of Jamaica's size and approximate shape in the tropical
trade wind belt. The pattern, as expected, approximates to but also improves the spatial coverage of recorded point
rainfall data for the months in question. With further refinements, a radar generated model would have considerable
utility in the analysis and prediction of water supply, currently critically low in Jamaica.   -Authors

Figure 2-6   Entries from *Geographical Abstracts*

article listed in a newspaper index, you can probably find a similar article
in your own area newspaper within a day or two of the listed date.

A citation index is another kind of index. If you know the name of an
expert or authority in the field you are researching, you can look up that
person's name in a citation index. As Figure 2-7 shows, under the author-

Cited author ⎯⎯

Cited reference, ⎯⎯
year and jour-
nal

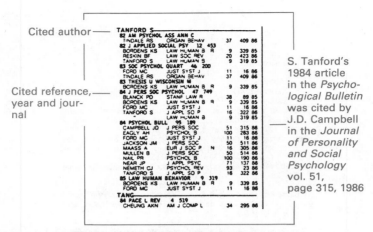

S. Tanford's
1984 article
in the *Psycho-
logical Bulletin*
was cited by
J.D. Campbell
in the *Journal
of Personality
and Social
Psychology*
vol. 51,
page 315, 1986

Figure 2-7 *Social Sciences Citation Index* entries

ity's name are listed the names of other writers who have cited the authority and the articles they have written. These articles will help you find recent information related to your topic and comments on what authorities have said. Citation indexes also have subject and source indexes. They are useful both for finding information and for evaluating the reliability of information. If a writer is frequently cited by others, he or she is probably considered an authority. But be careful; sometimes writers are cited because they are wrong! The major citation indexes are the *Science Citation Index*, the *Social Sciences Citation Index*, and the *Arts and Humanities Citation Index*.

In technical writing, it's often a good idea to look in indexes for the most recent titles; once you have about 10, it's time to find the articles themselves. You will save time if you gather information in blocks of about 10 items at a time. Ten sources give you enough possibilities so that at least one or two should be useful. At the same time, the list of 10 is short enough so you won't spend all your research time listing references without ever getting to the sources themselves. In many libraries these articles will be on microfilm or microfiche. Ask a librarian for help if you don't know how to use these tools.

Government Documents. More than 1400 libraries in the United States have been designated as depositories for literature published by the U.S. government. These publications cover many topics, and intended readers range from children to highly trained scientists. However, government publications are sometimes separated from regular library holdings and cataloged in a different way. Therefore, you need to find out if your library has government publications and how you can locate specific documents.

In addition to your library's classifications of government holdings, you can consult special books like the *Monthly Catalog of U.S. Government Publications, Index to Publications of the United States Congress* (also called the CIS/Index), and the *American Statistics Index.*

The three primary sources of technical and scientific documents are the National Aeronautics and Space Administration (NASA), the Department of Energy (DOE), and the Department of Defense (DOD). Documents from these agencies are available for sale through the National Technical Information Service (NTIS). Listings from NTIS can be found in regular indexes or in the *Government Reports Announcements and Index* (GRA&I), which is computerized in the data bases called DIALOG®, ORBIT, and BRS. Indexes will give you a scale of document costs and the address for ordering them. Allow two to four weeks to receive documents.

Using Company Sources   Companies provide technical and scientific information in two ways: (1) through published (therefore external, or public) documents, and (2) through unpublished (internal) company reports, proposals, letters, and memos. Published materials include annual reports, sales brochures and specification sheets, and manuals and instructions of all kinds. You can often find material such as annual reports in libraries, but you must obtain the rest by contacting the company directly with a letter of request.

Another source of published material is the trade journal—a publication focused on specialist or technician readers, often sent to them free, and supported by advertising aimed at those readers. While trade journals frequently are easier to read than the official journals of scientific and technical organizations, they may also be biased, so double-check information from those sources. Many trade journals are listed in indexes available at the library reference desk.

Internal company documents provide excellent sources of information on the job, but they may be unavailable to a student. These documents will be in a company's files of previous work; they include memos, reports, specifications, flow charts, and even transparencies and videotapes. You can, of course, write a letter requesting specific information. See Chapter 15 for help on writing request letters.

Computer Search

Your approach to a computer search for information will depend on (1) the computer facilities and data bases available to you, and (2) your budget. Before you begin a computer search, get an overview of your topic and pick out 10 or so key words or phrases. Do *not* ask for a computer search:

- if you need only two or three articles
- if your topic requires information more than 15 years old (most databases go back only to the 1970s)
- if the cost will be too high for you

In these situations, do a manual search instead.

The computer is a natural tool to use in finding information because it can use your key words to search through thousands of files in seconds—a process that might take you many hours in a manual search. However, since the computer doesn't think, it will respond only to what you tell it, and it won't "notice" information indexed under synonyms or under broader or narrower terms. Therefore, you should supplement any computer search with a manual search.

Usually, computers provide three information sources:

1. on-line library catalogs
2. on-line data bases
3. laser disk data bases

On-Line Library Catalogs   Many libraries have replaced or supplemented their traditional card catalogs with a computerized on-line catalog. Following instructions posted near the computer terminal, you type your request on a keyboard, and the information will be displayed on the terminal screen. You may be able to search by author, title, call number, or subject. When you find a likely source, you should copy the information onto a 3 × 5 index card. Besides standard catalog information, the computer may tell you the location of the book in a multibuilding library and whether the book is currently checked out. On-line catalogs are free to library users.

On-Line Data Bases   An on-line data base search corresponds to a manual search of periodical literature. Just as articles in journals, newspapers, and magazines are listed in indexes by subject, title, and author, so too are they listed in data bases. The key to finding what you want is to use the proper data base and to select good key words and phrases. You should talk to a librarian about the data bases available to you, the cost of an on-line search, and the procedures at your library for on-line searching.

Data Bases. Many of the most heavily used periodical indexes are now available on-line. Examples include the *General Science Index*, the *Business Periodicals Index*, the *Reader's Guide to Periodical Literature*, and the *New York Times Index*. In addition, an information retrieval service like DIALOG® has more than 150 available data bases in specialized areas. For example, DIALOG® includes AGRICOLA (provided by the U.S. Department of Agriculture Technical Information Systems), GEOARCHIVE (provided by Geosystems of London, England), Water Resource Abstracts (provided by the U.S. Department of the Interior), and MATHSCI® (provided by the American Mathematical Society).

Costs. Costs vary according to the data base you use and the time your search will take. Though search costs are high, the time you save may be worth the money because you might locate information that cannot otherwise be found. The current cost for a typical student search is about $15, but you must also pay printout costs if you have specific articles printed. Ask if your library or major department will pay for some or all of the cost. In a company library, the cost will usually be covered by your employer.

Procedures. To conduct an on-line data-base search, you will usually need to work with a librarian. Ask about the procedure to follow; you may have to fill out a form that requires you to outline your topic, list your key words, and detail your time frame. With that information the librarian will help you conduct the on-line search.

Laser Disk Data Bases One of the newest computer search strategies combines the huge data base of the on-line search with laser disk technology to give you cheaper and more convenient access to information. Here's how it works: The information in a heavily used data base is loaded onto one or more laser disks, which are then kept at the individual library and updated frequently. You use a personal computer at the library to search the disk. You follow the same general procedure to locate information as you would in an on-line search; the only difference is that the laser disk data base is now located at your library, and you need not pay the costs of connect time and long-distance phone lines.

For example, the laser disk data base called InfoTrac™ indexes articles in

- general magazines like *Smithsonian, Time,* and *New Republic*
- business journals like *Business Week, Credit and Financial Management,* and the *Journal of International Business Studies*
- government publications and laws
- major newspapers like the *New York Times* and the *Wall Street Journal*

These articles can be located by key words, and the system gives cross-references to help you find what you need. When you find a reference you want to investigate, you simply press a print button, and the bibliographic information is printed out for you.

The compact disk data base from Silver Platter indexes references from the Educational Resources Information Center, commonly called ERIC. In 1987 three compact disks in this index contained over 600,000 entries. Other new indexes on compact disks are PsycLIT for citations and psychological abstracts, MEDLINE® for medical citations, and DISCLO-SURE® for extracts from documents on market share, profits, and sales filed with the Securities and Exchange Commission. A useful feature of some of these data bases is their ability to print out the abstracts as well as the bibliographic information.

No matter what sources you use, you will probably find references to information that sounds perfect for your needs, only to discover that the information itself is not available from your library. Don't be discouraged. If you started searching early enough, you may be able to use Interlibrary Loan. If not, simply note on the back of your index card that the material isn't available and go on to another source. Remember that the word *search* is at the core of *research*, and searching means looking, and sometimes looking and looking.

CHECKLIST OF POTENTIAL WRITTEN SOURCES

Which of the following are likely sources of information for my specific assignment?

   _____ library card catalog
   _____ bibliographies
   _____ reference books
   _____ periodicals
   _____ government publications
   _____ on-line catalog
   _____ on-line search
   _____ laser disk data base

EXERCISE 2-5   Conducting a Search of Written Material

Directions: Consult your Planning Sheet (see section 1.6) to refresh your memory about your topic, scope, time frame, and time constraints. Consult the Search Process Plan you filled out in Exercise 2-4 to find your key words. Armed with that information and your source cards, visit your library to conduct a literature search for your report. Ask your instructor how many sources you should consult or use in your report. If this is a collaborative project, break up the tasks and assign one or two persons to each type of source: library, interview, survey, personal observation. Follow the general guidelines below.

1. If your time frame goes back five years or more, consult the library subject catalog using your key words and phrases. Copy all the key information about potential books on your source cards, one book to a card. Find about 10 sources and then look for the books. If you have problems, ask for help.
2. If your time frame is very current (within the last two years), or when you have found the relevant books, look for periodical information. Search manually or by computer, consulting
   • the most relevant indexes for your particular topic. Ask a librarian for help in locating the best index to use. Use your key words and phrases to locate articles. Copy all the key information on your source cards, one source to a card. Find about 10 sources, and then look for the articles.

or
- the most relevant data base for your topic. Ask a librarian for help searching on-line or by laser disk. Use your key words and phrases. Get a printout of the information or the articles themselves, or copy the information from the screen.

3. Consult government publications for information on topics like laws, statistics, regulations, and scientific and technical studies. If you're unsure about potential government publications, talk to a librarian from that area.

4. Turn in your source cards if your instructor requests them.

2.3 CONTACTING KNOWLEDGEABLE PEOPLE

"People in the know" (specialists, technicians, and managers of various kinds) can often provide you with the most current information, and because technology changes so quickly, the most current may be the best. As a working professional, you will have relatively easy access to knowledgeable people, either at your own company or outside through vendors or customers. But even as a student you can talk to experts if you are creative and persistent. This section details three ways of contacting experts for information. You may choose to use only one way or all three for a complex project.

Letters of Request

You can request information from a knowledgeable source with either a memo or a business letter. A memo is written to someone within the same organization, while a business letter is usually written to someone outside your organization. You should write a letter when

- you don't personally know your source
- you can't contact your source by telephone or in person
- you want to ensure either the accuracy or the formality of your request
- you can't get the information any other way

Note the last item in that list: "you can't get the information any other way." It's tempting to write for information as your first search strategy when, in fact, you should always do preliminary research to find out what is available in your college and local libraries and at local companies and dealerships. *Then* you write letters asking for very specific information that will fill the gaps.

Before you can write for information, you need to find the names and addresses of individuals, organizations, or companies that will be likely to answer your questions. When you are a working writer, your company can help you locate these sources, but while you are a student, you will have to find the names and addresses on your own. Here are some places you can look in a library:

FOR INDUSTRIES AND BUSINESSES:

- *National Directory of Addresses and Telephone Numbers.* This volume lists toll-free (800) numbers. Call a company and ask for the name of a customer or sales representative to whom you can write for specific information. Be sure to get the complete address, including department, mail stop, and zip code. You can also call 800-555-1212 and ask for the toll-free number of a specific company.
- *50,000 Leading U.S. Corporations.* This reference volume groups companies by general classifications such as "Ship and Boat Building and Repairing," "Guided Missiles and Space Vehicles and Parts," "Motorcycles, Bicycles, and Parts." Within each classification, the major companies are listed with their addresses and phone numbers.
- *Million Dollar Directory.* This three-volume reference lists companies in alphabetical order, by geographical location, and by the Standard Industrial Classification (SIC) number. A list of SIC numbers assigned to specific industries appears in the front. Company names, addresses, phone numbers, and division names are included. Similar information for larger companies appears in the *Billion Dollar Directory.*
- Standard and Poor's *Standard Corporation Descriptions.* This reference gives names and addresses as well as cross-references to subsidiaries.
- Standard and Poor's *Register of Corporations, Directors, and Executives.* This volume gives addresses, phone numbers, and products as well as listing key personnel.
- Thomas's *Register of American Manufacturers.* This reference gives company profiles with addresses, phone numbers, branch offices, asset ratings, and company officers.
- Ads in technical journals or trade magazines. Often the ads will give a name and address to write to for information, or they will give an 800 number you can call.
- Local business directories or yellow pages

FOR GOVERNMENT ORGANIZATIONS AND STUDIES:

- *United States Government Manual.* This reference gives information on agencies of the legislative, judicial, and executive branches. It includes a Sources of Information section and gives mailing addresses. It is revised every two years.
- *Congressional Directory.* This reference has biographies of members of Congress and information about committees and departments, including addresses and phone numbers.
- State almanacs or government directories

Note: Government names and addresses change with each election and administration; therefore, you need to check carefully to be sure the information is current.

Once you have the name and address of a likely source of information, write a letter of request. Chapter 15 will help you write a clear, concise letter of request and show you accepted forms to follow.

WRITING ASSIGNMENT #3   Writing Letters That Ask for Information

Directions: Using your library or company resources, find the names and addresses of 8 to 10 sources of information for your term project. Find the names of specific people whenever possible, but substitute job titles (such as Consumer Information Representative) or departments (such as Service Department) if necessary. Follow the suggestions and forms in Chapter 15 to draft a sample request letter and submit it to your instructor if requested. Then type and send a letter to each potential source. Consult your Plan Sheet to set cutoff dates for information.

Interviews

To most students the word *interview* triggers thoughts of job hunting and that clammy-palm ordeal they both desire and dread. But not all interviews are for jobs; one important way to tap the knowledge of specialists and technicians is through an informational interview, a meeting between two or more people for the purpose of sharing information. Most people are gracious about sharing their knowledge with students, and they also might be valuable contacts in the future.

Professionals in technical fields, whether they are writers, engineers, or scientists, rely heavily on personal contacts for information. Many interviews are informally conducted by telephone, while others are formal face-to-face meetings. As a working professional, you will find it advantageous to set up formal interviews with key sources early in a project; then, later, knowing your source, you can phone with a question or problem. As a student, you can supplement data gathered from written sources by conducting informational interviews; often you can gain current information and opinions from specialists—information that is simply not available in written form.

However, be aware of the difference between an expert's oral opinions given "off the cuff" in an interview and the carefully structured, considered, and reviewed written opinion that appears in a professional journal. Because a published book or professional article has been reviewed by specialist and peer editors before publication, it is generally a more reliable source than an opinion expressed in an interview.

Choosing someone to interview   You need a clear idea of the purpose for your interview before you can decide whom to contact. Once you've decided on the purpose, write it down in one or two sentences. The interview's purpose is related to the paper's overall purpose listed on your Planning Sheet, but it is also more specific. In other words, you should do your background reading and literature search first; then you'll have good questions to ask, and you'll be able to fit the answers into a coherent whole.

For example, if you are studying 10-speed bicycles to recommend one for the touring enthusiast, whom should you interview? The answer depends on what you need to know. You might interview

- a bicycle salesperson for information about what options are available
- a repair person to determine repair records of competing brands
- a bicyclist to get the knowledge and opinion of an experienced rider
- a manufacturer to learn about advances in frame or gear construction
- an engineer or quality control expert to learn about a part's reliability or safety

You can find names by jotting down people mentioned in your reading, by calling manufacturers and likely sources, or by talking to bicycle enthusiasts.

Planning the interview   Once you have chosen your source, write or phone to arrange a specific appointment. Since your informant is doing you a favor, you should accommodate yourself to his or her schedule. Try to plan for 30-60 minutes of interviewing time.

To use that time most efficiently, write out questions in advance. My students like to put three or four questions on a page, leaving plenty of space between each question. Design the questions to call for a response beyond a simple yes or no. Thus, instead of saying, "Do you think . . . ?", say "What do you think about . . . ?" or "Why do you think that . . . ?" Open-ended questions like those should get your source talking.

In your planning, consider tape recording the interview, but ask permission in advance because some interviewees will be uncomfortable if they are taped. A verbatim tape will let you fill in blanks in your notes and help you verify details of names and part numbers. If possible, use a battery-operated recorder so you don't have to depend on nearby electrical outlets. But do plan to take notes, and bring a supply of paper and pens for that purpose. Also bring paper big enough for your source to use for sketching diagrams or flow charts. That way you can bring that valuable information away with you, and you won't have to copy it from a blackboard or a scrap of paper.

Conducting the interview   Review your Planning Sheet (especially the Purpose section) before going to the interview. Arrive at your destination a few minutes early. Before you launch the interview itself, again ask permission to tape the conversation. Make your purpose for the interview clear at the outset. Opinion surveyor George Gallup (quoted in Brady 1977, 71) says, "When you start asking questions, the other person immediately wonders, 'Why does he want to know?' Unless your purpose is clear, he may be reluctant to talk."

Once you've established your purpose, begin with an appropriate question and listen carefully to the answer, writing down key words that will remind you of the points being made. Don't feel bound by the order of your questions; you may find that your informant skips from one topic to another. However, do use the questions to keep the flow of information coming and to keep the interview on track. Don't be afraid to admit that you don't understand something, and ask for examples and definitions of terms. Sometimes you'll find it useful to read back specific responses to confirm that you have clearly understood what was said. This technique assures the person being interviewed that you have recorded the conversation accurately.

As your alloted time nears an end, thank your source for his or her time and prepare to leave. A source who wants to prolong the interview will let you know, but usually an hour will be enough time for both of you.

Using interview material   As soon as possible after you leave the interview, fill in your notes. Your *short-term memory*, your mind's temporary storage, will be loaded to capacity, but if you immediately write down what you can remember, you'll be able to capture most of it. Within an hour or two, though, that information will be gone. If you taped the interview, you probably won't need to transcribe the whole thing. Experts tell us that only about 10 percent of what's said in an interview is important or relevant; therefore, the tape will be most helpful as a way of filling in your notes. When you write up the interview, be sure that you quote accurately.

As a courtesy to your source, send a brief thank-you letter along with a copy of the document you write based on the interview. Keep a record of the date and place of the interview as well as the name and title of your source. Include this information in the List of References or Sources Cited section of a formal report. (See Chapter 13.)

EXERCISE 2-6 Planning and Conducting an Interview to Provide Additional Information for a Feasibility or Comparative Study

Directions: Review the notes you took in your background reading and literature search to determine the specific information that would enhance your report. Choose one or more likely sources of information and set up an interview. Following the suggestions given in this book, ask specific questions. Complete your notes as soon as possible. Credit your source in your List of References or Sources Cited section. (See Chapter 13.)

EXERCISE 2-7   Planning and Conducting an Interview for a Separate Interview Paper

Directions:

1. Choose a person to interview who is working in the field you plan to enter.
2. Phone or write a letter to request an interview.
3. Following the directions in this chapter, complete your background reading.
4. Write out a series of questions; if required, get your instructor's approval of the questions.
5. Conduct the interview, recording it and taking careful notes.
6. Go over your notes immediately after the interview to fill in blank spaces.
7. See Chapter 14 for information on writing up the interview.

Surveys

If you are looking for information in breadth rather than depth, you may want to survey a group of informants. Surveys are useful in learning opinions rather than facts; think of the publicity given to preelection polls and the frequency of consumer surveys about brands of toothpaste or naturally flavored soda. In technical writing, surveys may help you establish such things as

- the need for a product (market surveys)
- criteria for judging an item
- the relative merits of one proposal or product over another (comparative surveys)
- the popularity of a product or procedure

Kinds of surveys   You can survey in two different ways: (1) with a written questionnaire mailed to potential sources, or (2) with telephone or personal interviews. Written questionnaires can be distributed over a wide geographical or social area and have the advantage of soliciting anonymous responses. Unfortunately, it's hard to get people to fill out questionnaires; response rates to mailed questionnaires typically are only 15 to 20 percent. In addition, you have the cost of printing and mailing. Personal surveys seem to work best for students, who usually have only limited time to gather information.

My students have conducted successful surveys in many different areas: they have, for example, contacted

- hospital personnel to determine success in using various surgical lasers or EKG machines
- companies to check preferences in copy machines or surge suppressors
- parents to find the preferred children's safety car seat or stroller and the reasons for the choice
- repair shops to test the reliability of washing machines, touring bicycles, or video cassette recorders

Survey questions   Effective surveys ask only a few questions, which can be answered quickly and easily. Answers fall into two groups: those that are quantitative (can be counted) and those that are qualitative (must be read and evaluated). Answers that are quantitative yield totals and percentages; they are good if you need to survey a large sample. You can also design questions so that answers can be read and tabulated by computer, a process that can save hours of your time. The four kinds of quantitative answers are

1. yes/no and either/or types with only two choices
2. multiple choice
3. ranking, in which the reader arranges items in order by preference
4. rating, in which the reader rates questions across a scale such as "very important," "important," or "unimportant"

Two kinds of questions yield qualitative responses that must be read and evaluated: short answer (fill in the blank) and essay questions. Even though these answers can't be counted easily, you may find that the details of the responses are more illuminating than the statistics. Whatever kinds of questions you use, design them carefully and pretest them on a small group to see if your questions are clearly understood.

Here is a sample of a brief student survey on children's car seats (Nelson 1984):

What brand of car seat(s) do you own? If you own more than one, please give names of both or all.
What do you like about the seat(s)?
What don't you like about the seat(s)?
Would you recommend the seat(s) you have to someone else? If not, why not?

Caution: Without formal training in how to construct surveys and how to choose representative samples, you should be cautious in using your survey data for any major conclusions. Remember that even expert surveyors come up with wrong predictions. However, you can use survey data as support for information you gather in other ways. Be sure to credit your survey in your Sources Consulted section, and perhaps include a copy of the survey questions in an Appendix.

CHECKLIST FOR CONTACTING KNOWLEDGEABLE PEOPLE

1. Can I enhance this document with information from people?
2. If so, whom shall I contact? What's the best way to find a source?
3. What method should I use to contact sources?
   _____ letters of request
   _____ interviews
   _____ surveys

EXERCISE 2-8 Designing and Administering an Effective Survey

Directions:

1. Decide what specific information you want from a survey.
2. Choose a target survey group (for example, by age, occupation, or use).
3. Using some of the key words from your reading, write four or five questions that will yield useful responses from your targeted survey group.
4. Administer the survey.
5. Tabulate or summarize the results and incorporate the data into your formal report.
6. Credit your survey in the List of References section of your report. Include a copy of the survey questions in the Appendix.

2.4 DEVELOPING INFORMATION BY PERSONAL INVESTIGATION

Most college students, whether in technical majors or not, have written laboratory reports for a science class. A lab report is a good example of writing up information developed on your own—that is, by personal investigation. In writing up a physics experiment, for example, you might be asked to

- state the object of the experiment
- describe the apparatus used
- describe how the experiment was performed
- state the method of deducing the results from the original data
- summarize the results
- give a physical interpretation of the results

All this writing stems directly from the work you have performed in the laboratory.

Other ways of developing information on your own but outside the laboratory are trip reports, site reports, analyses of collected data, and progress or status reports. These kinds of reports present what is called *primary* information (you develop the information yourself), whereas research reading produces *secondary* information (you compile the information from other sources).

If you are majoring in a technical subject, much of the writing you will do as a professional will be primary reporting on your work. Your main job will be to do the work: examine, investigate, experiment with different options, design, and so on. The writing will result from that work. That writing, though not your central concern, is very important; in a field like engineering, it enables you to share what you have done, and in the sciences, publication is your way of officially announcing your findings. Unfortunately, many technical professionals forget to document what they are doing; for example, they build or test prototypes without writing down what they did and how they solved problems. This lack of written documentation causes serious problems when valuable information is lost or when coworkers pursue paths already investigated and repeat errors that have already been corrected.

Personal investigation can be divided into three major activities: observation, examination, and experimentation.

Observation

In observation, you simply report what you see. For example, you might report on a process you have watched taking place, like the manufacturing of cheese. You might describe a physical object like a transformer, or the results of some action like the damage caused to a warehouse roof by a violent storm. In a time management study, you might observe the specific repetitive movements of workers in manufacturing plywood. If you are asked to report on potential sites for your company's office complex, you would observe and describe the sites—their size, physical features, elevation. Observation usually results in some kind of descriptive or procedural writing. These are explained in Chapters 11 and 18.

Examination

Although the word *examination* is sometimes used as a synonym for *observation*, examination implies more active involvement by the writer. When you examine something, you might take it apart in order to see how the parts work in relation to one another. You might study a bearing surface under a microscope to see the wear pattern. You might also take raw data, break it down, and analyze it in order to develop conclusions or recommendations. For example, you might study the vandalism reports in city parks over a six-month period to determine the types of vandalism and patterns of occurrence so you could recommend preventive measures.

Examination and analysis are different from observation because they involve evaluation of relationships or cause and effect. The results of such examination and analysis may be written up in a short, informal report or as a section of a longer, formal report.

Experimentation

Experimentation requires the most active involvement on the part of the writer: evaluating, operating, or testing objects, processes, or equipment. A software computer manual, for example, is best written by someone who tests the instructions by hands-on operation. An analysis of protein exchange at the capillary and tissue level is best written by someone who has participated in the laboratory study.

My students have experimented with a wide variety of tasks to develop information for technical writing projects. One student tested the strength and elasticity of various brands of monofilament fishing line. Another spent an afternoon in an electronics store testing various brands of stereo speakers within specific parameters, setting up distances and measuring responses. Still another gathered a whole box full of the kinds of cartons, bags, cans, and bottles that he would keep in a refrigerator. Taking this box to an appliance store, he stacked these items into various models of refrigerators to test the usefulness of shelf and storage space. Results of experiments like these can be used in the body of a report to support conclusions and recommendations.

Personal investigation—whether by observation, examination, or experimentation—will vary depending on your field and your assignment; therefore, it's hard to be specific about how you should proceed. You should, however, keep these general principles in mind:

1. Have a purpose, topic, and one major point written down.
2. Prepare for any personal investigation by doing background reading.
3. Keep your key words in mind as you proceed in order to maintain your focus.
4. Take careful, extensive notes while you are observing, examining, or experimenting.
5. Complete the analysis or calculations as soon as possible.
6. Write up your conclusions or observations while the material is still fresh in your mind.
7. Be sure to document the details, including times, places, equipment, and supplies.

CHECKLIST FOR PERSONAL INVESTIGATION

1. What kind of personal investigation do I need to do for this document?

_____ observation
_____ examination
_____ experimentation

2. How shall I proceed? Do I know my purpose? topic? one major point?
3. Is my background reading complete and have I decided on key words?
4. Do I have good notes to support my investigation?
5. What have I concluded from this investigation? What shall I recommend?
6. Have I written down details of time, place, equipment, supplies?

EXERCISE 2-9 Personal Investigation

Directions: Evaluate your collected data to see if you need additional information acquired by one of the methods of personal investigation. Review your key words to choose the method. Conduct the observation, examination, or experimentation and write up the results, following one of the informal report forms in Chapter 14.

REFERENCES

Applied science and technology index. 1987. Vol. 75, no. 2. February, Bronx, NY: H.W. Wilson.

Blakeman, Michael and David Fairbairn. 1986. _Geographical abstracts G Remote sensing, photogrammetry, and cartography._ Norwich, England: Geo Abstracts, Ltd.

Brady, John. 1977. _The craft of interviewing._ New York: Vintage, 71.

Library of Congress subject headings. 1986. 10th edition. Vol. 1. Washington: Library of Congress, 328.

Nelson, Pamela, 1984. _A technical report on children's car safety seats._ Formal report, San Jose State University.

Sheehy, Eugene P. 1986. _Guide to reference books._ 10th edition. Chicago: American Library Assn., 1197.

Social sciences citation index. 1987. Part 3 MURR-Z. Philadelphia: Institute for Scientific Information.

CHAPTER 3

Organizing for You—
The Writer

At last you have your information, or most of it. You have searched in the library, interviewed specialists, and written away for specification sheets. The stack of information on your desk is impressive, and according to your task sheet, you are on schedule. All your accomplishments give you a glow of satisfaction, but unfortunately, that glow begins to dim when you realize that you still face the hard part. Now you have to organize this mass of information into a logical whole and present it so you can fulfill your purpose and meet your readers' needs.

Any organizing task can be divided into two subtasks: (1) what you must do to put the parts in a coherent order before you write, and (2) what you must do to help your reader follow your organization. The first subtask is organizing for you—the writer, which will be covered in this chapter.

76

The second subtask, revealing your organization to the reader, is covered in Chapter 6.

When you organize, you do the thinking and arranging that will guide your writing. At this stage you can (and should) experiment with different ways of organizing the same material. You may also want to analyze a document that was written to perform the same task you are assigned. For example, on the job you are often given a report or memo written by someone else and told, "This will give you an idea how to do it." You need to study that report carefully to see if its organization meets the needs of your assignment. If it does, you can preserve the organization, enhancing it with your own ideas. If it doesn't, you'll have to try a new organizational pattern. You must, of course, follow company policy; nevertheless, when existing reports are used uncritically as patterns, bad organization and bad writing are perpetuated. Don't just blindly follow; be critical and ask questions. This chapter will help you to evaluate the organization of existing documents as well as organize material on your own. Model outlines are included here and throughout the book.

3.1 EVALUATING AND SORTING INFORMATION

Before you started gathering information, you may have spent a few minutes brainstorming to discover what you knew. Those words and phrases were randomly generated and may have appeared as a hodgepodge on the page. Nevertheless, from that brainstorming session, you probably found some key words and ideas that helped you focus your search for information. Using those key words, you searched in a variety of sources and gathered a mass of information—which may now appear just as much of a hodgepodge as a brainstormed list. But don't be discouraged, because now you are ready to evaluate and sort the information you have gathered.

The first step is to dig out your Planning Sheet from Chapter 1 and reread your stated purpose and the description of your intended reader. Ask yourself if either one has changed. For example, you may find that you need a narrower purpose than you'd originally thought. Now look at the topic you originally set. Is the scope of that topic still the same? Do you have good information that will allow you to cover more territory? Have you discovered that information on some aspects of the topic is not available? Adjust your scope if necessary. When you're satisfied with the purpose, reader, and scope, spend some time simply reading through the notes you've taken from books, the articles you've copied, and the information you've obtained through interviews, surveys, or examination.

As you read, you need to ask three questions:

1. Is this information useful or superfluous?
2. Is this information valid?
3. What am I missing?

If you have ever played any kind of card game, you have routinely asked yourself similar questions about the cards as you arranged your hand. Just as the rules of each card game determine the value and importance of each card, so the boundaries you have set for this assignment will determine the value and importance of each piece of information.

Is This Information Useful or Superfluous?

In many card games, you must judge the relative value of the cards you are given. Some cards will please you, and you'll begin planning ways to use them to your best advantage. Other cards will go in the discard pile.

In sorting information, you can follow the same principle. At this point, you can sort into two piles: information that will contribute to your document, and information that may be interesting but is extra and unneeded. Librarians who monitor student research say that about 30 percent of the information most students gather is not useful for the final document, but don't be surprised if an even greater percentage never appears in your final report. Information gathering is a cumulative process, and you may go astray in the early stages as you follow leads that appear interesting. As more sources are covered, though, your mind begins to fit the pieces together, and your focus becomes more concentrated. For example, if you were comparing different brands of stereo turntables, you might consider the manufacturer's warranty an important factor until you discover that all warranties are virtually identical. Then you must shift your focus to factors that do vary. Sometimes a source (an interview, perhaps) will open an avenue that you hadn't thought about and will shift your focus slightly or even radically. You also may have started a project with a preconceived idea of the "best" procedure or the "best" product. Students tell me they are frequently surprised by what they learn when they gather and analyze information. Be flexible in your thinking until all the facts are in, so you don't skew your information to fit your prejudgments.

Don't throw away notes that you put in the discard pile. You may want to include some of this material in an appendix to a long report or keep it in your file for later reference. You may even find as you begin to write that a discarded item can be used after all.

Is This Information Valid?

If you are dealt a joker in a card game that does not use jokers, you will quickly ask for a redeal. In evaluating information, the same principle operates. You need to examine each bit of information by asking:

- Are examples or evidence available to support this?
- Where did it come from?
- Is the source respected and unbiased?

How can you determine the validity of any piece of information? One way is to be sure that generalizations are supported by facts, statistics, examples, or expert opinions. You may have to drop back to the information-gathering step to find additional support. Another way is to examine the source. If the source of information is a person, you can look at the person's (1) credentials—that is, education, experience, and position, and (2) acceptance as an expert by others. You might check expertise by looking in a citation index (see Chapter 2) to see how often the person has been quoted by others. If the source is a journal article, you can estimate its validity by the kind of journal. Professional journals in all fields have panels of specialists who must approve submissions; in this case, publication itself may imply validity. If the article is in a trade journal, however, you may need to determine its validity by checking the writer's credentials, since trade journals do not generally use panels of specialists. Many journals give short biographical sketches of each author, so you can evaluate education and experience that way. If you are in doubt, ask professors in those fields.

What Am I Missing?

As you must discard in some card games, so too you can replenish your hand by drawing new cards. You always hope that those new cards will help fill the gaps in your hand. The information sorting and evaluating process will also show you what's missing and where your information is thin or unreliable. If you sort and evaluate soon enough, you'll still have a chance to ferret out the missing information before your writing deadline approaches. Find the information to fill gaps as soon as possible.

CHECKLIST FOR EVALUATING INFORMATION

1. Before sorting your information, review your Planning Sheet (see section 1.6) and ask:

    Has my purpose changed? If so, how?
    Have my intended readers changed?
    Should the scope of my topic be enlarged? diminished?

2. As you read through the information you have gathered, ask:

    Is this information relevant and useful?
    Is this information valid?
    What am I missing?

EXERCISE 3-1 Sorting Information: Cafeteria Chairs

Directions: You are a member of a student committee to advise your college on the purchase of new chairs for the student cafeteria. You have researched the field and acquired the following data. Now you must sort and evaluate the data to determine what is valid, what is useful, what is irrelevant, and what is missing.

Read the entire list below to get a sense of the whole. Then, in a group of three to four fellow students, sort each item into an appropriate category. You are not trying to organize the information, but simply to decide:

1. what is valid
2. what is useful
3. what is irrelevant
4. what is missing

Make sure your group is prepared to defend your choices to the class as a whole.

Chair A is made of hardwood and is available three days after ordering.

Chair D comes in pink/blue or yellow/green combinations.

Chair B is guaranteed for three years.

Chair C has a molded plastic seat with chrome legs.

Chair A is recommended by salespeople at three locations.

Chair D can be washed with soap and water.

Chair D costs $39.95 ($35.95 if purchased in dozens).

Chair C is used in more cafeterias than any other type of chair.

Chairs D and C are sold at Interstate Contract Furnishing Co., 1 mile away.

Chair D is advertised by a television talk show hostess.

Chair C has no warranty after 120 days.

Chair B has the high-tech look.

Chair A costs $69.95 and has a five-year warranty.

Chair B is made by a company owned by a Japanese conglomerate.

Chair D is called the "stay alert" chair.

Chair A costs $10.00 more if the seat and back are made of fabric.

Chair C has both arms and a padded back for comfort.

Chair D is made by a company that's been in business since 1952.

Chair B is recommended by doctors and insurance companies.

Chair B comes in any color and material desired.

Chair D has a chrome frame (wood at extra cost).

Chair C was designed by C. Bean for schools.

Chair B has optional cushions.

Chair D never needs to be replaced.

Chair B costs more than Chair C but looks better.

Chair A is made in Mexico.

Chair D outsold chairs A, B, and C combined last year.

Chair C is manufactured by a company owned by the brother of the vice president for student affairs.

EXERCISE 3-2   Sorting Information: Term Project

Directions: Examine the list of brainstormed ideas for your project, the proposal you wrote in Writing Assignment #2, and the notes, article copies, and other data you gathered in your information search. Read through that information, evaluating and sorting it into four categories by asking:

> Is this useful?
> Is this relevant?
> Is this valid?
> What am I missing?

Then go back to information gathering to find any information you may be missing.

3.2 BUILDING A SKELETON OUTLINE

Now that you have evaluated and sorted your information, you are ready to build a skeleton outline, or framework, for your paper. But first you have to learn how information can be organized.

Learn How Information Can Be Organized

Theorists tell us that information can be handled in three different ways: *randomly, sequentially,* and *hierarchically.* You use all three ways every day.

Random organization   You "draw straws" to determine who gets the extra piece of chocolate cake: it's a random choice. You enter a sweepstakes contest to win a trip to Hawaii: the prize is drawn at random. You shuffle the cards and deal a hand of poker: the cards are randomly distributed. You brainstorm ideas for a writing assignment: your ideas are randomly generated. In random organization you treat each item as equally important, and you want the arrangement to be by chance.

Sequential organization   To look up a phone number for the Cox Tire Service, you used the phone book, arranged alphabetically. Later, to write a check for a new tire, you used your checkbook, arranged numerically. On the way to school you saw a rainbow, with colors sequentially arranged. Tomorrow you will write a set of instructions for installing the tone arm on a stereo turntable that tell the reader what to do first, second, and third. In all the kinds of sequential ordering, you treat each item as equally important, but you choose *one* identifying characteristic (the first letter of the word, an assigned number, the location in a series), and you organize by that characteristic.

Hierarchical organization   This morning a revised company organizational chart was posted on the bulletin board where you work. When you studied it, you noted that a new management level had been added to your department and that your boss had moved up in the hierarchy. At the beginning of this term, you opened a new economics textbook to the table of contents to see how it was organized. You found the book divided into six key sections. Within each section, four to nine chapters covered divisions of the key points, and within each chapter three to eight subdivisions discussed details. You were looking at a hierarchical organization. In hierarchies, you no longer consider items equally important. Instead you assign a value to each item, and by that weighted value you determine the rank, level, or tier in which you place it.

Combinations   Often an arrangement uses both sequence and hierarchy. To understand how a combination works, think of a typical family tree. A simple family tree, like the one in Figure 3-1, is primarily hierarchical because it shows a father and a mother in key positions, with their children at a lower rank.

However, in time, the family tree grows and extends; the lower-ranked children marry and reproduce, and the tree also shows progression in chronological time. Figure 3-2 shows how the extended family tree combines hierarchical and sequential elements.

Now let's return to the topic of skeleton building. Think of your items of information as a pile of bones of different shapes and sizes. Some bones are long and sturdy, while others are small and delicate. Your job is to sort through the pile of bones and construct a skeleton that will fulfill your writing purpose. Remember, too, that there are all kinds of skeletons: you can't assume that you'll be building a human skeleton. Maybe your skeleton will be four-footed; maybe it will have a tail and no feet at all.

Because you are trying to help the reader see relationships, random arrangement doesn't work in technical writing. Instead, you use *either* sequential *or* hierarchical organization, or you combine sequential *and* hierarchical organization. Whatever arrangement you choose, the major points you select and the order in which you present them will provide

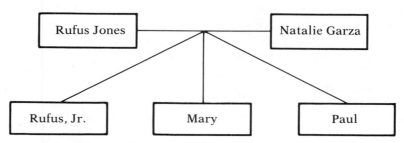

Figure 3-1   A simple family tree: hierarchical organization

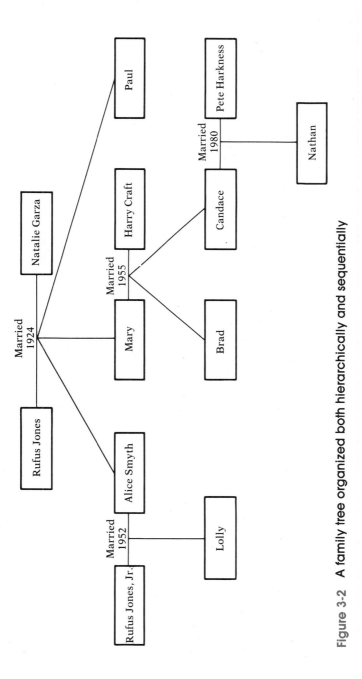

Figure 3-2   A family tree organized both hierarchically and sequentially

the skeleton outline or framework for your writing. Skeletons and frame-works are supporting structures; with that interior framework in place, the writing task is easy because you can simply flesh out the skeleton.

To build a skeleton outline, then, you must

- decide on the one major point you want to make
- choose the subpoints that support your major point
- arrange the subpoints in an effective method of organization

Decide on the One Major Point You Want to Make

Consult the purpose statement you wrote while you were planning, and use it to help you frame your major point. If possible, write that one major point in a single sentence or phrase. For example:

- Araxo Corporation's portable generator is the one most suitable for Jones Co. to use on construction sites in northern Utah.
- At the planning and budget review sessions scheduled for Nov. 14–18, 1989, managers should provide details on assigned staffing, department budget, and program requirements.
- In hand developing X-ray film, six steps are followed in darkroom conditions.
- When you read my qualifications, you will want to interview me for the position of junior statistician.

This major point may eventually shape the title of your report, form the conclusion of a memo, or become the overview statement that begins your written document. It may even be an *unstated* point (in a job application letter, you wouldn't, for example, state your desire for an interview as boldly as in the preceding example; nevertheless, that is your major point). At this step in the writing process, you should think of your major point as your goal and keep it in front of you as you begin choosing the main bones for your skeleton outline.

What if you have more than one major point? Suppose that your purpose is to describe a new digital blood pressure gauge and tell how to use it. What will be your major point? You could (1) divide your assignment into two sections with one major point for each section (how it looks; how to use it), (2) make how it looks the main focus and explain the use of each part as you describe the part, or (3) make its use the main focus and add description as you spell out the instructions. But notice that you make choices; you don't just let it happen.

CHECKLIST FOR CHOOSING THE MAIN POINT

1. After studying my purpose statement and my stack of useful, relevant information, what is the main point I want to make to my reader? Can I write it in one sentence?

2. Do I seem to have more than one more point? If so, what are the points?
3. Do I want to make one point more important than the others?

EXERCISE 3-3   Choosing the Main Point: Bicycles

Directions: If you brainstormed for a description of a 10-speed bicycle in Exercise 2-1, now examine that list. Determine what your one major point would be if you were going to write that description. In one sentence, write down the major point. Be sure to refer to your Planning Sheet from Chapter 1 before you write.

EXERCISE 3-4   Choosing the Main Point: Cafeteria Chairs

Directions: If you sorted the data on cafeteria chairs in Exercise 3-1, now examine your list of useful information. Determine what your one major point would be if you were going to write that study. Write the main point in a sentence and be prepared to defend your sentence.

EXERCISE 3-5   Choosing the Main Point: Term Project

See Exercise 2-2, 3-2, and Writing Assignment #2
Directions: Review your proposal, Planning Sheet, and collection of sorted data from Exercise 3-2. Then choose the main point you will want to make when you write up the results of your study. Write the main point in one sentence and be prepared to discuss it.

Choose the Subpoints That Support Your Major Point

Once you have decided on your main point, you are ready to choose supporting subpoints. To do that, you need to group your pieces of information—always remembering what you wrote on your Planning Sheet about your purpose, readers, and topic. When you group or cluster related ideas, you must keep your reader's short-term memory in mind. Short-term memory is the mind's temporary storage. You use short-term memory when you hold a new telephone number in mind long enough to dial it, when you listen to a speech, or when you read a sentence. Harvard psychologist George A. Miller (1967) says that the capacity of short-term memory is seven plus or minus two items. Thus, for easy understanding and remembering, three to five points is excellent, seven is acceptable, and nine is the maximum. Personally, I believe short-term memory is related to the number of fingers we have. I tell my students that they understand best when they can tick off on one hand the major points they're hearing or reading; it's harder when they have to use both hands; it becomes impossible when they have to take off their shoes!

In order to make communication easier, then, choose from two to nine subpoints and subsume your other points under them.

Besides the *number* of subpoints, you need to consider their *relationship*. Your two (or four, five, or seven) subpoints need to be equal in importance. They also need to be logically related and grammatically parallel. In this chapter I will talk about equal importance and logical relationship. In Chapter 6, I will talk about grammatical parallelism.

In a human skeleton, you would probably consider the skull and sternum to be more important than the toe bones, with the arm and leg bones perhaps equally important. Think about your items of information in the same way. Which individual items are equal in importance, rank, or degree? Once you have determined that, consider how the points are related. Both arm bones and leg bones, for example, provide movement (one kind of relationship you might want to develop). But if you were looking at supporting structures, then leg bones might be more logically related to the vertebrae that make up the backbone. However you approach your subject, logical relationship means that the major subpoints are tied to one another in a recognizable way.

For example, suppose that in the 10-speed bicycle exercise (see Exercises 2-1 and 3-3), your generated list of parts looked something like this:

wheels	handlebars	spokes	brake levers
seat	reflectors	tires	gearshift levers
derailleur	pedals	tubes	toe clamps
brakes	frame	forks	drivetrain

You need to determine some way of handling all those pieces of information that will make sense to your reader. If you discuss them in the order they are listed here, the reader will not see the relationships or understand what your main idea is. So, you study the list and, remembering that your purpose is to write a description, decide that your one major point is "A 10-speed bicycle has three major parts." You decide on three major parts somewhat arbitrarily, remembering short-term memory. Now you can test your main point by choosing subpoints.

When you choose your major subpoints, suppose you list these:

1. frame
2. drivetrain
3. reflectors

If you choose these three, your subpoints are not equally important. Reflectors are, of course, helpful in preventing accidents, but they are not equal in rank or degree to the frame and drivetrain.

You look again at your list and choose new major subpoints:

1. frame
2. drivetrain
3. brakes

Now ask yourself if these three are of equal importance. The key word is *major*. Are these the three *major* parts? Should you perhaps add a fourth part? If you do, what will it be? Which of these choices would be best?

- tubes
- tires
- wheels

In your choice of supporting subpoints, you need to see how the parts relate or work together. Wheels might be a larger term that could incorporate tires and tubes, just as drivetrain is a larger term that can incorporate derailleur.

Remember that as the writer you can choose your organization. But since you want to communicate that organization to a reader, you must think about relative importance and logical relationship and, beyond that, what the reader knows, does not know, and will do with the information.

EXERCISE 3-6   Choosing Subpoints: Bicycle

See Exercise 3-3.

Directions: Refer to your list of parts for the description of a 10-speed bicycle and your choice of the one major point. From your list, choose the most important subpoints that support your major point. Remember to keep the number of subpoints under nine so you don't overload your reader's short-term memory. Be prepared to discuss your choices in class.

EXERCISE 3-7   Choosing Subpoints: Cafeteria Chairs

See Exercise 3-1 and 3-4.

Directions: In Exercise 3-1 you designated one group of facts about cafeteria chairs as "useful information." From that list choose the key subpoints that will support your main point about cafeteria chairs, and be ready to explain why you chose the points you did. Remember to keep the number of subpoints under nine. You may want to do this exercise with a group of fellow students.

EXERCISE 3-8   Choosing Subpoints: Term Project

See Exercises 2-3, 3-2, and 3-5.

Directions: Review the sentence you wrote as the main point for your term project. Keeping in mind your purpose, reader, and topic, list from two to nine major points that will provide the framework for your paper when it is written. Submit the list to your instructor if requested.

Arrange the Subpoints in an Effective Method of Organization

When I talked about the ways of organizing information, I said that technical writers organize points in two major ways: sequentially and hierarchically. All the *methods of organization* described in this chapter can be classified in one or the other of these two categories or a combination of both. Figure 3-3 gives you an overview of these methods of organization.

You have probably already used many of these in writing without thinking of them as methods of organizing. Some methods discussed in the explanations that follow may be new to you. Read through each explanation to see how the material is organized and what the potential uses are. Study the example. Later you can come back to this section for help in choosing a method of organization that will fulfill your purpose and meet a specific reader's needs. When you organize, you need to be flexible, because often the same information can be handled in two or three different ways, each way emphasizing significant information differently. Your job is to choose the most effective method of organization based on purpose, reader, and topic.

Methods of sequential organization *Sequence* means "following one another in order." The two main types of sequential organization are chronological and spatial.

Chronological. In a chronological arrangement, points are arranged on the basis of time, from first to last. Chronological sequence is useful for

- work schedules
- minutes of meetings
- instructions
- test procedures
- process descriptions
- experiments
- historical reviews

The following outline for a process description is organized chronologically.

PURPOSE:	to explain how microwaves cook food
READERS:	generalists
ONE MAJOR POINT:	The microwave cooking process consists of three steps.

   I. Production of Microwaves
  II. Control and Distribution of Microwaves
 III. Absorption of Microwaves

This chronological arrangement works well because it follows the process: the microwaves must first be produced, then they are controlled and distributed, and finally, they are absorbed by the food. Subpoints under each of the three main points can explain each step.

SEQUENTIAL	HIERARCHICAL	COMBINED METHODS

Chronological
Spatial
	Comparison/Contrast
	Division or
	   classification
	Pro and con
		Problem-cause-solution
		Cause and effect
		Decreasing or increasing
		   order of importance
		General-to-specific
		   or specific-to-general

Figure 3-3   Methods of organization

Spatial. In a spatial outline, points are arranged as they appear physically, in relationship to either one another or the surroundings. For example, spatial arrangement may be top to bottom, left to right, outside to inside, or front to rear. This organization is useful for describing

- objects
- places
- test procedures
- processes

The following outline is arranged spatially; that is, the skin is viewed in cross section, and the discussion is oriented to the layers of the skin.

PURPOSE   to warn of the dangers of indoor tanning salons
READERS: generalists
ONE MAJOR POINT: The ultraviolet A rays produced in tanning salons
                          cause damage to the deepest layer of the skin.

1.0 Epidermis
  1.1 Penetrated by UVC rays
  1.2 Minimal damage because UVC rays are screened out of natural sunlight
2.0 Dermis
  2.1 Penetrated by the UVB rays in natural sunlight
  2.2 Damage
    2.2.1 Reddening of skin
    2.2.2 Thickening of skin
3.0 Subcutaneous Tissue
  3.1 Penetrated by the UVA rays produced in tanning salons
  3.2 Damage
    3.2.1 Harmful to blood vessels
    3.2.2 Can cause cataracts
    3.2.3 Causes skin swelling and injuries when mixed with chemicals or
          perfume on skin

Viewing the skin spatially from the top layer down helps achieve the purpose of warning of the dangers of tanning salons. The reader can see that the deepest skin layer sustains the most damage.

Methods of hierarchical organization Hierarchy implies choosing the most important points and assigning areas of responsibility or coverage to each of them. Three types of hierarchical organization are commonly used: comparison or contrast, division or classification, and pro and con.

Comparison or contrast. Points are chosen to show similarities or differences between elements of two or more subjects, or to explain a difficult or unfamiliar subject by relating or contrasting it to one simpler or already known. The most important ways to explain the comparison become the main points. The comparison or contrast method is useful for

- sales pitches
- recommendations
- analyses
- extended definitions

This sample outline analyzes two kinds of woodworkers' carbide-tipped saw blades, those manufactured for the industrial market and those manufactured for homeowners. Comparison or contrast is a logical method of organization.

PURPOSE: to analyze various carbide-tipped saw blades
READERS: operators
ONE MAJOR POINT: While industrial-quality blades cost more than homeowner blades, a comparison of materials used, manufacturing processes, and the sharpening process shows them to be a better buy for fine carpentry.

I. Materials
   A. Saw blade blanks
      1. Industrial blades
      2. Homeowner blades
   B. Carbide teeth
      1. Industrial blades
      2. Homeowner blades
II. Manufacturing Processes
   A. Machining
      1. Industrial blades
      2. Homeowner blades
      (1 and 2 will be the same for each following subpoint)
   B. Heat treating
   C. Surface grinding
   D. Tensioning
III. Sharpening Process
   A. First grinding
   B. Polishing—number of faces

An alternate method of organization using comparison/contrast would be:

I. Industrial Blades
   A. Materials
   B. Manufacturing Processes
   C. Sharpening Process
II. Homeowner Blades
   A. Materials
   B. Manufacturing Processes
   C. Sharpening Process
III. Conclusion

In either method of organization by comparison or contrast, the two kinds of blades are measured against each other. The first method stresses the criteria or points of comparison; when you want to measure something against judgment points, this is a good choice. The second method discusses first one type and then the other and requires a concluding section to make direct comparisons. This method is useful when you want to present all the information about one type in one place.

Division or classification. In division, points are determined by breaking a topic into its most important parts or functions. In classification, points are determined by grouping related items and assigning an umbrella term to each group. Division or classification is useful for

- analyses
- descriptions
- design discussions
- procedures

In this example, the writer wants to describe the design of a large computer information system workstation. Dividing the material into the component parts of the workstation makes it easier to understand.

PURPOSE: to describe the CompuCorp 1000 workstation
READERS: managers (potential purchasers), operators
ONE MAJOR POINT: The four components of the CompuCorp 1000 workstation work together to provide easy solo operation and the ability to connect to both local and distant networks.

1.0 Introduction giving overview of the workstation
2.0 The Four Major Components
   2.1 Processor
   2.2 Display
   2.3 Keyboard
   2.4 Screen Pointing Device
3.0 Network connections

Because the workstation is a complex product, the writer gives a broad view first and then breaks the material into small, manageable segments.

The exercises on cafeteria chairs (see Exercises 3-1, 3-4, and 3-7) provide a good example of classification. Here you examined the pieces of information, grouped them, and named each group—thus setting up a hierarchical outline with the group titles as key points.

Pro and con. In pro and con, points are determined to be either advantages or disadvantages. The presentation is one category at a time followed by conclusions and recommendations. Pro and con organization is useful for

- evaluations
- proposals
- justification
- persuasion (sales)
- recommendations

The partial outline below comes from a report that analyzes grand pianos in order to recommend one for purchase. Part of the analysis looks at the "action" of the piano.

PURPOSE OF THIS SECTION:	to evaluate chemically treated hammer heads
READERS:	generalists, managers, operators (potential consumers)
ONE MAJOR POINT:	The disadvantages of chemically treated hammer heads outweigh the advantages.

IV. Hammer heads
   A. Description
   B. Chemical treatment
     1. Advantages
       a. Shock resistance
       b. Resilience
       c. Shape retention
     2. Disadvantages
       a. Excessive hardening over time
       b. Production of harmonics
       c. Long-term metallic or tinny tone

Giving both the pro and con sides of an issue provides balance and lets the reader participate in the evaluation or analysis. In this case, the advantages are listed first, and the writer closes with the point she wants to make—the disadvantages. In another situation, disadvantages might be listed first. Do not assume that you should always put your main point at the end, however, for in most technical writing the opposite is true: readers want to know your main point or your conclusion at the beginning. Then, if they wish, they can read on to see how you arrived at that main point.

Combined methods Sometimes sequential and hierarchical organization are combined to produce the following four methods of organization:

1. problem-cause-solution
2. cause and effect (or effect and cause)
3. decreasing (or increasing) order of importance
4. general to specific (or specific to general)

Problem-cause-solution. When you develop an outline using the problem-cause-solution method, points are grouped by the category into which they fall. This kind of classification is useful for

- recommendations
- problem-solving
- analyses

The following outline is for a "recall" letter from a major automobile manufacturer. The method of organization explains the problem, discusses the cause, and then recommends the solution (bring the car in for a no cost refit).

PURPOSE: to recommend bringing the car in for a refit
READERS: generalists
ONE MAJOR POINT: The possible safety hazard is separation of an engine
                       mount caused by fatigue of the rubber cushion and
                       can be eliminated by installation of restraints.

1.0 Introduction explaining affected models
2.0 Explanation of problem
  2.1 Description of engine mounts
  2.2 Engine mount separation allows engine rotation
    2.2.1 Can hold throttle open
    2.2.2 Can disconnect power brake vacuum hose
    2.2.3 Can interfere with shift linkage
3.0 Cause of the problem
  3.1 Fatigue of rubber cushion in mount
  3.2 Separation of rubber cushion from mount
4.0 Solution
  4.1 Installation of restraints
  4.2 Procedure for arranging installation

Stating the problem first gets the reader's immediate attention. Once the reader recognizes what can happen to the car if the engine rotates, he or she will pay attention to the causes and will be more likely to follow the solution—to bring the car in for a refit. This organization is primarily hierarchical, but cause also precedes solution sequentially.

Cause and effect (or effect and cause). In cause and effect organization, points are chosen hierarchically; that is, on the basis of which is more important, the cause or the effect. Once the supporting points are chosen, they are presented sequentially. This method of organization is useful for

- accident reports
- predictions or plans
- analyses

In this example, the writer wants to use an airplane maintenance problem to push for better maintenance procedures. This part of the outline describes an "incident" that could have led to an accident. Effects are described first, then causes.

PURPOSE: to describe a near accident and discover why it occurred
READERS: operators (maintenance), technicians, managers
ONE MAJOR POINT: The plane's problems in flight were caused by a tiny piece of potting compound not discovered for four months due to inadequate maintenance procedures.

2.0 Effects in flight and actions taken
    2.1 First flight
        2.1.1 Plane is unstable—pitches up.
        2.1.2 Autopilot system is checked; some parts replaced.
    2.2 Second flight
        2.2.1 Plane is sloppy in pitch.
        2.2.2 Autopilot sytem is checked; other parts replaced.
    2.3 Third flight
        2.3.1 Pitch problems continue. Trim problems noted.
        2.3.2 Pressure check of system. Small hole found.
    2.4 Fourth flight
        2.4.1 Pitch problems continue. Other trim problems noted.
        2.4.2 Entire autopilot system disassembled for visual inspection and testing.
3.0 Cause of problem
    3.1 Piece of potting compound 1 cc in diameter found in venturi.
    3.2 Obstruction causes decreased ram air pressure to trim and autopilot.

Notice in this outline that the main divisions are the effects and the cause, a hierarchical organization. But within the effects section, the order is sequential: first flight problems and subsequent maintenance, second flight problems, and so on. Arranging the material this way stresses the time lapses and helps the writer make his point about inadequate maintenance procedures.

Decreasing (or increasing) order of importance. When organizing by decreasing order of importance, you first choose points by their importance (hierarchy) and then present them in sequence from the most to the least important. Decreasing order of importance gets the reader's attention with the key point and makes a strong initial impression. In increasing order of importance, you again choose the most important points (hierarchy), but when you present them, you do it from least to most important. Increasing order of importance works well with a potentially hostile reader because you can meet the reader on common ground and then use the pattern to win him or her over to your way of thinking. Order of importance arrangements are useful for

- conclusions and recommendations
- scheduling reports
- purchasing reports
- explanations
- status reports
- persuasion and argumentation
- problem-solving

In the following example the writer uses decreasing order of importance, starting with the most critical feature and moving down to those less critical.

PURPOSE OF THIS SECTION: to explain the importance of safety in choosing one surgical laser over another
READERS: managers, technicians, operators
ONE MAJOR POINT: Surgical laser safety depends on specific features and procedures.

3.0 Criteria for Evaluation
  3.1 Safety
    3.1.1 Safety Features
      • Keylock master switch
      • Emergency shut-off control
      • Simple meters and scales
      • Test controls
      • Audible and visual warning signals
      • Shields and shutter systems
      • Short focal lens
    3.1.2 Safety Procedures
      • Recalibration before each use
      • Use of focusing guides with short focal length and easy focal spot control
      • Use of wet towels near incisions
      • Avoidance of combustible anesthetic gases
  3.2 Versatility

Notice in the Safety Features section that the first two points are the most critical: they prevent unauthorized use and permit the system to be shut off in an emergency. Although the other points help the surgeon observe safety, they are less critical.

General to specific (or specific to general). When you organize from general to specific (the funnel method) or specific to general (the horn method), you classify points by degree of detail and then present them sequentially from broadest to narrowest or narrowest to broadest. In general to specific organization, the broad picture comes before the details. In specific to general organization, the opposite is true. This method is also called simple to complex (complex to simple) or deductive (inductive) reasoning. It is useful for

- teaching or tutorials
- reviews
- recommendations
- evaluations
- introductions

In the following outline, the writer plans an introduction that will move from the specific example to the general point. The same material could be arranged from the general point to the specific example.

    I. Introduction
       A. Tucson residents discard 9,500 tons of food annually.
         1. Plate scrapings account for one-third.
         2. Spoiled lettuce, partially eaten apples, rancid cheese, leftover macaroni, and other forms of waste account for two-thirds.
       B. Tucson residents discard 15 percent of the food they buy.
       C. Studies show similar patterns in Wisconsin and California.
       D. U.S. discards enough food every year to feed all of Canada.
    II. Report on Consumer Food Waste

The details here catch the reader's attention; as the introduction proceeds, the points become more general—moving from Tucson to other individual states to the United States as a whole. This organization is especially good for introductions when readers are generalists.

    Remember that these methods of organization can be combined in any one writing project. An individual paragraph usually follows a single method of organization, but a long report or proposal might combine several methods.

CHECKLIST FOR ORGANIZING INFORMATION

    1. What is the main point?
    2. What are the main supporting points?
    3. Have I considered short-term memory by using fewer than nine points?
    4. What method of organization best supports my purpose and meets the needs of my targeted reader?

Sequential	*Hierarchical*	*Combinations*
chronological	comparison or contrast	problem-cause-solution
		cause and effect (effect and cause)
spatial	division or classification	decreasing (or increasing) order of importance
	pro and con	general to specific (specific to general)

EXERCISE 3-9   Choosing a Method of Organization by Purpose

Directions: Listed below are purposes for specific writing assignments. For each one, choose an appropriate method of organization from the checklist. If you think more than one method is appropriate, indicate your alternate choice also.

1. to evaluate the potential closing of ZBD's Denver manufacturing plant
2. to explain the function of the autopilot in an airplane
3. to explain why the thermocouple in the toaster failed
4. to teach the reader about lasers
5. to describe the organizational structure of ZBD Mfg. Co.
6. to recommend which leasing service should be selected to provide fleet automobiles for ZBD branch offices
7. to report on an automobile accident
8. to explain the responsibilities of the vice president of sales at ZBD Mfg. Co.
9. to instruct the reader in how to set the switches for the warehouse burglar alarm
10. to explain the priorities in scheduling shutdown of the generator

EXERCISE 3-10   Choosing a Method of Organization: Bicycles

See Exercise 3-6.
Directions: Refer to the subpoints you chose to describe a 10-speed bicycle. Now arrange them in one of the methods of organization explained in this chapter. Review your assigned writing purpose to help choose the method and name your method of organization.

EXERCISE 3-11   Choosing a Method of Organization: Cafeteria Chairs

See Exercise 3-7.
Directions: Refer to the subpoints you chose to determine the criteria for purchasing new chairs for the cafeteria. Now arrange your subpoints in a

logical method of organization. Review your stated purpose and name the method of organization.

EXERCISE 3-12   Organizing: A Comprehensive Exercise

Directions: The following table presents information that might be contained in a report on the storage conditions of paper stock at a printing company. It does not provide the details, only the major points. Remembering that information can be organized in more than one way, use this information to write three simple outlines, showing three different ways to organize these facts. Be prepared to discuss how each method of organization would be better for particular purposes or groups of readers.

CONDITION OF PAPER STOCK AFTER STORAGE

Kind of Paper	Warehouse A	Warehouse B	Warehouse C	Warehouse D
coated	poor	good	good	poor
book	good	good	poor	fair
tag	poor	fair	poor	fair
newsprint	fair	poor	poor	poor

3.3 FLESHING OUT THE SKELETON OUTLINE

Add Details to the Subpoints

Now that you have chosen your main points and arranged them in an appropriate method of organization, you are ready to add the details to each main point. You need to do two things:

1. Make sure that the details support their particular point.
2. Arrange each subset of details in the method of organization that will work for that section.

Details need not follow the same organizational method as main points. For example, the eight chapters in Part 1 of this book follow a chronological method of organization: in the process of writing, planning comes before gathering information. But the sections that make up the gathering information chapter follow a division-and-classification method of organization: section 2.2, Searching Written Sources, is one way of gathering information, and section 2.3, Contacting Knowledgeable People, is another way, but there is no reason that one of these steps must precede the other.

As you add details to your major points, you will also discover where you may need more information, and you can drop back to that step. It's easier to find additional information now than it will be when you begin the actual writing.

Choose an Outline System

Now is also a good time to choose an appropriate outline system. So far all your organizational efforts have been on rough or working outlines with simple numbering and indentation. Rough outlines will work for short documents, but before review of a long document, you may want to set up a formal numbering system. The two systems in Figure 3-4 are commonly used. Your company or instructor may favor one system over the other; increasingly, companies are using the decimal system, and military documents use it exclusively.

In either system, if your instructor or company style guide permits, you might want to use "bullets" instead of numbers after the third or fourth level of subordination. Bullets are usually either filled-in circles or squares. On the typewriter you can make bullets by typing a lowercase *o* and filling it in. Using bullets will simplify the document. Compare:

4.2.1.1	4.2.1.1
4.2.1.1.1	• ———————————
4.2.1.1.2	• ———————————
4.2.1.2	4.2.1.2

Decimal System	*Roman Numeral System*
1.0	I.
1.1	A.
1.2	B.
1.2.1	1.
1.2.2	2.
2.0	II.
2.1	A.
2.2	B.
2.3	C.
2.3.1	1.
2.3.1.1	a.
2.3.1.1.1	(1)
2.3.1.1.2	(2)
2.3.1.1.2.1	(a)
2.3.1.1.2.2	(b)
2.3.1.2	b.
2.3.2	2.
3.0	III.

Figure 3-4   Common outline systems

Bullets are especially useful when you do not want to indicate any particular order in a list or when the list will not need to be referenced later in the text.

This textbook uses a modified form of the decimal system. Chapters are numbered, and the major divisions are denoted by decimals, but subdivisions are indicated by type size and style. The book also includes many examples of bulleted—as well as numbered—lists.

You need to remember three important things when you add a formal numbering system to your working outline:

1. Numbering systems are useless unless they are combined with clear headings.
2. You should always have at least two subpoints for each major point. If you end up with only one, study your organization. Perhaps it needs revision, or your subpoint needs renaming. Perhaps the subpoint is an explanation of the main point, not a division of it.
3. The numbering system in an outline is less important than the actual relationship of ideas. The numbers are primarily an aid to the reader, but they can also help you see the relationships among your ideas as you write the draft.

CHECKLIST FOR COMPLETING AN OUTLINE

1. As I add details to each point, have I
   _____ chosen those that really do support that point?
   _____ arranged them in a method of organization that is appropriate for that section?
2. If I have any gaps in my data, do I know where to find the missing information?
3. Have I chosen an outline system that is appropriate for my purpose and reader? How many levels of detail does it include? Do I want to use bullets at the third or fourth level?

EXERCISE 3-13   Organizing Outlines: Turntable Purchase Recommendation

Directions: The following table summarizes the information in a report comparing four stereo turntables and judging each on four criteria. Using this information, choose a method of organizing the information for the recommendation section of the report. Then *outline* your proposed organization of this information.

PURPOSE: to recommend the purchase of one of the four stereo turntables
READERS: generalists

Table 3-1   TOTAL POINTS FOR STEREO TURNTABLES

Product	Drive mechanism	Tone arm	Operational criteria	Misc. criteria	Total points
Hitachi L 303	19	18	16	20	73
JVC QL-L2	32	16	16	24	88
Sony Ps-F17	28	20	16	21	85
Technics S1-J2	32	16	16	22	86
Total number of points possible					100

Table 3-1 lists the total number of points acquired by each turntable in this study.

WRITING ASSIGNMENT #4   Writing an Outline for a Term Project

See Exercises 3-5 and 3-8.
Directions: If you have completed Exercises 3-5 and 3-8, you already have the major points of the outline for your term project. For this writing assignment, review those exercises to make sure that your choices of main point and subpoints still stand. Then add the necessary details and choose a formal outline system. This outline will be the basis for your paper. Your instructor may ask you to prepare several copies for a workshop review by fellow students.

WRITING ASSIGNMENT #5   Writing Status Reports

Directions: If you are engaged in a term project, you should keep your instructor appraised of your progress with biweekly or weekly status reports. The completion of the organizing step is a good time to begin reporting the status of your project. Following the form in Chapter 14 or one suggested by your instructor, write a status report in memo form.

REFERENCES

Miller, George A. 1967. *The psychology of communication.* New York: Basic Books.

CHAPTER 4

Reviewing Your Organization

4.1 UNDERSTANDING WHY REVIEW IS NECESSARY

4.2 FOUR TYPES OF REVIEW

Just as construction projects have building inspectors, so writing projects have reviewers. Thus, after you finish organizing, you'll want to take a second or third look at what you have done. This review should be done after you have been away from your outline for some time. Ideally, you should let the outline rest a week or so before you review, but realistically, you may have to settle for letting it rest (and letting you rest) for a day or several hours. This chapter explains why you need to review at this stage and describes four possible kinds of review.

4.1 UNDERSTANDING WHY REVIEW IS NECESSARY

When you're ready to review, think about what the word *review* really means—"to look again." While you were assembling and shaping your information into a skeleton outline, the arrangement was open to change. Once you finished your outline, though, you may have felt that the organization you chose was the only possible one. Now, in review, you can take a second look when the ideas are cold and you have achieved a certain distance.

Almost all information can be arranged more than one way. For example, Terri, a physics major, has been studying 8-inch reflecting and refracting telescopes. Her purpose is to find the one that will best meet the needs of an amateur (her reader) in the market for a deep-sky photographic and observational instrument. When she sorts her gathered information, this is what she finds:

All the telescopes evaluated had the following common features:

- aperture
- f/ratio
- optical system
- stellar magnitude limit
- useful visual magnitude limit
- aberration

The telescopes differed in these features:

- optical surface accuracy
- oculars (eyepieces)   These features were important.
- size and weight
- cost

- gears   These features were not important.
- tube construction

The main point: the Celestron C8 is the best choice.

How should Terri organize her material? She could set up an outline this way:

    I. Conclusion
   II. Common Features of Telescopes
 III. Differing Features of Telescopes
      A. Important
      B. Unimportant

Or she could do it this way:

    I. Conclusion
   II. Features (or Criteria) for Discrimination
      A. Optical surface accuracy
      B. Oculars

     C. Size and weight

     D. Cost

  III. Common Features (or Criteria)

     (omitting in this outline gears and tube construction since they are not critical)

Or this way:

    I. Important Features

   II. Unimportant Features

  III. Common Features

  IV. Conclusion

These are only three possibilities; there are several others. At the review stage, Terri must apply a series of questions to the organization she has chosen to see if it works—if it is appropriate for the reader and the reader's needs and if it will fulfill her purpose. If it doesn't work, she should try another method.

    In organization review, then, you want to ensure that whatever arrangement you chose is the best one. Remember that you've created a working outline—one that will guide you when you begin to write. Cooled though the ideas may be, your outline is not embedded in cement. You can change it now by looping back to the organizing process, and you can change it still more when you begin to write. What's important is the relationship of the ideas, the achievement of your purpose, and the clarity with which you give the readers the information they need.

4.2 FOUR TYPES OF REVIEW

Four kinds of review are possible at this stage: self-review, peer review, multiple author review, and instructor or management review. How elaborate your organizational review must be depends on the complexity of your assignment: for a simple letter or memo, self-review or a quick look by one other person may be enough. For a long proposal or report, though, most students and working writers appreciate feedback from three or four peers in a review workshop. Ironically, very short documents like resumés—which have multiple sections—also benefit from group review, because short and packed documents are very hard to write. The more sections the writing has, the more elaborate the review should be because it is difficult to keep the relationship of several sections in mind. Multiple author review is necessary when more than one person is developing the organization. The fourth kind of review, by your instructor or manager, can be extremely helpful because this person can view your organization in the light of considerable experience. Instructors or managers can also verify that you are doing the right thing—that is, fulfilling the assignment.

Self-Review

Many times you will do the reviewing yourself. Check your Planning Sheet (see section 1.6) and compare it with your assignment. Check the Planning Sheet against the outline. Answer each question below:

1. Have I fulfilled my writing purpose?
2. Do I have a single main point?
3. Do the subpoints support the main point?

(Here you can ask other questions too. Are significant points emphasized? Are there two to nine subpoints—no more than nine? Have I chosen the right organizational pattern for my purpose? Are the subpoints arranged in the best sequence? Are the details logically organized under each subpoint?)

4. Do I have a numbering system that helps the reader understand the organization? Does every division have at least two subpoints?

5. Is the main point I started out with the main point I still want to make? (Sometimes as you work through your material, you find that it adds up in ways you hadn't anticipated. That means you have literally "written to learn"; you have discovered something you didn't know before you began and that's good. But if so, now is the time to shape the new major point and reshape your working organization. You'll find it easier to change your organization now than when you begin writing.)

6. How does my outline compare to existing documents? If your company has a format or outline for this type of report, it may help you to look at it. Also see Part 3 of this book for sample formats for particular documents.

Peer Review

If possible, you should have your outline reviewed by fellow workers or fellow students. They can see what you have written from another point of view and help you clarify your organization. The questions listed under self-review can also be used for peer review.

Small groups of three to four people are best for peer review. If possible, submit your outline to them to read in advance. During the review session, listen carefully to suggestions you receive and take notes, resisting the urge to interrupt the reviewers to defend yourself. Later you can adapt your organization to include those suggestions that improve what you have written.

While you may find the idea of peer review intimidating at first, you'll soon see how helpful it can be. Because fellow students will be unfamiliar with the material, they can tell you if your organization makes sense. They can spot holes in your information and sometimes suggest alternate methods of organization. Students often tell me that their review sessions were among the most valuable parts of the class. After all, review and reorganization of your outline can save you many hours of time in the actual writing of your draft—which is the next step in the process.

Review of the Work of Multiple Authors

When you work on a large and complex project, you may be only one of several authors. Usually each author is assigned a piece of the writing pie and is responsible for organizing that piece. The lead writer may assign the sections and also rework the various methods of organization so they are compatible. At the review stage, you may be meeting with the other authors to work out an overall organization, or you may be meeting with the lead writer to see how your contribution meshes with the rest.

    Just as you can't buy a single pie that contains one slice of cherry, one of strawberry, one of chocolate, one of pecan, and one of coconut cream (interesting as that might be), so you can't have a single document that contains six radically different methods of organization. With multiple authors, everybody has to compromise, and this review stage is a good time to alter your organization as needed.

Instructor or Manager Review

Instructors or managers will approach your outline from a slightly different perspective than will your peers. With their depth of experience and their vision of a broader picture (such as how your contribution fits into the whole project), instructors or managers can offer sound advice about organization before you begin to write. And remember that the more help you can get at this point, the easier the writing will be and the more likely it is to be successful.

CHECKLIST FOR REVIEWING THE ORGANIZATION

Should I use
_____ self-review?
_____ peer review?
_____ multiple author review?
_____ instructor review?

My review questions should include the following:

1. What is the writing purpose?
2. What is the single main point?
3. Do the subpoints support that main point? Are any subpoints missing?
4. How many subpoints are there? Are they equal in importance? Should some subpoints be combined and others be separated?

5. What is the main method of organization? What is the method of organization of the subpoints? Is this the right organization to choose? If not, how should the order be rearranged?
6. Do the details support each subpoint?
7. Is the numbering system sound? Does each division have at least two subpoints?
8. Is the writing task fulfilled?

EXERCISE 4-1 Organization Review: Self-Review

Directions: After letting your outline for Exercise 3-10, 3-11, or 3-13 cool, review it yourself. Answer each question in the checklist in writing, commenting on problems. Then revise your outline as needed.

EXERCISE 4-2 Organization Review: One-on-One

Directions: Pair up with another student and exchange Exercise 3-10, 3-11, or 3-13. Review the outline using the checklist, and add questions you would like the writer to answer. Write your comments on the outline and submit to your instructor or to the writer, as directed.

EXERCISE 4-3 Organization Review: Group Review

Directions: Join a group of three to four fellow students. Supply each member of the group with a copy of the outline for your project (Writing Assignment #4). As you review each outline, answer the questions on the checklist. Write your comments on the outline. Submit your review pages to the instructor or to the writer, as directed.

EXERCISE 4-4 Organization Review: Instructor Review

Directions: Submit the revised outline for your term project to your instructor for review. Attach a page with three or four questions about your organization that you'd like the instructor to address.

CHAPTER 5

Writing the Draft

Once you have reviewed and revised your outline, you are ready to plunge into writing the first draft. For many writers, staring at that first blank page is as frightening as standing atop the high dive and staring at the blue water far below. What's more, starting the actual writing can be as difficult for professionals as it is for students. In one study of an engineering group, for example, Paradis, Dobrin, and Miller (1983) learned that of the total time spent on writing, 14 percent was spent procrastinating!

Fortunately, though, writing does not require the same kind of all-or-nothing commitment as does stepping off the high dive. Writers usually think of the main part of a document as containing an introduction, a body, and a conclusion, and almost every piece of technical writing is composed of separate sections within those larger parts. These parts need

108

not always be written in the order they will appear in the final document. This means that you can ease into the writing process much as you might ease into chilly water with your toes first. For example, you might want to begin with the easiest sections of the body first: those for which you have the most information or those with a sequential method of organization. Often, the definitions and descriptions, because they are so specific, can be written most easily. Then you can go on to more difficult sections because your outline has given you the total picture and will help hold the separate pieces together.

This chapter is divided into four activities that you can carry out in writing the draft of a technical document, using the activities listed on the preceding page.

Are you surprised that preparing an introduction comes last in the list? If you are writing a short memo or letter, you might want to change the sequence and write the introduction first. For a long or complicated document, though, you may find it easier to construct the body first, write out your conclusions next, and then drop back to write the introduction when you already know where you are going and how you are getting there.

5.1 PREPARING TO WRITE

Revise and Follow Planning Sheets and Schedules

In Chapter 1 you learned the details of planning a writing project, and I encouraged you to fill out a Planning Sheet like the one in section 1.6 for each writing project. As you complete short writing assignments, you can see how much time they take and how accurate your estimates on the Planning Sheet are. If you are involved in a long-term writing project, you need to review your original Planning Sheet before you write the draft to see if you are on schedule. If you are, congratulate yourself for good planning. If you have fallen behind, analyze the reasons. Remember that your original time estimates were just that—guesses—and maybe you needed more time than you thought. On the other hand, perhaps other assignments intruded on your schedule. In any case, think through the current status of your project and revise your schedule as needed. This revision will help you in two ways: (1) it will give you up-to-date information for any status or progress reports you have to write, and (2) it will prepare you psychologically to begin writing.

Now is also a good time to review the other items on your Planning Sheet and revise those that may have changed. Pay special attention to the purpose you checked on the Planning Sheet, your written purpose statement, and the main point you wrote as you organized. Make sure these are still your purpose and main point because you'll need them to write a thesis statement.

Keep a Time Log and a Document Log

Planning practical time allotments is difficult for writers, but you can make that planning easier on the next project if you keep track of your writing time on this one. The best way is to log your time as you work. Many professionals keep a log on a desk calendar—marking at the end of each day the number of hours or minutes spent on a particular project or blocking off time segments. Entries can be as simple as those in Figure 5-1. Other writers keep logs in an engineering or laboratory notebook or on a computer, using software programs designed for this purpose. Logs like these can help you charge for your time as well as help with future planning.

You also need to log your information sources for later documentation. When I talk to students at midpoints in the writing process, I often see them carrying notebooks stuffed with odd-sized scraps of paper and leaking information sources. Such students usually can't find what they are looking for, and they often forget to write down their exact sources. Unfortunately, at the typing stage these students must backtrack to figure out where specific points of information came from.

You can avoid this common problem by planning now how you will keep track of source material. You can do this most easily by using your stack of 3 × 5 index cards with bibliographic and identification information on each card. (Let's hope that you *did* put each source of information on a separate card, and that you put down complete information—author, title, volume number, date, pages, publisher, city, and so on; see Chapter 2.) One way to keep track of information as you write is to alphabetize those cards by author's last name, then number the cards consecutively. Then when you are writing and want to document information, you can simply note in your draft either the page number of the source and the card number or the author's name, page number, and year of publication. These two methods work well if you plan to use the *author-date* or *author-page* documentation system (see Chapter 13). Another method is to number the index cards in the order you use them in writing your draft. Then when you want to document information, you write down those consecutive numbers. This method works well with the *number* system of documentation. Of course, if you change the order of documentation when you revise, you will also need to change the numbering. Your instructor may specify a documentation method or you may choose one now. See Chapter 13 for specifics on documentation.

Learn to Write in Segments

A segment is one of the parts into which something naturally separates or is divided. Once you've written your thesis statement (see section 5.2), you can take advantage of the segmental nature of technical writing by beginning to write with the segment that comes easiest to you. That segment

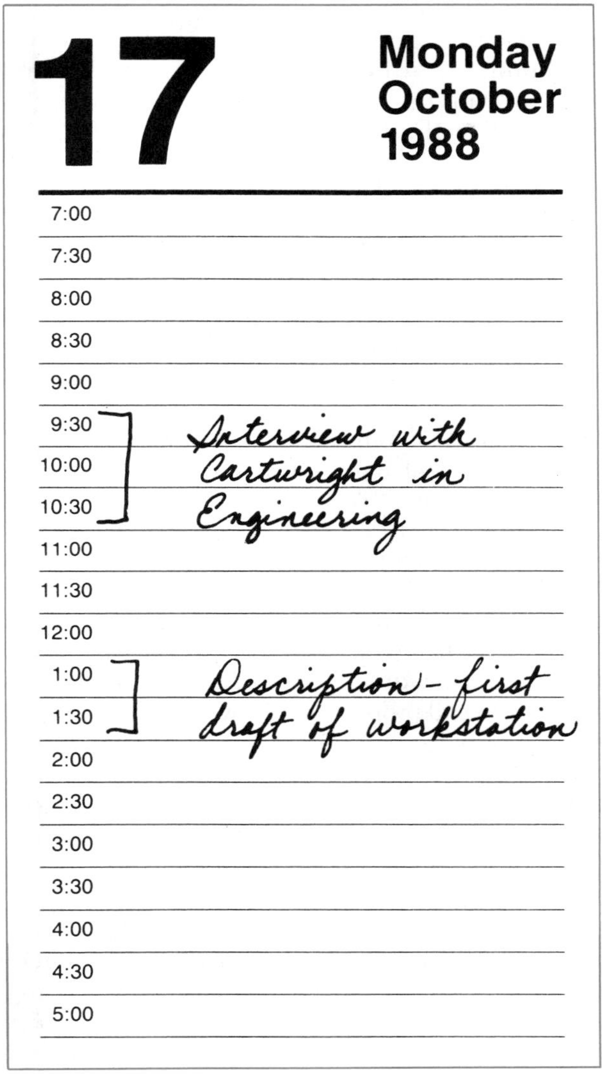

17	**Monday October 1988**

7:00	
7:30	
8:00	
8:30	
9:00	
9:30	*Interview with*
10:00	*Cartwright in*
10:30	*Engineering*
11:00	
11:30	
12:00	
1:00	*Description – first*
1:30	*draft of workstation*
2:00	
2:30	
3:00	
3:30	
4:00	
4:30	
5:00	

Figure 5-1   **Planning Log**

may or may not be the introduction. The introduction should provide an overview, and it's hard to have the perspective for an overview until you've written and assembled all the segments. But no law says that just because the introduction is read first, it must be written first.

Think about the segments of an orange. Each orange segment contains what is essential to making up the orange, but a whole orange consists of eight to ten segments put together and then covered with skin. Any piece of technical writing is something like an orange because it has independent yet related parts. Those parts may be paragraphs in a letter or memo, but often they are divisions of a longer piece, each with its own subheading. Put together and unified with a thesis statement and introduction, those individual parts make up the larger whole. Together, the thesis statement and the outline are like the skin of the orange; they hold the segments together.

To make writing easier, consider the following strategies:

1. Choose a comfortable place and time of day to begin writing. The place should be free of distractions and have a surface big enough so you can spread out your notes and papers. The time should give you about two uninterrupted hours. Many people write most easily in the morning when they are fresh; others are more alert in the evening. Choose the time that's best for you.

2. Whether you write in longhand, on a typewriter, or on a computer, begin with the segment you know best and are most comfortable with. Write fairly quickly, concentrating on ideas and content. Don't worry about spelling, punctuation, or perfect expression; just leave plenty of empty space on the page. Remember that this is a draft; later you can revise, add, or subtract, but now you want to get ideas on paper.

3. Write everything you have to say about that particular topic. You may feel that you're writing too much, but you can always trim later. Remember how much more quickly a reader can read than you can write. An average person can read more than 250 words a minute; a hard-working writer may only write between 250 and 500 words in 30 minutes!

4. At the end of about two hours, or when you finish a section, stop writing. Most people can only concentrate productively for two hours or so and then begin to tire. When you stop, spend a few minutes thinking about what you'll write next, and jot down a word or two to jog your memory when you begin again. Rereading what you have written is also a good way to get back into the writing mood.

5. Different kinds of writing may require slightly different strategies:
   - In a *development project* in engineering, your writing reports what you learned during the project. These drafts are best written incrementally (that is, in small segments while the project is still going on). You can write each section quickly at the end of a project phase, and by the project's end, you will have the paper already written in draft form. See section 14.3 on informal reports.
   - In a *research paper* or *lab report*, the paper's structure is more rigid: introduction, materials and methods, results, and discussion. Here you might want to begin with the introduction in order to state the problem first, and then continue with the easiest section, which is probably materials and methods. Chapter 22 explains the format and organization of research reports; Chapter 14 discusses lab reports.

- In a set of *instructions* or *procedures,* the method of organization is sequential, so you can begin with the first step and then proceed in order. However, you can put off writing any overview statements or introductions until after you have written out all the steps. See Chapter 17 for instructions and Chapter 18 for procedures.
- In a *research, feasibility,* or *comparative project,* the writing itself may be a learning process—leading you to new ideas and to insights about the data. In this case, once you have a purpose statement and an outline, you can fairly easily launch into the body of the paper; then write conclusions and recommendations; and then write the introduction and abstract or summary. Chapter 21 covers feasibility and comparative reports and Chapter 22, research reports.

CHECKLIST FOR PREPARING TO WRITE THE DRAFT

1. Have I reviewed the Planning Sheet and revised it if necessary?
2. Have I established a time log for a long report?
3. What procedure shall I follow to keep track of documentation?
4. What segment will be easiest for me to begin writing?

EXERCISE 5-1 Setting Up a Plan for Writing

Directions: Study your Planning Sheet and revise it as necessary. Then study the major points from your outline to determine where you will begin writing. List those points on the Planning Sheet in order from easiest to hardest.

5.2 CONSTRUCTING THE BODY

The main part of a piece of technical writing—whether it is a one-page letter or a 45-page proposal—has three sections: the introduction, the body, and the conclusion. But unlike a typical essay or short story, these parts are not always written in that order. In fact, in a long report or proposal, they may not even appear in that order: the conclusion may come first, then the introduction, and then the body.

Thus, when you begin to write, you may want to begin with the section that contains specific details: the body. The controlling factor in the body is the thesis sentence, which you'll want to write first.

Combine the Purpose and Major Point into a Thesis

Early in your planning you determined the purpose of your specific writing assignment. That purpose may have been to explain how something works, to evaluate the feasibility of a new procedure, or to propose a specific course of action. Whatever the purpose, I encouraged you to write it down so that it was clear to you and would be clear to your readers. When you began to organize, you decided on the one major point you wanted to make. Once you wrote down that single major point, it became a reference for organizing the rest of your material. Each subpoint in your outline contributed to moving you and your reader toward recognition and acceptance of the one major point.

For example, suppose that in the 10-speed bicycle exercise (3-3) you decided that your main purpose was *to describe a 10-speed bicycle*. A secondary purpose was *to persuade the reader to buy a Diamond bicycle*. Then, as you organized in section 3.2, you decided that your one major point was, "A 10-speed bicycle has four major parts: the frame, the drivetrain, the wheels, and the brakes." Now, as you begin to write, you need to combine the purpose and major point into a thesis statement. What will it say? Before you can answer that question, you need to think about what a thesis is.

Definition of a thesis   A *thesis* is the main idea you want your reader to reach because it provides the answer to the problem or question being investigated. In technical writing you can have one of two different kinds of thesis:

1. If you are simply presenting information or explaining a complex idea, the thesis is a *statement of content*. The 10-speed bicycle description requires this first kind of thesis. Since your primary purpose is to describe, you are presenting information; thus, your one major point can also be your thesis: "A 10-speed bicycle has four major parts: the frame, the drivetrain, the wheels, and the brakes."

2. If you are analyzing data and drawing conclusions, the thesis is a *proposition to be defended*. This kind of thesis statement is more common. Here the writer adopts a specific view, defends that view by logical argument, and offers a valid solution. The following section describes these two kinds of thesis statements more fully and gives advice about how to write them.

Ideally, both kinds of thesis

- tell the purpose
- announce the topic and limit its scope
- make the main point
- map the order of supporting statements (or tell how many there are)

In addition, the thesis that is a proposition to be defended

- indicates the writer's point of view—often, in fact, taking a strong and persuasive stand

You should always write a thesis statement for yourself because it will guide and focus your writing. Also, most of the time you'll want to state the thesis within your document as an aid to the reader's understanding. Occasionally, however, the thesis will be implied but not directly stated.

How to write a thesis statement Typically, you will write a thesis that is a *statement of content* when your purpose is to define, describe an object or process, instruct, or present narrative or objective information. As the writer, you do not take a strong stand, and your tone is objective. For example:

If your purpose is	*And your one major point is*	*Your thesis sentence can be*
to define a cave	A cave is a natural opening in the ground that extends beyond the zone of light and is large enough to permit human entry.	the same as your one major point, with—perhaps— an elaboration to indicate the order of supporting explanations
to describe how a laser works	The basic process of laser action involves four steps: excitation, spontaneous emission, stimulated emission, amplification.	the same as your one major point, since the four steps map the order of the supporting statements
to report how suspect Jones was arrested (a narration of events)	Suspect Jones was arrested by Reporting Officer Smith on May 14, 1989, in response to observed driving behavior, physical indications, and results of a field sobriety test.	the same as the one major point, which outlines the sequence of events

You write a thesis that is a *proposition to be defended* when your purpose is to interpret, evaluate, judge, recommend, propose, sell, or persuade. As the writer, you do take a stand, and you construct your thesis to spell out both your point of view and the support for your stand. For example:

If your purpose is	And your main point is	The thesis sentence might read
to propose a remote sensing survey of Portsmouth Harbor, NJ	My company should win the survey contract.	Allied Marine Survey is best equipped to conduct a remote sensing survey of Portsmouth Harbor for sunken vessels because of its previous success with magnetrometer and sidescan sonar, its recent acquisition of improved sensing equipment, and its experienced staff.

(Note that the thesis implies the main point, clearly states the writer's point of view, and supports that point of view with details. This kind of thesis could appear in an introduction and/or in a conclusion.)

to evaluate the effects of stabilizer buildup on operating costs of 20–30,000 gallon swimming pools	The stabilized form of chlorine in trichloro-s-triazinetrione is more costly to use over the long run.	The stabilized form of chlorine in trichloro-s-triazinetrione, while initially appealing because the chlorine remains in the water longer, is costlier in the long run because the pool water must be partially drained at intervals to eliminate the stabilizer concentrations.

(Note that the thesis in this case *assumes* the purpose, clearly states the main point and the writer's point of view, and then outlines the support for that point of view. Such a thesis might go in the conclusion section, though the conclusions will probably be presented at the beginning of the report.)

| to recommend purchase of a four-track cassette recorder for a home recording studio | The Tascam 244 is the best recorder of the four evaluated. | The Tascam 244 four-track cassette recorder is recommended for use in a home recording studio because it best satisfies the six evaluation criteria and combines all the features necessary for recording, processing, and mixing on a high-quality four-track cassette tape. |

(Note that the thesis goes far beyond the main point by clarifying the topic and by setting out main supporting areas: the evaluation criteria and the features. Such a thesis would probably appear in the recommendation section, which might well appear at the beginning of the report.)

In summary, then, the first kind of thesis, the *statement of content,* often appears in the introduction. The second kind of thesis, *the proposition to be defended,* often appears in the conclusions or recommendations. For the second kind of thesis, you might also include a separate purpose statement in the introduction.

What is the most important as you begin to write is that you know where you're going. Writing a clear thesis statement will help you construct meaningful paragraphs.

CHECKLIST FOR WRITING A THESIS

1. What is my purpose? What is my major point?
2. Should my thesis be
   _____ a statement of content?
   _____ a proposition to be defended?

3. When written down, does my thesis
   _____ tell the purpose?
   _____ announce the topic and limit its scope?
   _____ make the main point?
   _____ map the order of supporting statements?
4. If it is a proposition to be defended, does my thesis
   _____ indicate my point of view as the writer?

EXERCISE 5-2   Writing a Thesis about Cafeteria Chairs

Directions: Suppose you are a member of a student committee to advise your college on the purchase of new chairs for the student cafeteria (Exercises 3-1, 3-4, and 3-7). In this case you determined what you wanted from a chair, studied the available chairs, and are ready to recommend purchase of one brand. Your thesis will be a combination of purpose and one major point. The purpose is to recommend the purchase of one brand. Meet with your small group again and, based on your findings in Exercise 3-4, list your one main point. Then together write a thesis statement that combines the purpose and main point. Be prepared to discuss the kind of thesis you chose and its construction.

EXERCISE 5-3   Writing a Clear Thesis Statement

Directions: Choose one of the following:

1. If you conducted a research study or a comparative product study, combine your purpose statement and one major point into a thesis. Write the thesis in one sentence and submit it for review if requested. Use it to guide your writing of the report.
2. If you conducted an interview for a separate paper, combine your purpose statement and your one major point into a thesis. Write the thesis in one sentence and submit it for review if requested. Use it to guide your writing of the interview report.
3. If you worked on a collaborative feasibility study, meet together to write your thesis based on your purpose statement and your one major point. Use it to guide all group members in writing their individual portions of the report.

Construct Meaningful Paragraphs

With a good thesis sentence, you have a controlling focus for your writing and are ready to write the paragraphs that will support that thesis. Fortunately, you also have a well-organized outline to help you construct those paragraphs. Paragraphs are the building blocks that make up any major point. Most paragraphs contain a topic sentence and support sentences. Before you begin writing, refresh your memory about those terms.

Definition of a paragraph A *paragraph* is a group of related sentences, complete in itself but also usually part of a larger whole. Paragraphs are set off as independent items with white space: an extra line of white space in block form or an indented first line. You can write good paragraphs if you remember three things:

1. A paragraph should contain only one central idea, which will be stated in a topic sentence.
2. The rest of the sentences should support or explain the topic sentence.
3. Each paragraph should be logically related to the paragraphs before and after it.

Definition of a topic sentence A *topic sentence* contains the main point of a paragraph. For example:

> Computer Output Microfilm (COM) recorders can use either of two methods of film development: wet processing or dry processing.

With this topic sentence, what do you expect the paragraph to say? You expect it to discuss the two methods of film development—either explaining each one or comparing advantages and disadvantages of each one. Just as the thesis statement controls the whole document, so the topic sentence controls the paragraph. Technical readers usually must read quickly, and they depend on the topic sentence to tell them the main point. This in turn tells them whether or not they need to read the rest of the paragraph carefully. For this reason, most paragraphs in technical material place the topic sentence at the beginning—as the first or second sentence.

Support sentences Once you have stated the main point in the topic sentence, you should support that idea in the rest of the paragraph. Support can come through giving examples or statistics, through defining or describing, through comparing or contrasting, or through logical argument. Just as a document or section of a document should have a method of organization, so should each paragraph. Methods of organization are discussed in detail in section 3.2; you might want to review that material before you begin to write.

The important thing to remember is that support sentences should be specific, related to one another and the topic sentence, and sensibly organized. Look at the way this student writer (Clayton 1985) supported the topic sentence on film development:

> Computer Output Microfilm (COM) recorders can use either of two methods of film development: wet processing or dry processing. Wet processing involves printing the character images onto a silver-based film, which is then passed through five to nine chemical tanks. These tanks develop the film and wash off the chemical residues. Dry processing involves

printing the character images onto a heat-sensitive film that needs no development and is immediately available for viewing. The advantage of wet processing is that the film can be stored for a long period of time in dark or light areas, whereas dry-processed film can be stored for a long time only if kept in a dark place. This can be inconvenient. On the other hand, dry processing requires no additional supplies—only the film. Wet processing requires chemicals, filters, and developing supplies in addition to the film; therefore, wet processing is more expensive in terms of cost of supplies.

Note that the paragraph first defines wet and dry processing, then gives the advantages and disadvantages of each process—and both explanations support the topic sentence.

Paragraph length   How long should a paragraph be? To answer that question, you need to keep your reader's short-term memory in mind. Short-term memory is the number of things you can hold in your mind at one time in order to process the information. The magic number is seven plus or minus two items. Thus, if your topic sentence is one point, and the reader also has to keep in memory the general point of the previous paragraph, you are left with three to seven items of information you can add. Typically, each item of information is in a separate sentence, so the number of information items will dictate the paragraph length.

In shorter documents like letters and memos, paragraphs can also be shorter. But don't make a paragraph too short; a series of very short three-line paragraphs is hard to read because the topic becomes fragmented and the continuity is lost. With many very short paragraphs, the reader will try to keep them all in short-term memory in order to see the relationships. Remember too that most of your paragraphs will naturally set their own limits; the sense of the paragraph, the complexity of the document, and the intended reader's background all control how long a paragraph should be.

Paragraph indentations, or the white space setting them off, give the reader's eyes and brain a chance to relax by subtly saying, "We're changing subjects here." As a writer you can also supply eye and brain relief by using headings, lists, and graphics to break up the solid blocks of print on a page. Chapter 6 tells you more about headings, and Chapter 12 gives detailed information on graphics.

Write Straightforward Sentences

When you are writing a draft, it's more important to get your ideas on paper than it is to perfect each sentence. Later you can revise for clarity. If you need either help in revising or practice in writing good sentences, you'll find many exercises in Chapter 8 and in the Punctuation section of the Handbook Appendix. At the draft writing stage, simply keep these things in mind:

- Help the reader understand in every way you can. After all, you are the expert; your reader depends on you to explain and clarify.
- Keep your purpose and your thesis clearly in mind. If you aim at the purpose and thesis as you write, your writing will be focused and clear.
- Write as simply and clearly as you can. Don't use fancy words just to impress the reader, and define any terms that your reader might not know.
- Get at the writing; don't put it off. You will learn something from the act of writing itself. E. B. White (quoted in Elledge 1984, 233) said, "I always write a thing first and think about it afterward, which is not a bad procedure, because the easiest way to have consecutive thoughts is to start putting them down." You have already thought about your subject for a long time. Now you need to see those consecutive thoughts on paper.

CHECKLIST FOR WRITING PARAGRAPHS AND SENTENCES

1. Are my paragraphs self-contained units making one major point?
2. Do my content paragraphs each have a topic sentence?
3. Do the rest of the sentences support that topic sentence?
4. Are the support sentences organized in a sensible way?

EXERCISE 5-4 Finding and Evaluating Paragraphs

Directions: Search textbooks, technical reports, and magazine or journal articles to find two examples of well-written paragraphs on technical subjects. Photocopy the examples or clip and attach them to a sheet of paper. After each paragraph, write a short evaluation, telling what's good about the paragraph. If any improvements are needed, indicate them. Be prepared to discuss your findings in class.

EXERCISE 5-5 Evaluating Your Own Paragraphs

Directions: Choose two paragraphs at random from a paper you are working on. Bring the paragraphs to class and exchange with another student. Analyze your fellow student's paragraphs to see if each one has

1. a topic sentence
2. clear supporting sentences
3. those sentences organized to make the writer's point

Write comments on the paper and submit it to your instructor if requested.

5.3 WRITING A CONCLUSION

Once you have written the central portion—or body—of a document, you have prepared yourself and your reader to reach some conclusions based on that information. You may think of technical writing as totally non-judgmental and objective, but remember that usually you are writing from your knowledge and experience to readers who know less about the subject than you do. Therefore, readers look to you for (1) factual knowledge (in the body), (2) analysis and interpretration of that knowledge (in a conclusions section), and (3) recommendations based on your interpretation (either in the conclusions or in a separate recommendations section). In other words, you often must not only take a stand, but recommend action based on that stand.

Be aware that writers use the word *conclusion* in two ways: (1) as the final section of the main part, or body, of a document, and (2) as the results section of a report, which includes logical interpretations and inferences drawn from the data in the body. In short documents like letters, memos, and informal reports, the logical conclusions may go in the final section. But in long reports and proposals, they are more often placed at the beginning.

Definition of Conclusions

When you write the body, you provide facts and even analysis, and the reader is free to interpret those facts, form judgments, and reach conclusions. But most readers want that job done for them; they expect the writer (who after all has been working on this) to weigh and evaluate the facts and the results of any analysis. Readers want to know what is important, what the relationships are, what the results are, and what their significance is. Such interpretation usually takes place in the *conclusions* section of a technical report or the *discussion* section of a scientific paper.

For example, to arrive at conclusions about awarding a contract, the writer might compare the submitted proposals on such established criteria as experience with similar projects, time estimates for completion, expertise of project directors, and projected costs. Conclusions must often consider many requirements—and cost, though important, is not always the deciding factor.

Such conclusions might look like this:

Four proposals were submitted for the remote sensing survey of Portsmouth Harbor, by

Allied Marine Survey
Excel Corporation
PMS Co.
Smith Oceanic Survey

Of these four, Allied Marine presents the best record of experience, especially with magnetometer and sidescan sonar. PMS and Excel Corp. have recently added new staff members, but their experience is in shallow water and therefore only partially applicable. All four proposals accept the three-month time limit, with October 1, 1989, as the completion date. Project costs by each company are

Allied Marine Survey	$67,000
Excel Corp.	72,000
PMS Co.	65,000
Smith Oceanic	67,500

Notice that while this conclusion is a summary that pulls together the key facts of the report, it also evaluates those facts and begins to guide the reader's thinking. This kind of conclusion is especially appropriate with a thesis of evaluation or judgment.

Another conclusion might be a simple summary of the key facts. See Figures 18-1 and 18-2 for examples.

Definition of Recommendations

In many documents the writer will move beyond interpretation of evidence to make an actual recommendation, telling readers what they should believe or what action they should take. Recommendations answer the question, "What should I do about it?" A recommendation is also an opinion, but because it is based on fact and arrived at through reasoning, it is an informed judgment—much more valid than an unsupported emotional claim.

Recommendations can be either in a separate section or combined with conclusions. Sometimes recommendations appear first to be supported by the conclusions; at other times the conclusions come first and lead into the recommendations. The method you choose will be determined by your purpose (what's the main thing you are trying to do?) and your reader (what's most important to that reader?).

In the Portsmouth Harbor study, the recommendation might look like this:

Allied Marine Survey should win the contract for the remote sensing survey of Portsmouth Harbor. It has the best record of experience in deep water, and its bid is second lowest of the four bids at $67,000.

Requirements for Concluding and Recommending

If you are going to be successful in persuading readers to believe or act as you wish, you must meet these four requirements:

1. Provide a solid basis of fact in the body, including statistics, examples, evidence, and expert opinion.

2. Use logical reasoning to evaluate the evidence. Often you must establish criteria (standards) as a basis for judgment, and then test alternative products or courses of action against those criteria. To help set up criteria, you can ask questions like these:

Will it work?
Can it be done?
Will it yield the results I want?
How long will it take?
Can I (or the company) afford it?
Is this product or way of doing it better than others?

3. Be fair. To be fair, you must present *all* evidence, even that which promotes the other side. If you suppress alternatives, your readers won't trust your judgment. In the evidence weighing process, therefore, you must consider *in writing* the implications of alternatives.

4. Use a positive, confident tone (see section 1.4 for more on tone). You cannot hedge or be half-hearted in your judgments, because readers won't trust that either. Sticking your neck out with a firm decision may be difficult, but you should do it. Remember, though, that this is your *informed* judgment, so your tone should be serious and unemotional as well as positive and confident.

Following is the Conclusions and Recommendations section of a student report (Jarrett 1984) that analyzed stethoscopes in order to find the one that best met these requirements:

1. was priced at $50 or less
2. included two chest pieces, a bell, and a diaphragm
3. had screw-out eartips and optional, flexible eartips
4. was lightweight
5. had metal side pieces (binaurals) that could be adjusted to the angle of the ear canal
6. had a non-chill bell

Conclusions and Recommendations

Based on this study, the Deluxe-Dual Head Stethoscope by Marshall is the best purchase. This stethoscope received high marks in all six categories.

The Lightweight Stethoscope by Littman is rated as the second choice. Although this stethoscope rated high in four categories, it does not offer the optional flexible eartips. However, if the consumer is able to purchase these separately, this stethoscope would be a good choice.

The Double Head Color Coded Nurse Stethoscope by Ritter-Tycos is rated third. The cost of this stethoscope exceeds the $50 limit, and the optional, flexible eartips are not offered. However, if the consumer wishes to pay the cost of the stethoscope and to purchase the flexible eartips separately, this stethoscope is also a good choice.

The Sprague-Rappaport Stethoscope by JAC is rated as the fourth and last choice. The chestpieces on this stethoscope are heavy, resulting in an overall heavy instrument. The bell does not have a non-chill ring, which is important for patient comfort. This stethoscope is not recommended for purchase.

After reading these statements, would you be willing to accept the author's conclusions and recommendations? Do they seem to be based on fact? Does she explain why she made the choices she did? In this case, the conclusions and recommendations are combined, but sometimes they will appear in separate sections.

Placement of Conclusions and Recommendations

Where do you place conclusions and recommendations? The answer depends on your purpose and intended readers as well as on where the document will appear. In a journal article, for example, conclusions and recommendations usually appear at the end, after the body, but in a report for industry, they are often better placed before the introduction, where busy managers, technicians, or specialists can easily find them. Exact placement in a document depends on the primary reader. Remember that not all documents require recommendations, and a few—like instructions—do not even require conclusions. For placement of conclusions and recommendations in specific forms of technical writing, see Chapters 14 to 22 in Part 3.

CHECKLIST FOR WRITING CONCLUSIONS AND RECOMMENDATIONS

1. Are my conclusions supported in the body of the document?
2. Are the conclusions fair?
3. Is the tone appropriate?
4. Are any recommendations based on the conclusions?

EXERCISE 5-6   Evaluating Conclusions and Recommendations

Directions: Study the following purpose statements and conclusion and recommendation sections from student reports. Decide whether the conclusions fulfill the stated purpose and, if not, where the problems are. Be prepared to discuss your analysis in class.

EXAMPLE 1: Multivitamins

Purpose
This report on multivitamins will inform the reader about the need for adequate vitamin intake in nutrition and the causes of inadequate nutrition. It will also illustrate the functions of each vitamin by its letter, its sources,

and its synonyms. The reader will walk away with a more complete knowledge of vitamins and their necessity in the diet.

Conclusions and Recommendations

The Myadec multivitamin is highest in milligrams per vitamin. I recommend this one in comparison to Centrum and Theragram-M. Vitamins are as important for children as for adults. Individuals have a greater requirement for vitamins during growth, during periods of hard work, during pregnancy, lactation, and menstruation. Conditions accompanied by fever and stress also increase the body's requirement for vitamins. Older people, whose food intake may be less, may need vitamin supplements to maintain good health.

Know the manufacturer whose vitamins you purchase. Without the assurance of high quality-control standards during manufacturing, vitamin potency may be worth less than what appears on the label. Seasons and temperatures have little effect on the body's need for nutrients. Excessive consumption of vitamins, however, has not proven to have any particular benefit. Be sure that you are taking the amount of vitamin supplement that is right for you— winter or summer. Ask your doctor.

A careful selection of foods within a properly balanced diet and the proper cooking of foods should place all the vitamins at your command. Unfortunately, many people eat "favorite" foods, snack foods, or imitation foods to excess. These people may need a vitamin supplement.

EXAMPLE 2: Table Saws

Purpose

This is a comparative study on non-commercial 10-inch table saws. The different features of four saws are discussed in terms of quality and performance for use in a small woodworking shop.

Conclusions

The Shopsmith Mark Five is the best of the four saws compared in this report. The Shopsmith Mark Five is five tools in one and uses only one motor, one stand, and one work table. It does everything that five individual tools do, and has been engineered with accuracy and precision. The Shopsmith Mark Five is not only a 10-inch table saw, but it also includes:

1. a vertical drill press
2. a horizontal boring machine
3. a 34-inch lathe
4. a 12-inch disc sander

These are all standard items when purchasing the single unit.

The next best saw is the Powermatic Model 66. This saw is excellent for both non-commercial and light commercial use. The DeWalt Model 7756 and Rockwell Model 34-710 lack the necessary features to produce high-quality wood products.

Recommendations
1. The Shopsmith Mark Five is the best choice for our needs and preferences.
2. The Powermatic Model 66 is perfect for light commercial use because of its construction.
3. The DeWalt Model 7756 and Rockwell 34-710 are not recommended because they are not as well made as the other brands.
    The unit price for these saws can be found in Table 1.6 (Zertuche 1982).

EXERCISE 5-7   Writing Conclusions and Recommendations for a Formal Report

Directions: When you have finished the body portion of your report, write conclusions and recommendations that are supported by the facts and analyses in the body. Make sure that the conclusions and recommendations are understandable even to a person who has not yet read the report itself, and that they are justified. Submit the conclusions and recommendations for review.

5.4   PREPARING AN INTRODUCTION

You might want to put off writing the introduction of a long paper until you have finished writing the body and the conclusions and recommendations. When the paper itself is written, you will have a good idea what to include in your introduction.

In a way, an introduction is an overview because it tells readers where they are going, the word *introduction* coming from two Latin words meaning "to lead inside." Short papers generally do not have a separate introduction, but the opening paragraph will include introductory material such as the purpose and topic. In a very short paper, purpose and topic can be combined in a single thesis sentence. Introductions to longer papers should include

- a full statement of the topic
- the purpose (reasons for conducting the study) }
- the scope (how much is covered)   } Often these are combined in a thesis sentence.
- the plan (method of development) }
- definitions of key terms (with equivalent acronyms or abbreviations)
- a brief history or explanation of theory
- a description of the intended reader
- documentation of source material

In a very long or complex document, you might also include mini-introductions—that is, smaller introductions to individual sections. In this book, for example, each chapter has a mini-introduction, as do many major sections within chapters.

The following introduction is from a student report that compares surgical lasers for possible purchase by a hospital (Leonard 1985). Study the way the writer has organized the introduction and note the information that is included.

> This report is an evaluation of state of the art Nd:YAG (neodymium: yttrium aluminum garnet) surgical lasers suitable for Allied Hospital and available in 1985. Allied Hospital is a private, metropolitan hospital, which presently owns a mobile carbon dioxide gas laser and a fixed argon laser.
>
> In this report four surgical lasers are compared on the basis of five criteria:
>
> versatility
> cost
> safety
> mobility
> ease of operation
>
> I do not intend for this report to be the final word on available models. Rather, I hope to provide useful information and an objective recommendation that will be a valuable resource in final purchasing decisions by Allied Hospital staff.
>
> I encourage hospital staff to inquire professional opinions of the specific models from the physicians and facilities listed in the appendix. I consider this information vital to a purchasing decision, but, understandably, it is not openly distributed to the public.
>
> The report is directed to the reader who may have medical or other science-related experience but who is not an expert in the field of laser surgery. It is intended to aid Allied Hospital specifically, but much of the information contained could be useful to other facilities and to anyone with an interest in lasers and their use in surgery. For the convenience of the reader, terms defined in the Glossary, page 56, appear in capital type the first time they are used in a section of the report.
>
> The report includes four major sections:
>
> - an explanation of what surgical lasers are and how they work
> - a description of the criteria used to choose a suitable laser and how those criteria are applied
> - a technical description of the four models evaluated
> - a recommendation for Allied Hospital.

In summary, an introduction should give a broad overview of the subject without going into detail. As a technical writer you need to review your Planning Sheet before you write the introduction so that you are sensitive to the needs and knowledge of your primary and secondary readers.

CHECKLIST FOR WRITING INTRODUCTIONS

Does my introduction include

_____ a full statement of the topic?

_____ the purpose?

_____ the scope?

_____ the plan?

_____ definitions of key terms (with equivalent acronyms or abbreviations)?

_____ a brief history or explanation of theory?

_____ a description of the intended reader?

_____ documentation of source material?

EXERCISE 5-8   Evaluating Introductions

Directions: The following introductions are from two student reports that analyze and compare consumer products. Evaluate each one against the checklist to see if it is complete, and write a brief evaluation based on your analysis. Be prepared to discuss your evaluation in class.

INTRODUCTION TO PAPER #1 (Vargas 1984)

The most recent development in audio equipment is the compact digital disc player, or CDP. Co-developed by Sony and the Phillips Corporation, the CDP uses a laser beam to "read" musical information off a microscopically encoded plastic disc. The laser then transmits the digital binary code of computers through a converter, changing it into the standard audio signal which is heard through speakers.

High Fidelity Magazine calls the compact disc player "the most fundamental change in audio technology in more than eighty years." *Time Magazine* claims ". . . the difference between digital and analog sound is, in its own way, as striking as the distinction between mono and stereo." Since the release of the first CDP in the United States in October of 1982, many audio equipment manufacturers and record companies are entering the compact disc market.

Because of its relative newness, objective information is scarce on CDPs. Several audio magazines, such as *Stereo Review, High Fidelity,* and *Audio,* have laboratory tested some models, comparing specifications and features. Many of those models, however, have already been discontinued in favor of the "second and third generation" of players. Glossy information sheets accompany many of the CDPs, but none make comparisons. They function mostly as advertisements for their own product.

Purpose of Report

The intention of this report is to present objective information on a range of players as a guide for purchasing a compact disc player. The sales pitch from a commissioned salesperson can be persuasive, but as a basis for making a choice, it is the wrong medium. The salesperson will suggest buying the brand in stock, or will say theirs is "one of the best" on the market. That type of meaningless phrase can only mislead or confuse the buyer.

Instead, this report presents facts about a variety of different brands and models, pointing out strong and weak points of each. No preference is given to house brands, favorite brands, in-stock merchandise, or special order items.

Intended Audience

The prospective buyer of a CDP does not have to be an avid audiophile, or even know much about stereo equipment. The buyer merely wants a playback system which will produce better sound than his present system is capable of; therefore, no special knowledge of audio jargon is necessary. The report is written for the reader who has minimal experience with stereo equipment. Some terms specific to audio equipment are necessarily included (see section of specifications), but they are fully defined and explained in the text.

Scope

Limited time and resources did not permit the extensive scope of research possible in a consumer magazine. Nine players were examined, brands chosen to represent the lower, middle, and higher price ranges available. The following is a list of CDPs examined in this report:

Brand	Model	Price
MGA	DP-103	$ 600.
Hitachi	DA-800	699.
Marantz	CD-73	799.
Sony	CDP-400	799.
Sony	CDP-610	850.
NEC	CD-803	975.
NEC	CD-705	1000.
Kyocera	DA-01	1050.
Sony	CDP-701	1500.

The criteria used to evaluate the models are listed below:

specifications	shock resistance
features	error correction
warranties	price

Attractive design/appearance, ease of use, and color were not used as comparison criteria since these are mainly subjective items. The buyer, however, will most likely take note of them when shopping for a CDP.

Sources

Traditional sources of information consulted were back issues of audio magazines and reference books on stereo equipment. Also consulted were sales brochures from manufacturers and salespeople with "hands-on" experience using compact disc players. Personal experience with the CDPs was considered if other sources were incomplete, contradictory, or unclear.

INTRODUCTION TO PAPER #2 (Bernucci 1981)

When one is in the market for a new sewing machine, the available information on the different models is usually obtained from the individual stores that carry them. The dealer will demonstrate the machine to the customer and elaborate on its special features; however, it is difficult to obtain information from the dealer about the weaknesses and flaws of his product. This report supplies that information to the consumer. It does not attempt to show how to operate the machine because that information is readily available from the store.

The *Elna SU 68* is first discussed at length because it is the most well-designed machine. In pointing out its features, similar features of other models are mentioned and compared. A discussion of the *Riccar 808 E* comprises the second part of this report because it is considered the next best buy following the Elna. It is not discussed at great length in this part because many of its features are mentioned in relation to the Elna in the first part of the report. This is also true of the *Pfaff 1222E*, which rates third in the comparison. The *Singer Touch-Tronic* is discussed at the end and at greater length because it has incorporated a micro-computer into its design, and a longer explanation must be given to inform the consumer how this works. The reader will find a recapitulation of all the major points in the conclusion of this report.

EXERCISE 5-9 Writing an Introduction for a Major Paper

Directions: When you have finished drafting both the body of your report and your conclusions, write an introduction that will cover as many of the items on the checklist as appropriate. Submit it for review if requested.

WRITING ASSIGNMENT #6 Writing the Draft of a Term Project Report

Directions: Combine the introduction, body, and conclusions or recommendations into a draft of your project report.

REFERENCES

Bernucci, Julie. 1981. *Comparative study of top-of-the-line sewing machines: Elna SU 68, Riccar 808E, Pfaff 1222E, Singer Touch-Tronic 2010*. Formal report, San Jose State University.

Clayton, LaVada. 1985. *Technical evaluation of COM recorders*. Formal report, San Jose State University.

Elledge, Scott. 1984. *E. B. White: A biography*. New York: Norton, 233.

Jarrett, Nancy. 1984. *A comparative study of four brands of stethoscopes to determine one suitable for home health nurses.* Formal report, San Jose State University.

Leonard, David. 1985. *An evaluation of Nd:YAG surgical lasers for Allied Hospital.* Formal report, San Jose State University.

Paradis, James, David Dobrin, and Richard Miller. 1983. *Writing at the Exxon Intermediates Technology Division: A study in organizational communication.* Cambridge, MA: Massachusetts Institute of Technology.

Vargas, Jeffrey. 1984. *A comparative study of nine compact digital disc players.* Formal report, San Jose State University.

Zertuche, Ralph. 1982. *The selection and purchase of a new non-commercial 10-inch table saw.* Formal report, San Jose State University.

CHAPTER 6

Revealing the Organization to Your Readers

You've been working hard planning, gathering information, organizing, and writing. Now, as you begin to polish your first draft, you should consider how you can communicate your organization to your reader. Unfortunately, readers don't have the same experience with the material you do; you are now an expert, but they may find the information totally new. Think of your draft as a dense forest without a path. Your job is to lead your readers through that forest in such a way that they find the important features—the giant specimen tree, the swamp, the rare species of fern— that you have already noticed. Your readers have made their way through document forests before, so they have certain expectations, and you'll communicate best when you meet those expectations. That means you provide meaningful signposts, you don't overload their memory with too much detail, and you tell them where they're going and review for them where they've been. In other words, you mark the path you want them to follow.

You can reveal the organization to your readers in five different ways, by using the activities listed on the preceding page.

6.1 COMPOSING A MEANINGFUL TITLE OR REFERENCE LINE

What a Meaningful Title Is

The first clue you give a reader about the purpose and content of any piece of writing is the title or reference line. For readers of technical literature, that title or reference line needs to tell both what information is covered and how that information is handled. Before your readers plunge into the thicket of your writing, they deserve to know what's in store for them, and you cannot assume that they are familiar with your subject. For example, here are the titles of two technical articles and two books. What subject do you think each one covers?

> "How to Find the Right Software" (Did you guess computer programs for stockmarket investors?)
>
> *The Two Faces of Management: An American Approach to Leadership in Business and Politics* (Did you guess points of similarity between management in industry and management in government?)
>
> "RAM Mods: Horns of Plenty" (Did you guess modified Cessna airplanes?)
>
> *What Color is Your Parachute?* (Did you guess a book on job hunting?)

Whether you are searching in a library catalog, a computer data base, or a periodical index, you depend on clues to help you find what you want. In titles or the reference lines in memos, letters, and reports, the primary clues are *key content words* and *organizational indicators*. Key content words tell you about the subject of the document, while organizational indicators tell you the method of organization and the purpose. Look again at the titles listed above.

In the first example, the words "how to" provide a valuable clue, indicating that you will get instructions. The key word "software" identifies the subject. But software (computer programs) can be used for many things. In this title, the key content words that would identify the specific use are missing. A title "How Stock Market Investors Can Find the Right Software" would supply that missing information.

In example two, *The Two Faces of Management: An American Approach to Leadership in Business and Politics,* notice all the key words in the title:

"management," "American," "leadership," "business," "politics." The combination of all these key words gives a good general idea of the subject matter. In addition, the words "two faces" imply a comparison, as does the parallel structure of "business and politics."

Example three, "RAM Mods: Horns of Plenty," could be very confusing to a general reader because it assumes familiarity with a name and an abbreviation. If you know computer terminology, you probably thought RAM meant "random access memory." In this case, however, RAM is the name of an airplane company that MODifies airplanes, and the article describes such a plane. The title is catchy and attention-getting, but it does not clearly indicate either the subject or the approach. To be fair, though, this title appeared in an aviation magazine, whose readers would be more likely than most to recognize the name and the abbreviation.

Example four, *What Color Is Your Parachute?* is a fanciful, attention-getting title, but it gives no clue to the subject. Recognizing this, the author supplies a secondary title, *A Practical Manual for Job-Hunters and Career-Changers.* Notice that the secondary title contains both an organizational indicator, "practical manual," and key content words, "job-hunters," "career-changers." The secondary title is the *real* title. The problem with secondary titles, however, is that they are not always indexed, and readers searching for job-hunting information might not be led to this book. On the other hand, shelved in a bookstore with other job-hunting books, the catchy title does attract book buyers (the generalist readers). In general, attention-getting titles are better for generalist readers than for others in the technical spectrum, who want information and are impatient with fancy words.

If you are writing a memo or a letter, you obviously do not have a title, but you do have a subject or reference line to guide your reader. If your memo deals with a proposal for a comparative study of music synthesizers, don't write a reference line like this:

Subject: Term Project

Instead, write one that tells the reader the task, the form, and the key content. Write one like this:

Subject: Proposal for a Comparative Study of Four Music Synthesizers

How to Compose a Meaningful Title

Notice that you can take two specific steps to create a meaningful title. You can:

1. *use key content words,* which are words that tell the reader about the topic or content of the document. These words should be as specific and concrete as possible. For example, if you are writing about automobiles, "Ford" is more specific than "Car," and "Escort" is more specific than "Ford." (See Chapter 8 for more information about concrete words.)

2. *use organizational indicators*, which are words that tell the reader what the method of organization and purpose are. Such words include:

How to . . .
An Explanation of . . .
An Analysis of . . .
A Description of . . .
A Definition of . . .
The Process of . . . (or How _____ Works)
A Recommendation for . . .
A Comparison of . . .
An Evaluation of . . .
A Justification of . . .
A Report on . . .
The Plan for . . .
The Results of . . .

EXERCISE 6-1 Writing Meaningful Titles Based on Abstracts

Directions: Read the following abstracts of technical articles and write a meaningful title for each. Remember to choose words that show both the organization and the key content.

Abstract 1: A student example (Neher 1984)

> This study, using a set of six evaluation criteria, compares four-track cassette recorders in order to recommend one or more for use in a home recording studio. The products compared are Aria Studiotrack IIII, Clarion XD-5, Cutec MR402, Fostex 250, and Tascam 244. The evaluation criteria used include the following: tape speed, pitch control, noise reduction, equalization, rack mount, and price. The study includes (1) a technical definition of a four-track cassette recorder, (2) a description of the principal components of a multitrack recorder, (3) a process description of the multitrack sound-recording process, (4) an explanation of the evaluation criteria and the point evaluation system, (5) individual product evaluations, (6) product comparisons, and (7) conclusions and recommendations. The study recommends the Tascam 244 for use in a home recording studio because it best satisfies the criteria set up for the study and combines all the features necessary for recording, processing, and mixing a high-quality four-track cassette tape.

Abstract 2: A student example (Leonard 1985)

> This report evaluates four Nd:YAG surgical lasers for use at Allied Hospital: Lasers for Medicine: FiberLase 100, Cooper LaserSonics: Model 8000, MBB-Angewandte Technology: mediLas 2, Laserscope: OMNIplus. Allied Hospital is a private hospital in a large urban community. Knowledge of surgical laser action, application, and market competition is

needed to choose the right laser. The evaluation is based on five criteria: versatility, cost, safety, mobility, and ease of operation. Each model is analyzed by these criteria; the comparison emphasizes key differences among the models. The FiberLase 100 is recommended for Allied Hospital. It costs about $20,000 less than its nearest competitor, the LaserSonics Model 8000. Its audible fault indicators and self-calibration system combine to give it the best overall safety features of any model. Additionally, its broad delivery options and push button input panel give it creditable versatility and easy operation.

Abstract 3: A professional example (Thomas and Campbell 1985)

Assembly, operation, and disassembly of the Battelle Large Volume Water Sampler (BLVWS) are described in detail. Step by step instructions of assembly, general operation and disassembly are provided to allow an operator completely unfamiliar with the sampler to successfully apply the BLVWS to his research sampling needs. The sampler permits concentration of both particulate and dissolved radionuclides from large volumes of ocean and fresh water. The water sample passes through a filtration section for particle removal then through sorption or ion exchange beds where species of interest are removed. The sampler components which contact the water being sampled are constructed of polyvinylchloride (PVC). The sampler has been successfully applied to many sampling needs over the past fifteen years. 9 references. 8 figures.

Abstract 4: A professional example (Verosub, Lott, and Hart 1985)

The technical aspects of using observations of animal behavior reported by the general public to predict earthquakes are discussed. The possibilities and problems inherent in two alternative approaches are explored. One approach would be to obtain daily reports of egg and milk production on farms along faults. The other approach would be to encourage a selected group of volunteer observers to report all observations of apparently abnormal animal behavior.

EXERCISE 6-2 Evaluating Titles

Directions: Bring to class three examples of misleading, ambiguous, or unclear titles from magazines, newspapers, or books that you have read this term. Be prepared to discuss what's wrong with them and how they might be improved.

EXERCISE 6-3 Writing a Meaningful Title

Directions: Write a meaningful title for your term project paper. Test the accuracy of your title by asking two fellow students to write three to four sentences explaining (1) the exact topic, and (2) the purpose and the method of organization of your report. If their responses are not accurate, rewrite your title.

6.2 MAKING SUBPOINTS GRAMMATICALLY PARALLEL

One of the easiest and best ways to lead the reader through your organization is to write grammatically parallel subpoints. Readers look for patterns; thus, whenever you want to show that points are equal or related, you should state those points in the same way. In your written document, those parallel points might become section headings, the first sentences in a series of paragraphs, sentences within paragraphs, or even the parts of an individual sentence.

A Definition of Grammatical Parallelism

Grammatical parallelism means using the same structure in sentences or parts of sentences for points that are alike, similar, or direct opposites. Parallelism gives writing balance and proportion, and it creates a pattern for the reader by saying, "Look at these things in the same way." As you polish your first draft, you should use parallelism (1) when you want to give the same importance to two or more items, as in a list, an outline, or a sentence, and (2) when you want the reader to see the way two or more items are related.

How Parallelism Helps the Reader Understand

Parallelism helps the reader understand because it

- communicates a pattern
- clearly indicates when one point has been completed and another begins
- emphasizes the relationship between ideas

Parallelism is important to all writing, not just technical writing. Good examples of grammatical parallelism have become classics in our language. Here are two effective examples by nontechnical writers. In 1940, British Prime Minister Winston Churchill said:

> . . . we shall fight on the beaches, we shall fight on the landing ground, we shall fight in the fields and the streets, we shall fight in the hills; we shall never surrender.

Churchill spoke these words of defiance at a time when British forces had suffered a major defeat at Dunkirk and when the whole world expected the Nazis to invade the island of Britain. Notice the way each part of the sentence repeats the key words, "we shall fight." Notice also how Churchill shows the fighting moving inland, from the beaches to the hills. Each point is equivalent, however, and stresses the "we shall fight" portion. Especially effective is Churchill's closing clause, which is both equal to and a summary of the others. It uses the same construction "we shall" but switches to the negative "never surrender."

Novelist Virginia Woolf used parallelism for irony when she wrote in 1929: "One cannot think well, love well, sleep well if one has not dined well." Notice how the repetition of the structure here makes thinking, loving, and sleeping all equal and all dependent on dining. Since we do not usually equate these actions, the parallelism adds a humorous and ironic touch.

Technical writers use parallelism not for drama or humor, but for clarity. Notice the repetition of structure in the following overview statement from the 1982 IBM Information Development Guideline called *Designing Task-Oriented Libraries for Programming Products.*

> *Operation* consists of the following subtasks:
> - starting a program
> - checking and controlling a program
> - recording the status of a program and its data
> - reacting to abnormal events
> - stopping a program
>
> The following sections describe each subtask in greater detail.

The five subtasks are clarified by placement in a list, by marking with a bullet, and especially by parallelism—each subtask begins with an "ing" word.

The following example is from an operator's manual (Varian 1986). Notice how the parallel structure of the sentences is designed to emphasize the difference between two temperature ranges and to direct the operator's attention to the colored areas on the dial face.

> Operating ranges are shown on the dial face. When the cryopump is at room temperature, the needle should be in the silver-colored area. When the cryopump is at proper operating temperature, the needle is in the blue-colored area (i.e., the temperature of the second stage is then below 20K).

In contrast, here is a confusing job description for a software engineer:

> Duties: Performing concept definition, software design and development, and to produce system software.

Can you determine what the job requirements are? Try listing each item, one under the other, to determine how many parts the job has:

- performing concept definition
- software design
- software development
- to produce system software

These are all job functions, but they seem to be different because they are different in form. The writer has given you one verbal, two nouns, and one infinitive. Try to clarify the sentence by using the same grammatical structure for each one. You may have to try several possibilities. Since the structure describes job functions, try verbs:

- define concepts
- design software
- develop software
- produce system software

This may not work because it's unclear what the writer means by "to produce system software." You might have to talk to the writer to find out. If you have guessed right, you could restate the job description like this:

> Job requirements for a software engineer:
> define concepts
> design and develop software
> produce system software

Another option might be to use nouns:

> concept definition
> software design and development
> system software production

Still another choice might be verbals:

> defining concepts
> designing and developing software
> producing system software

Whatever option you choose, notice that in each case you can make the points understandable and equal by using grammatical parallelism. Remember too that the order you choose for items will influence the reader in a subtle way. You may choose to arrange items from most important to least important (descending order) or from least important to most important (ascending order). If no other order is important to you, follow the principle of emphasis by position: the most important position is last, and the next most important position is first.

In summary, here's how to make subpoints grammatically parallel:

1. List them one under the other (even if later you will put them in a sentence).
2. Examine the list. Sometimes the order of items should be changed for clarity, especially if some are singular and others are plural.
3. Use the same grammatical structure to state each one. The structure can use single words, phrases, or clauses. Here are examples of each:

SINGLE WORDS: The four operations are etching, masking, screening, and scanning.

PHRASES: Three typical printing presses are listed below:
blanket-to-blanket press

in-line open press
common impression cylinder
press

Note: Usually technical writers will not break sentences with only three parts into a list unless they want to call attention to the individual parts.

CLAUSES IN A LIST:	During the Jan. 1–June 1 period: 5,230 rolls were processed 742 tons of waste were baled 12,470 pieces of freight were handled
CLAUSES IN A SENTENCE:	Facepieces fall into three types: the full-face mask, which covers the entire face including the eyes; the half mask, which covers only the nose and mouth and fits under the chin; and the quarter mask, which covers only the nose and mouth and fits on the chin.

Note: These two sentences both use colons to introduce the lists. For more information on punctuation with colons, see the Classy Punctuation section in the Appendix. To see how parallelism operates on a larger scale, look at the chapter, section, and subsection titles in the Table of Contents of this book.

EXERCISE 6-4   Writing Parallel Sentences

Directions: Complete the following sentences as truthfully as possible by adding a series of parallel words, phrases, or clauses.

1. _____ are four units of measure in the metric system.
2. A driver can increase a car's gas mileage by (give three techniques).
3. I can get a job as a technical writer if I (suggest three courses of action).
4. You may find it necessary to (suggest two possible activities) in order to find a parking space near campus.
5. If you want to write a good resumé, you should (list four steps).
6. People who (list several undesirable forms of behavior) make poor employees.
7. Changes this campus needs urgently are (list three improvements).
8. The procedure for starting my car is as follows: (list at least four steps).
9. Three alternatives to oil as a home heating fuel are (list them).
10. (List four courses of action) will reduce air pollution.

EXERCISE 6-5   Designing Parallel Patterns

Directions: Revise each group of sentences below into a single sentence by condensing and by putting groups of similar ideas into parallel patterns. For example, instead of presenting a series of related ideas like this:

> Several engineering teams have designed family-sized electric automobiles. Some teams built electric cars. Now they are testing those vehicles.

Try arranging them into a grammatically parallel series like this:

> Several engineering teams have designed and built and are now testing family-sized electric automobiles.

1. The vehicle is a three-door auto.
   It is a cooperative effort.
   General Electric participated in the effort.
   So did Chrysler Corporation.
   Also included was Globe-Union, Inc.
2. The ETV-1 is intended for short trips.
   It can be used for commuting.
   Grocery hauling is one aim.
3. At a constant 45 mph the ETV-1 provides an expected 97-mile range.
   Sixty-nine miles is the figure dropped to with the SAE J227a driving cycle, which more realistically includes stops and starts.
4. The auto weighs 3300 pounds.
   The vehicle has a half-ton battery pack.
   A 600-pound payload can be carried by it.
5. Sixty miles at 60 mph is a good typical driving range for a practical electric vehicle.
   Today's electric cars can't claim an average velocity much better than 8 mph, however.
   An electric car should go 122 miles at 35 mph.
6. Vacuum-fluorescent displays have found considerably more applications recently.
   They are low cost.
   Power requirements are modest.
   High readability is a good feature.

EXERCISE 6-6   Rewriting for Parallelism

Directions: Rewrite the following to make the equivalent items parallel in structure. Then meet with three to four fellow students to review your rewritten sentences. Discuss each sentence and be prepared to defend your rewrites to the class.

1. The purpose of this section is to provide information about
   - our technical staff
   - performing other similar projects, and
   - qualifications.
2. The equipment she worked with included camera use, developing negatives, metal plates, light composer, and rub-on lettering.
3. They found the weave of the silk open enough to allow paint to pass through easily, and it was also found that the strands were so fine that they left no permanent impression on the paint.
4. This knowledge must include, but not be limited to
   > reading music quickly,
   > being familiar with most musical instruments,
   > an understanding of the range of each instrument,
   > a knowledge of music literature and history, and
   > he must be familiar with all kinds of music notation and proof-reading marks.
5. Some of the many things they check for are listed below.
   - The amount of wire allowed to be stripped of insulation should equal the wire width or be less.
   - The component should be flush or level, with no more than 1/16 angle.
   - When the wire insulation melts into the solder, it contaminates the solder joint and must be reapplied.
   - Cracked components are never allowed to be used.
   - Metal should never touch an open trace; this will cause a short in the unit.
6. The chemicals are kept in two separate tanks and with connecting rods are shot into a container through a gunlike dispenser around the product.
7. Painted on the front side of the vase is Hermes carrying the infant Dionysus, and on the reverse the three Muses are painted.
8. III. Technical Definition of Microwave Ovens
   > A. What is a microwave?
   > B. How microwaves operate within the oven chamber.
   > C. Operation of microwave ovens.

6.3 WRITING GOOD TRANSITIONS

Because technical documents are often composed of a number of discrete parts, you can help your reader see the relationships among those parts by writing good transitions. Transitions are like the string holding beads in a necklace. Individual beads may be very beautiful, but until they are strung, they are neither a necklace nor are they very useful. The same principle holds true for sentences and paragraphs. Individual sentences or paragraphs may be clear and well written, but until they are fastened to one another in a meaningful way, they will not make sense to the reader. You join sentences, paragraphs, and sections with transitions.

Definition of a Transition

A transition is a word, phrase, sentence, or paragraph that moves the reader from one idea to another. Transitions work by simultaneously looking backward to the previous idea and forward to the new idea. You can link both sentences and paragraphs by transitional words or by repetition of key words. You can also link paragraphs by sentences or short paragraphs that both summarize the previous ideas and introduce the new ideas.

Transitions between Paragraphs

The following paragraphs explain the process of color separation in printing (International Paper Company 1981, 79-81). Notice how the fifth paragraph sums up the previous information and then prepares the reader for a discussion of color reproduction.

> The process of color separation is analogous to the process of seeing by the eye, but the printing process introduces new concepts. The original is photographed using three filters, each corresponding in color and light transmission to one of the additive primaries.
>
> Placing a red filter over the lens produces a negative recording of all the red light reflected or transmitted from the subject. This is known as the red separation negative. When a positive or print is made from this negative, the silver in the film will correspond to areas which did *not* contain red but contained the other two colors of light, which are blue and green. In effect, the negative has subtracted the red light from the scene and the positive is a recording of the blue and green in the scene which is called *cyan*. The positive is the *cyan printer*.
>
> Photography through the green filter produces a negative recording of the green in the original. The positive is a recording of the other additive primaries, red and blue, which is called *magenta*. The positive is the *magenta printer*.
>
> The blue filter produces a negative recording of all the blue in the subject. The positive records the red and green which when combined as additive colors produce *yellow*. This positive is the *yellow printer*.
>
> These three colors, cyan, magenta, and yellow, are called *subtracting primaries* because each represents two additive primaries left after one primary has been subtracted from white light. These are the colors of the process inks of process color reproduction.
>
> When the three positives are combined and printed, the result should be a reasonable reproduction of the original. Unfortunately, it is not. The colors, outside of yellow and red, are dirty and muddied. There is too

transitional
paragraph

much yellow in the reds and greens and too much red in the blues and purples. This is not a flaw in the theory but is due to deficiencies in the colors of the pigments used in the inks. . . .

Transitions between Sentences

Transitional words like "for example" and "therefore" are simpler links; to be most effective, you should put such words near the beginning of the sentence. In the following example (Miller 1979, sec. 8, 45), notice all the transitional words (italics added).

> In the event divers are carried into a trawl from which they cannot readily extricate themselves, they must cut an exit through the web. *Usually* trawls have heavier web in the aft portion (cod end). *Therefore* an escape should be cut forward in the top of the trawl body and a 3-foot-long diagonal slit should be made in the trawl. *Another* similar slit should be made at 90 degrees to and beginnning at the upstream end of the first slit. The water current should *then* fold a triangular flap of webbing back out of the way, leaving a triangular escape hole. The diver's buddy should assist the trapped diver through the opening to free any gear which snags on meshes.

Transitional words and phrases show relationships. Table 6-1 shows how transitions function and gives examples of common transitional words and phrases.

Table 6-1   TRANSITIONAL WORDS AND PHRASES

Function	Examples
add	also, in addition, second (etc.), moreover
enumerate	first, second
compare/contrast	likewise, similarly, however, on the other hand
illustrate	for example, that is, for instance
summarize	in conclusion, in summary
tell place	below, above, on the right
tell time	later, then, by noon
show cause	therefore, consequently, thus

Key Words as Transitions

Another kind of transition is repetition of key words, a technique that keeps readers on track by refreshing their short-term memory. To tie sentences or paragraphs together, you repeat the key term at the beginning of the new clause, sentence, or paragraph. Such repetition reminds the reader what has gone before and says that more on this topic is coming. In technical writing it's important to use the same word to refer to an object or an idea from sentence to sentence and paragraph to paragraph; clarity is more important than variety. Thus, if you call something a "shaft" in one sentence, don't call it a "rod" in the next.

In the following paragraph (Neher 1984), notice how the italicized words link the ideas. The key words "engineer," "performers," "record," and "tapes" are picked up and repeated (in varied forms) throughout the paragraph to provide the links (italics added).

> Once the recording equipment is set up and properly adjusted, the master recording is made. The *engineer* turns on the recorder and signals the *performers* to begin. During the *performance*, the *engineer* monitors the *record* levels and tone adjustments to make sure they stay within acceptable bounds. When the *performance* is over, the *engineer* stops the *tape* and rewinds it. If, after playing back the *tape*, the *engineer* and the *performers* are satisfied with its quality, the *tape* is ready to be mixed down. Otherwise, the *recording* process may need to be repeated, or individual *performers* may do their parts over (overdubbing).

To avoid the stuttering of constant repetition, sometimes you must replace a key noun with an appropriate pronoun. If you use a pronoun, however, be sure the original noun is not too far away. Often it's better to add an adjective to the key noun instead of using a pronoun. The following paragraph (Lissaman 1980) uses both pronouns and adjectives with key nouns (italics added).

> Among the greatest reservoirs of solar energy are the ocean gyres, or circulating current systems, generated by the prevailing global winds. The energy density in *these currents* is very high, because *they* are created by a twofold intensification of the direct sunlight: first, by the transformation of the sun's heat into wind, then by transfer of the surface wind stress into the wind-driven currents. *This* intricate energy *transfer* is part of the mighty global heat engine. In the Northern Hemisphere, the gyres rotate clockwise, and because of the Earth's spin, are strongest on the western shores of the oceans.

6.4 CLARIFYING THE ORGANIZATION WITH VISUAL MARKERS

Advantages of Visual Markers

Readers look at a page as a whole visual picture before they begin reading the words printed there. Most of the time they do this unconsciously, but they always do it, and the impression made by that visual picture either helps or hinders their understanding. Think of how you feel when you open a report or textbook of solid, tightly packed words—page after gray page with no visual breaks. In contrast, how do you respond to a page with a picture or a drawing, with a list, with subheadings in boldface or colored type? The more visually interesting the page, the more likely you are to respond positively—even before you start to read. Readers will respond the same way to your pages, and you always want to make your readers respond positively.

A good rule of thumb for helping your reader to follow and respond to your organization is to break your writing at least every half page with some kind of visual marker. That visual marker can be as simple as a paragraph indentation or the white space surrounding a list, or it can be a heading, a block around a key word, or underlined key phrases.

Kinds of Visual Markers

Here are some common visual markers that will help your reader understand and follow your organization.

White space   You use white space every time you skip a line between paragraphs or skip an extra space between sentences. But you should also think of white space as the frame around a list or around a subheading. White space helps define the boundary between one idea and the next, helps relieve the monotony of gray print, and helps communicate by resting the eye. You can think of white space as a frame, and you should plan for it on the page as you begin writing. Don't be afraid to use white space. Even a sizable chunk of it is permissible before a major heading. In a document packed with information—like a resumé—white space is especially important. Use it to lead the reader's eye to the points you want to emphasize.

Headings   Most companies have a style guide that governs the kind and placement of subheadings to show their position in the outline hierarchy. If you are planning the layout for a company document, you will want to consult that style guide. If you must choose the kind and placement of subheadings yourself, here is a typical arrangement:

SECTION 1
SECTION HEADING

1.1 FIRST-ORDER HEADING
1.1.1 Second-Order Heading
1.1.1.1 <u>Third-Order Heading</u>. Text continues on the same line.

Note that the section heading is all capitalized, bold, and centered on the page. The text begins three lines below the section heading. First-order headings are numbered, all capitalized, bold, standalone, and flush to the left margin. The heading number consists of two parts separated by a period. Second-order headings stand alone flush to the left margin, have initial capitals only, and are bold. Articles, prepositions, and conjunctions are not capitalized. Third-order headings are run in with the text, use initial capitals, are bold or underscored, and end with a period.

Technical documents often use a decimal numbering system to show the various levels of headings in an outline, because readers can see at a glance where a numbered heading (like 4.3.2) fits into the whole document. However, too many decimal places can be intimidating to generalist readers, so some documents use a modified system, combining numbers with heading size to indicate the hierarchy. This book, for example, uses only two decimal places (this is section 6.4) and uses type size and style to show subdivisions.

Unless your company style demands many decimal places to show levels of subordination, consider using paragraphs or bullets beyond the fourth level of subordination. The fourth level means four layers down from your major point; beyond that you begin to deal with very fine discrimination. For example,

```
2.0 _____
  2.3 _____
    2.3.1 _____
        • _____
        • _____
        • _____
```

Remember that the text must be comprehensible without the headings. In other words, the headings are not a part of the text but are really an embedded outline. Remember too that maintaining parallelism within each order of the headings will help the reader see relationships.

Print highlights   Bold-faced print, varied type sizes, colored inks, underlining, blocks or rectangles around key words or phrases—all kinds of printers' techniques can be used to visually enhance key subpoints so they will "leap out" at the reader. As with white space, you may be so accustomed to seeing bold or larger type for subheadings, you seldom think about their effect. Yet when you scan a page searching for key points, what do you see? You see those terms or words or subheadings that are visually different from the rest.

Word processing software for computers has made this kind of visual enhancement of subheadings easier and cheaper to produce. Many companies can now prepare finished copy on their own computers with no need for typesetters and printers. Some word processing programs let writers choose among a variety of type sizes, styles, and fonts, or even combine graphics and text in the same document. But don't overdo a good thing. If you have too much variety or too many centers of interest on a page, your text will resemble a bowl of confetti, and it will lose coherence and clarity.

Using color for headings is the easiest way to provide visual contrast, but because of the increased cost, it may not be practical.

Paragraph indentation, bullets, and numbers   Not every bit of information in a paper or a report needs a subheading. (Remember the rule of seven plus or minus two; you don't want to overload your reader's short-term memory.) Therefore, as you work toward your less important subpoints, you can let paragraphs themselves serve as subpoint markers. You can use paragraphs that way because of what a paragraph is—a unit dealing with a single idea.

Other minor point markers include bullets, which can be made on the typewriter by filling in a lowercase or uppercase letter "o." Bullets, used more and more frequently in technical writing, are especially helpful when you want to list points that are equal in importance, but you don't want to indicate a sequence as you would with numbers.

You should use numbers, of course, if you do want to show a sequence. You should also use numbers to show a hierarchical arrangement, as in a formal outline. A good general rule for more than three subpoints of equal importance is to put them in a list. The physical arrangement of the list itself provides a visual clue to your reader about the items. Also, because lists are short, you will be using white space surrounding the list. Unless your company style guide calls for punctuation at the end of each item in a list, omit such punctuation even if the list completes a sentence.

To refresh your memory about the possibilities for visual markers, leaf through this text, noting how type size, color, numbers, lists, and paragraph indentations all contribute to helping you understand how the book is organized.

EXERCISE 6-7   Using Visual Markers

Directions: The following information is adapted from a section of Underwriters Laboratories *Standard for Safety: Microwave Cooking Appliances* (1981). Except for paragraphing, the visual markers in the original have been removed. Read the material carefully, seeking to understand the intent of the writer. Then determine what visual markers you would include and insert them. Be prepared to discuss in class.

40.1 The door and door interlock system, in conjunction with the complete appliance, shall be capable of completing 100,000 cycles of operation (opening and closing the door) for a household microwave cooking appliance and 200,000 cycles for a commercial microwave cooking appliance. During and at the conclusion of the test the performance of the system shall meet the following criteria: There shall be no electrical or mechanical malfunction that could result in the likelihood of fire, electric shock, personal injury, or excessive radiation emission. Only parts that fail safe and prevent completion of the test may be replaced or repaired. There shall be no loosening or shifting of adjustments or parts that could result in the likelihood of fire, electric shock, personal injury, or excessive radiation emission. Radiation emission shall not exceed 5 mW/cm² at any point 5 cm or more from the external surface of the enclosure.

40.2 Compliance with the requirements of paragraph 40.1 is to be determined by subjecting 2 complete samples of the appliance to the test described in paragraphs 40.3–40.5

40.3 The appliance is to be connected to a voltage source as described under the input test section (paragraph 35.2). Any timers incorporated in the appliance are to be by-passed to facilitate uninterrupted operation of the appliance. The door interlock switches shall be operational during the test except that external controls (such as externally operated switches wired in parallel with the interlock switches) may be provided in order to by-pass each door interlock while microwave radiation emission measurements are being taken. The external controls shall not affect nor take part in the operation of the appliance in any other way.

40.4 The microwave cooking appliance door is to be cycled by an automatic device attached to the door handle or in the operating area of the door handle. The door closure force is to be predetermined so as to simulate intended operation. In no case shall the cycling mechanism apply a closure force greater than 20 lbf (89 N) unless agreed to by the manufacturer. The door is to be swung open from the closed position to any angle within 135 degrees to 180 degrees, or to within 10 degrees of its maximum travel, whichever is the smaller angle. A sliding door shall be opened to within 1 inch (25 mm) of its full travel. The cycling rate is not specified, but there shall be sufficient on time to permit generation of microwave energy that may be delayed by warm-up time of the magnetron, a time-delay circuit, and so forth.

40.5 The appliance shall be operated in accordance with the following sequence: An initial microwave radiation emission measurement shall be made as described in Section 34. A fresh coating of corn oil is then to be applied to all door sealing surfaces, and any microwave absorbing load (such as dry brick load) shall be placed and maintained in the oven cavity. The appliance is to be energized and the door operated for 10,000 cycles. Following the 10,000 cycles of operation, the cycling load is to be removed. Microwave radiation emission measurements are to be made as described in paragraphs 34.2–34.8. A microwave cooking appliance incorporating a resistance element in the cavity, a self-cleaning feature, or if intended for use in conjunction with another heating appliance shall, in addition, be operated as described in the temperature test until temperatures stabilize. Following the operation, microwave radiation emission measurements are to be repeated. After the microwave radiation emission measurement is made, the door sealing surfaces are to be cleaned of the corn oil with a soft cloth. The above procedure is to be repeated until a total of 100,000 cycles of operation are accumulated on a household appliance, and 200,000 cycles of operation on a commercial appliance.

EXERCISE 6-8   **Reorganizing and Using Visual Markers**

Directions: The following process description "Combine Harvester and Thresher" (1967) is intended for a general reader—not a farmer or a mechanical engineer. Its purpose is to explain how the combine harvester and thresher works. Follow these steps:

1. Read the whole process description.
2. Write down the single major point.

3. List the major subpoints.
4. Group minor points under the proper subpoints.
5. Rewrite and include the visual markers you would use. Indicate any drawings or figures that would be helpful.

You may find this exercise easier to do if it is a group activity with two or three fellow students.

> The combine harvester is a combination of a grain harvesting machine and a threshing machine. The grain is cut, threshed and cleaned in one operation. The machine may be self-propelled or towed by a tractor.
> The wheatstalks are cut by an oscillating knife while the revolving reel pushes them back towards the knife and auger (feed screw). Grain flattened by wind or rain is raised by the spring prongs on the reel. The cut wheat is conveyed into the machine by the auger and reaches the threshing cylinder which rubs the grain out of the heads against a "concave." The grains of wheat, together with the chaff and short fragments of straw, fall through the interstices between the bars of the concave and into the cleaning "shoe." Some of the grain is carried along with the straw, is stopped by check flaps, and is shaken out of it on shaking screens on the straw rack of the machine. The straw drops out of the back of the machine and is left in a windrow for later baling, or is baled directly by a baling attachment or press, or is scattered over the ground by a fan-like straw spreader. The grain shaken out of the straw is also delivered to the cleaning shoe. In the shoe the grain is separated from the chaff and cleaned by sieves and a blast of air. The chaff and fragments of straw are thrown out from the back of the machine. The grains of wheat fall through the sieves and into the cleangrain auger (screw conveyor) which conveys them to an elevator and on into the storage tank or into bags. Any heads of wheat which fail to go through the sieves and are thrown backwards by the air blast fall short in comparison with the lighter chaff and drop into a return auger which, via an elevator, returns them to the threshing cylinder. Correct adjustment of the air blast—by throttling the intake of the fan and by altering the setting of baffles—is important in determining the degree of cleaning of the grain and the magnitude of the grain losses that occur.

6.5 PROVIDING OVERVIEWS

A good way to help your readers stay on the path you have planned is to

- tell them at the beginning where they are going
- remind them at midpoints where they have been and what's ahead
- review at the end the path they have followed

Technical writers use specific locational guides called overviews to carry out these tasks. Among the overviews are the table of contents, list of illustrations, introduction, summaries (at the end of a document and at the end of a section), abstract, and executive summary. In a long document, even the body can contain overviews where necessary. Obviously, not all writing tasks need these specialized guides. If you are writing a one-page letter or a three-page memo, you can provide all the necessary organizational signposts through titles or reference lines, parallelism, transitions, and visual markers. But as your writing task becomes longer and more complex in documents like reports and proposals, readers will need overview statements. Here's a brief look at each one.

Table of Contents

Throughout Chapter 3, you organized your material more and more carefully in working toward a formal outline. One big advantage of having a carefully constructed outline is that you can easily turn it into a table of contents for a long document such as a formal report or proposal. If your subheadings are equal in importance, logically related, and grammatically parallel, you need only add the page numbers to have a good table of contents. The table of contents shows your reader the main points and the method of organization. It also allows easy access to individual sections of the report. Since technical reports usually do not have an index, a clear table of contents is especially important. Study the table of contents in the student comparative report in Chapter 21. Notice how the table of contents reproduces the outline and shows where to find the information.

In a very long proposal, report, or book-length manual, you can also put minilistings of contents at the beginning of each chapter or major section. As the table of contents provides an overview of the whole, so these listings give an overview of the individual sections. This book has such minilistings at the beginning of each chapter.

List of Illustrations

Like a table of contents, a list of illustrations helps the reader see at a glance what is included in the document, especially if you have taken care to illustrate all your key points by some kind of graphics. "List of Illustrations" is a general heading that often includes separate lists—one for tables and another for figures. One expert in rocket fuels (a specialist reader) told me that he won't even read proposals or reports that don't have a list of illustrations at the front because the data he wants is best presented in graphic form, and he wants to be able to find it quickly (Brown 1986). Chapter 12 explains how to use graphics to enhance your writing, and Chapter 21 explains how to compile the list of illustrations.

Introduction

Another kind of overview (or perhaps *pre*view) is provided by the introduction, which should give information about the topic, the purpose, the scope, and the way in which the information will be developed. Introductions are discussed in section 5.4.

Summary

Papers of analysis or persuasion usually end with conclusions and recommendations. If the subject is complex, many writers include a summary of the report before such conclusions and recommendations to remind readers where they've been—to review the facts and add up the main points. Papers that simply present information without conclusions or recommendations can also end with a brief factual summary. In the same way, major sections of a long document or chapters of a book can end with intermediate summaries: paragraphs that review for the reader what was covered in that section or chapter. Some summaries are written as outlines or checklists rather than as paragraphs. This book uses such checklists at the end of each chapter to sum up and apply the chapter contents.

Again, a summary should be written after the body of the paper itself is written. It should be concise and should contain only information that appeared in the document itself. Summaries can be included as a part of conclusions and recommendations. See section 5.3 for examples.

Abstract

An abstract is really a condensation or summary of a longer document. It has a different name because of its specialized purpose and its special location in a technical document. Like any summary, an abstract's purpose is to give an overview of the key information in a document. However, that overview is used by readers to determine if they want to read the report itself. Abstracts appear in two places:

- separated from the report itself in a listing or journal of abstracts. Readers scan abstracts of technical reports, choose the ones in which they are interested, and order copies of those reports.
- separated from the report itself, but attached to it on a cover sheet, on the title page, or on the first page after the title page. Readers determine from the abstract whether the report is of interest to them and in general what the report covers.

Abstracts are written for the *primary* reader of the technical report (usually a specialist or technician), and they must be understandable without reference to the report itself. They are not considered part of the main report, so information in an abstract is also repeated in the text. Abstracts are called either *descriptive* or *informational* depending on their purpose and content. See Chapter 19 for explanations and examples of these two types.

Executive Summary

An executive summary is another kind of overview. Like an informational abstract, it condenses the key information in a document, including major facts, conclusions, and recommendations. However, the intended reader is usually a nontechnical person like a manager; therefore, the executive summary often contains background information, explanations of technical terms, and discussion that will help the reader make decisions or see the entire picture. The executive summary is considered part of the main text, and many documents include both an abstract and an executive summary. See Chapter 19 for examples of executive summaries and information on how to write them.

CHECKLIST FOR REVEALING YOUR ORGANIZATION TO THE READER

1. Does my title or reference line contain both key content words and organizational indicators?
2. Are any points that are of equal importance also grammatically parallel?
3. Have I included strong transitional words, phrases, sentences, and paragraphs between sections?
4. Have I broken up each page with some kind of visual marker?
5. Have I used

   _____ white space?
   _____ print highlights?
   _____ headings?
   _____ paragraph indentation, bullets, and numbers?
6. What overviews are appropriate for this document?

   _____ table of contents
   _____ list of illustrations
   _____ introduction
   _____ summary
   _____ abstract
   _____ executive summary
   _____ overview for long, complex sections of the body

EXERCISE 6-9   Writing a Table of Contents and a List of Illustrations

Directions: When you have finished writing your term project paper, turn your formal outline into a good table of contents by adding page numbers. After determining the placement of the figures and of the tables in your report, write a list of illustrations. Attach these overviews, along with your title page, to your report. See Chapter 21 for examples.

REFERENCES

Brown, Robert. 1986. United Technologies. Conversation with author. 15 February.

Churchill, Winston. 1940. Speech to House of Commons. 6 June.

Combine Harvester and Thresher. 1967. *The Way Things Work*. New York: Simon and Schuster, 432–3.

IBM Information Development Guideline. 1982. *Designing task-oriented libraries for programming products*, no. ZC28-2525-2, 77.

International Paper Co. 1981. *Pocket pal: A graphics arts production handbook*. 12th ed. New York: International Paper Co., 79–81.

Leonard, David. 1985. *An evaluation of Nd:YAG surgical lasers for Allied Hospital*. Formal report, San Jose State University.

Lissaman, P. B. S. 1980. Tapping the ocean's vast energy with undersea turbines. *Popular Science*, Sept., 72.

Miller, James W., ed. 1979. *NOAA diving manual*. 2d ed. Washington: National Oceanic and Atmospheric Adm. Section 8–45.

Neher, Jonathan. 1984. *A comparative study of four-track cassette recorders*. Formal report, San Jose State University.

Thomas, V. W. and R. M. Campbell. 1985. Abstract of Assembly, operation, and disassembly manual for the Battelle large volume water sampler (BLVWS). *Energy Research Abstracts* 10 (15 April), 1605.

Underwriters Laboratories. 1981. *Standard for safety: microwave cooking appliances*, UL 923. Northbrook, IL: Underwriters Laboratories, 38-9.

Varian Vacuum Products Div. 1986. *Cryopump operator's manual for CS8FA*, no. 87-400437. Santa Clara, CA: Varian Associates.

Verosub, K. L., D. F. Lott, and B. L. Hart. 1985. Abstract of Use of volunteer observers to detect abnormal animal behavior prior to earthquakes. *Energy Research Abstracts* 10 (15 April), 1617.

Woolf, Virginia. 1929. *A room of one's own*. New York: Harcourt, 18.

CHAPTER 7

Reviewing and Revising the Document

One test of your skill in planning comes when you have finished writing your first draft. The test is whether you have enough time before your due date to (1) carefully review what you have written, (2) revise it as needed, and (3) complete and edit the final draft. Students are often surprised at how long these steps take—and sometimes must work through the night at less than top form to finish a major paper. Both professional writers and professionals who write (engineers, scientists, and managers, for example) realize the importance of careful review. They schedule review and revision time and push themselves hard to prepare a first draft early enough to allow time for multiple reviews and careful revision.

　　No matter how hard you have worked on the draft, your memo, letter, or report will profit from revising. That's why it's called a draft; it is a

156

beginning or preliminary version, and as such, is subject to revision. Revision may be needed in any or all of these areas:

- choice of purpose, reader, topic, tone, and form
- technical or factual accuracy
- organization

Editing at the sentence level for good diction, appropriate syntax, and correct punctuation is covered in detail in Chapter 8 and the punctuation section of the Appendix.

7.1 THE NEED FOR REVIEW

Suppose that for the past year you have been building a single-engine airplane from a kit. You have put in countless hours building the frame, assembling the fuselage, building the engine, and attending to the details of wiring, control surfaces, and instruments. After you tighten the last screw, will you hop in the seat, rev the engine to its maximum, and take off at top speed down the runway? Probably not. Instead, you will run a series of tests, checking out each major component to see if it is doing its job and working properly with all the other components. You might even have it checked by an experienced mechanic—because only after careful checking will you feel ready for the sky. Your life might hinge on the outcome of those tests.

The writing situation is very similar. Even a short memo or letter has taken concentrated effort on your part. Before you release it, you want to ensure that each part is doing its job and working with the other parts to fulfill your purpose. A long proposal or report, like an airplane, has many parts—all the more reason for careful review and, if necessary, revision. Before you started writing the draft, you reviewed your outline to be sure that the organization you had chosen was the best way to fulfill your purpose and meet your reader's needs. Now you want to look again at what you have written to see if your document is technically accurate, clear, and correct—because your reputation as a writer might hinge on the outcome of this second look.

If possible, let the draft cool for a day or two before you begin reviewing. You need to make the difficult shift from being a writer to being a reviewer. When the material is still fresh in your mind, it's difficult to put yourself in a reader's place, but once you've been away from it for awhile, you can be more objective.

7.2 TYPES OF REVIEW

In both the classroom and the professional world, several types of review are possible, and for any written piece you may choose one or more of these reviews. The longer and more complex the writing assignment is, the more likely you'll be to use more than one type of review. The three basic types are self-review, peer review, and instructor or manager review.

Self-Review

Whatever other types of review you might use, you're sure to have to review your own document at each draft and then again in its final form. To help objectify your review, look at your original Planning Sheet and remind yourself of your intended purpose, readers, topic, tone, form, and style. Then check your document to see if your goals have shifted or if you have done what you intended to do. Follow the three phases of review discussed in section 7.4, revising as needed after each phase. Don't be afraid to read your document aloud. You yourself should provide most of the changes.

Peer Review

Most companies provide for peer review because colleagues can closely approximate your intended readers. In addition, your peers often have a stake in what you're writing, so they will review carefully. Peer review can be done internally or externally. In either case, it's important to give the manuscript to the reviewers in advance, so they have time to read it carefully before you meet.

Internal peer review on the job involves members of your department; in the classroom it involves fellow students in your writing class. In both cases, the reviewers will know something about what you're doing, and they may even be experts who can help with technical accuracy.

External peer review on the job involves people from other departments who have an interest in what you have written. They are concerned with what you say because it affects what they do. When they read, they can check for clarity as well as technical accuracy. When you are a student, external reviewers may be family members or friends whose critical abilities you trust. Often they won't know the subject matter at all, but they can provide a valuable check on clarity. If you are writing to generalists and your external reviewers don't understand, you know you need to revise. Peer review can be carried out one-on-one or in small groups. Often small groups work best because of the group dynamic: that is, one comment triggers another, or one observation encourages another person's new idea. Peer reviewers work best when they have questions like those listed under the phases of review in section 7.4.

Instructor or Manager Review

As a professional, you might well use all three types of review, the last reviewer being your manager. As the person who probably assigned you this writing task, he or she is in a unique position to assess how effectively you have performed. Sometimes managers see only the final copy, but more often they help in its preparation by reviewing for technical accuracy and clarity. You should also think of your instructor as a manager and seek interim reviews when you suspect problems. Most instructors will be glad to share their experience with you if you don't wait until the last minute to ask.

7.3 TECHNIQUES FOR REVIEWING A TEAM-WRITTEN DOCUMENT

When several writers work on portions of a long and complex document, their various writing styles and approaches must be knit together before or at the reviewing stage so the document will have unity and coherence. Typically, a lead writer or editor will cut, paste, and revise to blend the various portions. Then, perhaps, all the writers will participate in the review. If this is the situation, you must force yourself to be objective, viewing your own writing only for its contribution to the whole. Many writers are defensive about their writing and find criticism hard to accept. Yet think how much easier it is to be criticized at the draft stage, when you can revise the document before it is cast in its final form, than it is to be blamed for faults in the final document.

As a critic of someone else's writing, remember to be positive as well as negative. All of us are vulnerable when we expose our writing for criticism; sometimes you will accomplish the most by focusing on the best parts of a document and then revising the rest to bring it up to that standard.

Whether you are writing as an individual or part of a team, it's important to have your draft written well before the due date. That's why early planning and frequent reference to your schedule are so important. The only way to benefit from a review is to have time for a thorough evaluation—and time for the revision that will follow.

7.4 THE THREE PHASES OF REVIEW

Whether you are reviewing your own writing or someone else's, you can do a better job if you break the review into three phases, as shown graphically in Figure 7-1.

That means you should read through the document three different times, each time looking for different things. By concentrating on one thing at a time, you can be more thorough. Content review and organization review are covered in this chapter because you will probably need to revise after each one. The third review—editing—is covered in Chapter 8, along with exercises to sharpen your editing skills.

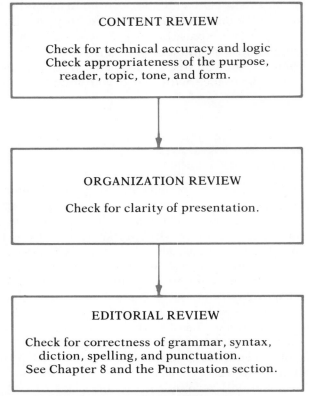

Figure 7-1   Flow chart of the three phases of review

Content Review

Phase 1 is a review of *content*; you are checking for technical accuracy and logic and the appropriateness of the purpose, reader, topic, tone, and form. Check your notes or your original sources to verify the accuracy of numbers, facts, dates, names, quotations, and conclusions. Check all the graphics for accurate presentation of data. Review documentation references to ensure completeness and correctness. Ask these questions about the content:

- Is the purpose clear and appropriate?
- Is the thesis clear and is it supported by evidence?
- Does the thesis support the purpose for writing?
- Are the facts, figures, and computations correct?
- Is all necessary information at hand?
- Is the information appropriate for the intended reader or readers?
- How can the tone be described?

- Is that tone appropriate for this document?
- Does this document need more or different graphics to make the point?
- Can anything superfluous be eliminated?
- Do the conclusions follow from the data presented?
- Do the recommendations make sense in light of the conclusions?
- Does the form of this document best convey the content to the reader?

Make any changes or corrections indicated by the Phase 1 review.

Organization Review

Phase 2 is a review of *organization;* you are checking for clarity of presentation, knowing that it does no good to be accurate if what is written is unclear. Start this review by reading only the headings in order; then read the subheadings in order. Verify that the numbering system is consistent, that the levels of headings follow an accepted format, and that the headings are grammatically parallel. Make sure that the headings in the table of contents are the same as those in the document. Ask if your method of organization clearly communicates the thesis.

After you have reviewed headings and subheadings, ask these questions about organization and clarity:

- Does this document make sense? If not, why?
- Is the major point clear? What causes confusion?

Now study paragraphs for organization and coherence. Ask:

- Do the subheadings and the topic sentences support the thesis?
- Do the sentences in each paragraph support that paragraph's topic sentence?
- Is each section or paragraph linked logically and smoothly to the one before and the one after it?
- Can the intended readers find the information they want? Is it easy to find information by skimming? Do most pages highlight key words and phrases so they are easily found?
- Does this document need more or different definitions, examples, descriptions, instructions, or details explaining who, what, where, why, when, or how?

When you complete the Phase 2 review, make any changes or corrections called for. You may have to move or interchange sections or paragraphs to achieve greater clarity. You probably already worked on this when you reviewed and revised your outline, but you may now have to organize once more. You can use the comments from your document review to guide you in reorganizing. If you are working from a typed or handwritten draft, chop sections apart with scissors if necessary and reassemble the draft with tape. If you are working on a computer, add, delete, and move blocks of data as needed. If at all possible, type or print out a clean draft before the Phase 3, or editorial, review.

Editorial Review

Phase 3 is a review of various *editorial* matters, including word choice, sentence construction, punctuation, capitalization, and spelling. These are covered in Chapter 8 and the Handbook in the Appendix.

CHECKLIST FOR CONTENT AND ORGANIZATION REVIEW

Phase 1: Content Review
    1. Do I have accurate
        _____ figures?
        _____ facts?
        _____ computations?
        _____ dates?
        _____ names?
        _____ quotations?
        _____ conclusions?
        _____ graphics
        _____ references?
    2. Does the thesis support the writing purpose? Is that purpose clear and appropriate?
    3. Is the thesis clear and supported?
    4. Is all information included?
    5. Is the information appropriate for the intended reader?
    6. Is the tone appropriate?
    7. Are conclusions supported by data?
    8. Do recommendations follow from conclusions?
    9. Have I chosen an appropriate form?

Phase 2: Organization Review
    1. Are headings and subheadings
        _____ accurate?
        _____ parallel?
        _____ correctly numbered?
    2. Do paragraphs concentrate on one main idea?
    3. Are topic sentences supported with specifics?
    4. Are paragraphs and sentences logically joined with transitions?
    5. Have I highlighted key words and phrases to improve reader comprehension?

EXERCISE 7-1 Self-Review

Directions: After letting your first draft cool for some time, review it yourself in two separate phases. Use the checklists and be meticulous in marking questions, unclear passages, and possible errors. Revise as needed.

EXERCISE 7-2 One-on-One Review

Directions: Pair up with another student and exchange papers. Review the document in two separate phases, making notes and suggestions on the manuscript. Apply the specific questions given in section 7.4 and summarized in the checklists. Note positive as well as negative things. When you get your own paper back, analyze the comments and revise as needed.

EXERCISE 7-3 Peer Review

Directions: Join a group of three to four fellow students. Submit a copy of your document to each group member. Review their documents in two phases, making notes on the documents and answering the questions from section 7.4 in writing. Make positive as well as negative comments. When your own document is reviewed, take notes on the comments and use them to revise.

EXERCISE 7-4 Instructor Review

Directions: Select a particularly complex or difficult part of your paper and give it to your instructor for review. Also give your instructor a set of comments, questions, or concerns to bear in mind while reading this section of your paper.

CHAPTER 8

Editing

The last review you give a document is an editorial review; you are checking for correctness of grammar, diction, syntax, spelling, and punctuation. You may need to improve

- word choice. Your vocabulary must be simple, clear, and appropriate. If not, you'll need to revise.
- sentence construction. Your sentences must say what you mean efficiently and correctly.
- punctuation. Each sentence should be punctuated to *help* the reader understand.
- mechanics of capitalization and spelling.

If you followed the suggestions in Chapter 7 and sought review of your writing, or if your instructor assigned peer review in or out of the classroom, you probably have rewritten and revised for content and organization. Now you will examine individual sentences—even words—to see if you can sharpen their focus. In order to do that, you may need to brush up on your editing skills. Sharpening your editing skills will also help you comment intelligently on other people's writing.

You can use this chapter as a guide to editing, a resource for information, and a place to practice those writing skills in which you are weak. This chapter assumes a basic understanding of grammar. If you need help in basics, consult a college composition handbook or tutors at your school. Ask your instructor for advice. The appendix contains a glossary of grammatical terms, a guide to punctuation, capitalization, spelling, and the treatment of numbers, and exercises to enhance those particular skills.

8.1 EDITING BY EYE AND BY COMPUTER

You can edit a document either by eye or by computer—sometimes you will want to do both.

Editing by Eye

Students often find editing at the sentence level difficult because they know their own writing too well. Here are some ways you can edit your own sentences more easily:

- Read the document aloud to yourself or someone else. Hearing the words often helps you spot awkward constructions, overly long sentences, misused words, or gaps in logic.
- Force yourself to read slowly by covering the bottom part of the page with a card and moving the card only when you complete each sentence.
- Read the document backward one sentence at a time, starting with the last sentence. This technique forces you to study individual sentences rather than skim whole paragraphs or sections.
- Have a trustworthy friend or colleague read the draft aloud to you while you follow on another copy. If the reader stumbles while reading, see if the problem is an unclear or awkward sentence.

In sections 8.2 and 8.3 and in the punctuation section of the Appendix you will find many explanations and exercises to help you improve your writing skills at the sentence level. When you find problems during your editorial review, you may want to read the appropriate section and do some practice exercises. For very basic problems, consult any standard college handbook or ask your instructor for help.

Specifically, in the editorial review you should look for problems in

- grammar

  Are all the sentences complete, with a subject and verb and a single thought?

  Do subjects agree with verbs? Do pronouns agree with their noun antecedents? *Are* there antecedents for pronouns?

  Are tenses accurate and consistent?

- syntax (word usage)

  Are sentences pruned of unnecessary words and inappropriate expletives like "there are" and "it is"?

  Are most sentences in active rather than passive voice? Are passive sentences that way for a reason?

  Is the action of a sentence lodged in the verb?

  Is each sentence arranged to put the most important information either at the beginning or the end?

- diction (word choice)

  Are the words specific and concrete?

  Is each word accurate? Check especially for words commonly confused.

  Do the words avoid sexist bias?

  Is any jargon used? If so, is it appropriate for the intended reader? Can simpler terms be substituted? Do terms need to be defined?

  Can pompous words be eliminated to improve the tone?

- spelling

  Is each word correctly spelled? When in doubt, check it out!

- punctuation

  Do the punctuation marks help the reader understand by showing the relationship of one part of the sentence to another?

  Does the punctuation follow generally accepted professional standards?

During the editorial review, use your dictionary and the relevant sections in this chapter and the Appendix to help you correct and revise words, sentences, and punctuation. After editing and revision, you are ready to submit your document for a second review or to produce your final copy. See Chapters 14 to 22 in Part 3 for help in preparing the final copy of specific documents.

Editing by Computer

Another way to edit a document is with editing software (a computer program). The most common editing function is checking spelling, but other programs highlight such things as sentences in the passive voice, nominalizations (nouns made from verbs), long words, and long sentences. Often the programs will highlight a suspect word and offer alternatives. Helpful as these computer programs are, though, the final responsibility rests with you, the writer. Spelling checkers won't flag correctly spelled words that are misused, nor will other programs discriminate between acceptable and unacceptable passive voice sentences. You can use computer reviews most effectively if you know the standards for good writing yourself. Don't rely on the computer to do the editing for you.

Another caution: Don't try to learn a computer program for word processing during the same week your final report is due. If you plan to use a word processor, practice on it first with short documents like memos, letters, and short reports.

CHECKLIST FOR EDITORIAL REVIEW

1. Check grammar.
_____ Do all sentences have subjects, verbs, and a complete thought?
_____ Do subjects and verbs agree in number?
_____ Do pronouns and antecedents agree in number?
_____ Are tenses accurate and consistent?
2. Check syntax.
_____ Are sentences lean and clean?
_____ Are most sentences in active voice? Are passive sentences justifiable?
_____ Is the action of the sentence in the verb?
_____ Does the most important information appear either at the end or the beginning of the sentences?
3. Check diction.
_____ Are words specific, concrete, and accurate?
_____ Are words neither sexist nor biased?
_____ Is jargon appropriate to the reader or defined?
_____ Are words the simplest that accurately project the meaning?
4. Check spelling.
_____ Is each word correctly spelled?
5. Check numbers.
_____ Is numeral and word use of numbers consistent?
6. Check punctuation.
_____ Do my punctuation choices help the reader understand the relationship of ideas?
_____ Does the punctuation follow accepted professional standards?
7. Check mechanics.
_____ Do mechanics and capitalization follow accepted standards?

8.2 USING CLEAR WORDS: DICTION

Writing is intended to communicate—to move facts and ideas from the mind of the writer to the mind of the reader. Too often in the technical and business world, though, that stream of communication is polluted: muddy or clogged with garbage and junk. Instead of communicating, the writing confuses; instead of educating readers, it enrages them. The lack of clarity in government writing is so serious that laws have been passed against it. In 1982 the California legislature, for example, passed a bill stating that:

> Each department, commission, office or other administrative agency of state government shall write each document which it produces in plain, straightforward language, avoiding technical terms as much as possible, and using a coherent and easily readable style.

The idea of passing a law to eliminate muddy writing may seem absurd. However, to the reader looking for information and to the writer trying to make a point, muddy writing is not a joke but an obstacle. You should remove the mud from your sentences during editing.

Before you can edit for clear sentences, you need to know what kinds of errors you are looking for and how you can eliminate each one. Muddy writing is a result of poor word choice and poor sentence construction. A convenient term to describe word choice is *diction:* the choice of words and the force and accuracy with which they are used. The term for sentence construction is *syntax:* the pattern, structure, or arrangement of words in a sentence. An easy way to remember the difference is

Diction = Word Choice
Syntax = Word Order

Use Concrete Words

As a child, did you ever play a game of locating yourself in space? You were sitting, perhaps, on the bed in your room, doodling on a sheet of paper, and you began there with your name and the place, as shown in Figure 8-1. With each level you moved outward from a place easily understood and easily visualized (your room) to larger places that were harder and harder to see and understand. You were creating something similar to a "ladder of abstraction."

On a ladder of abstraction, the bottom rung holds the most specific term. This might be a model name or a part number. Thus, the bottom rung has the most *concrete* word: a specific place, object, action, or person. Concrete words help readers understand because they create pictures in the mind's eye. For example, if you are writing about fruit, a concrete term would be Golden Delicious, a specific kind of apple. With each level that you move up the ladder, the term becomes less concrete and more general until you reach a rung where the term becomes too inclusive to be visualized at all. Now you have an *abstract* word: a word naming a general condition, quality, concept, or act.

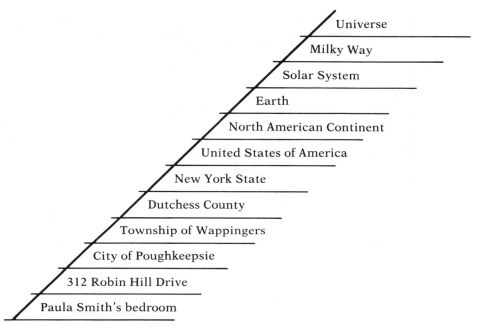

Figure 8-1 **Locating oneself in space**

Put the Golden Delicious apple on a simple ladder of abstraction, as shown in Figure 8-2.

As you move up the ladder, the words become more general and less specific. When you read the word "sustenance," do you *see* anything in your mind's eye? At what level on this ladder do you stop being able to visualize an object?

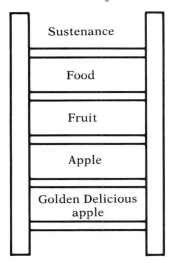

Figure 8-2 **A simple abstraction ladder**

Objects (named by nouns) are not the only words that can be placed on a ladder of abstraction. Verbs (action words) can also be concrete or abstract. Consider the verb "perform." What visual image do you have of this word? Do you see an actor on a stage? A pianist at a grand piano? A quarterback passing the football? Or do you see a 5-liter V-8 EFI engine accelerating smoothly and rapidly? "Perform" could be the verb you'd choose in each instance above, but you'd help your reader by choosing a verb to describe that specific action.

It's tempting to say that in technical writing you should always use words at the bottom of the abstraction ladder; in other words, that you should use only concrete words. But it's not that simple, because you will also need abstract words like "economy," "industry," "perseverance," and "liabilities." When you edit for diction, you can, however, achieve clarity if you:

1. Choose the most concrete words (those at the lowest level on the abstraction ladder) whenever possible.

   Mazda RX7, not *automobile*
   rheostat, not *resistor*

2. Clarify abstract words by giving concrete examples—several of them if you wish to indicate a range.

   Your *communications* [abstract word] must be clear and organized whether you are writing *request letters, status reports,* or *user manuals* [concrete words].

If you choose general or abstract words when concrete or specific terms are available, you will be guilty of vague or ambiguous usage. In other words, if your reader is not sure of your meaning, you have failed to communicate.

EXERCISE 8-1 Concrete and Abstract Words: Adding Specifics

Directions: Each of the sentences below is vague because it uses too many general and abstract words. First, circle the general words and phrases. Then, replace them with words and phrases that are more concrete or specific. Make up your own details. Discuss and revise your new sentences with a small group of fellow students and be prepared to present them to the class.

1. The substantial part of the body of the report rests on the information expected to be received from these companies.
2. They think they would not have time for them in the future.
3. The firm's excessive charges over a long period of time caused us to cancel their contract recently.

4. After flying at altitude for a long time, many pilots experience hypoxia.
5. Recent research indicates that substantial numbers of immigrants from the Far East experience feelings of disorientation and culture shock.
6. Adding a weak acid changed the color of the solution.

Avoid Clichés and Vogue Words

Clichés have been called "pre-fab" phrases. They drop into our minds without thought because we have heard them so many times. Once they were fresh and appropriate, but overuse has dulled their meaning. A cliché muddies a sentence by keeping the reader from focusing on a specific message.

Some writers of business letters seem to be still wedged in the 19th century. They use formal, archaic expressions that obscure their intended message. Study the list below and think of an equivalent in plain English.

Pursuant to your letter of May 3 . . .
Enclosed please find . . .
In re your letter of November 15 . . .
I enclose herewith . . .
Thanking you in advance, I remain . . .
Please be advised that . . .
I will sincerely appreciate . . .

Your business letters and memos will be shorter and cleaner if you replace pre-fab words with words of your own choosing.

If letter writers seem drawn to old-fashioned clichés, technical report writers often favor vogue words and phrases. Today many vogue or "trendy" words come from the computer industry, and they have quickly become overused. In the computer world, terms like "interface," "input," and "user-friendly" have a specific meaning and a specific place. But when words like these are used to describe interpersonal relationships, or interactions in another technology, they mislead and misinform.

Following are some vogue words that have become clichés in the last few years. Add others that you read or hear on the job or in the marketplace, and resolve not to use these terms unless they specifically apply.

glitch	state of the art	interface
parameters	utilizing	window
access	debug	bottom line
input	facilitate	networking

Many technical writers use clichés that are not vogue words at all, but simply old and tired expressions. Phrases like the following should be replaced whenever they appear in your first draft. To assure yourself that you do know other ways of stating the idea, write an equivalent phrase for each cliché.

each and every	no sooner said than accom-
in the final analysis	plished
needless to say	pave the way
last but not least	method in his madness
to make a long story short	better late than never
few and far between	to all intents and purposes
easier said than done	get down to brass tacks

EXERCISE 8-2   Clichés: Choosing Fresh and Accurate Expressions

Directions: Rewrite each sentence to eliminate clichés or vogue words. Make up details.

1. I enclose herewith sufficient postage to pay for shipping the new catalog.
2. Please be advised that we no longer manufacture photovoltaic collectors.
3. If you can access that information promptly and get back to me, I will do everything in my power to meet the schedule.
4. Needless to say, the relative humidity of the desert is low; average rainfall for a year is less than 125 mm.
5. The Antarctic region, to all intents and purposes, is a high ice plateau surrounded by mountains.
6. Each and every person on earth is utterly dependent on water for survival.
7. Recovered meteors are few and far between because most burn up through friction as they enter the atmosphere.
8. If the earth's polar ice caps and glaciers were to melt, the mean sea level would rise by about 200 feet. That additional water would pave the way for mass destruction because half the world's population would be submerged.
9. In the final analysis, the earth is watery; almost three-fourths of the surface is covered by water.

Discriminate between Words That Are Easily Confused

Many of the words in your vocabulary have come to you orally; you've heard them on radio or television broadcasts, but you haven't seen them in print. Therefore, you don't have a visual image of the word, and you may have only a hazy notion of its spelling or its meaning. It's easy under those conditions to confuse words that sound alike. Likewise, words may have visual similarities that don't correspond to similarities of meaning. The only way to ensure accuracy is to make constant use of the dictionary.

For example, what is the difference between "replace" and "reinstall"? Both words mean to put something back, but what is put back differs. If you *replace* a faucet, you take one out and put a different one its place. If you *reinstall* a faucet, you put the same one in again (perhaps repaired, but still the same one). Can you discriminate in this way between "respectfully and respectively"? Between "universally" and "generally"?

EXERCISE 8-3   Discriminating between Words That Are Easily Confused

Directions: To sharpen your understanding of the following paired words, look up each one in the dictionary. Do not just write down the definition. Instead, write a sentence that explains the *difference* between the two words. Then use each word correctly in a sentence.

1. adjacent/contiguous
2. advise/inform
3. affect/effect
4. alternative/choice
5. among/between
6. amount/number
7. anxious/eager
8. appreciate/understand
9. assure/insure/ensure
10. bimonthly/semimonthly
11. capital/capitol
12. censor/censure
13. complement/compliment
14. continual/continuous
15. credible/creditable
16. deteriorate/degenerate
17. disinterested/uninterested
18. few/less
19. filtrate/filter
20. further/farther
21. idea/ideal
22. infer/imply
23. its/it's
24. liable/likely
25. maximum/optimum
26. oral/verbal
27. perfect/unique
28. principal/principle
29. reaction/opinion

EXERCISE 8-4   Checking for Accurate Diction

Directions: Edit the following sentences to change any inaccurate words or terms. Check your dictionary to be sure of the meanings. Be prepared to justify your changes.

1. Caution: Take care in handling the film since the resistance to damage has now been increased.
2. Work assignments outside the noted shift hours will be coordinated on an individual basis with a supervisor.
3. Management has carefully studied the sight for the new building but has not yet decided to build.
4. The conference speaker had a real flare for words.
5. The amount of people in the room far exceeded those allowed by the fire regulations.

6. He inferred to me that I was in line for promotion.
7. The critic complained that Tri-Level Company did not "go after its goal in a humbling way."
8. The list of special features is infinitesimal.
9. I want to be appraised continuously of your progress.
10. The assembly base requires a four-legged tripod.
11. Because the President and other high-ranking officials were preoccupied with the scandal, affairs of both foreign and domestic importance weren't able to flourish.
12. Are freedom today is lessening as the price of gas becomes more expensive.
13. There are millions of microcomputers in use today in both the home and the workplace. Of these, perhaps as little as 5 percent are protected by electrical stabilizers known as surge protectors.
14. I have been anxious all day long to tell you the good news.

Recognize Denotative and Connotative Meanings

The *denotative* meaning of a word is its direct, objective, and neutral meaning. The word "mother," for example, denotes simply "female parent." The *connotative* meaning of a word includes associated meanings that are indirect, subjective, and often emotionally loaded. When you hear the word "mother," what are your immediate associations? Probably you first think of your own mother, and then your responses are triggered by your association with your mother—maybe warmth, caring, sharing, helping or perhaps nagging, meanness, and smothering.

Theoretically, technical writing is objective and denotative, but in fact, much of it must be persuasive in forms like proposals, memos, letters, and reports. Thus the *connotative* meanings of words become important because they affect the reader's attitudes. Often when the *tone* of a letter or memo is criticized, the offensive tone is caused by connotations of particular words. For example, a "white-collar worker" might be offended at being called a "worker," while a "blue-collar worker" would probably accept the title. "Rank and file" implies lower-level unionized workers and may have a negative connotation, while "front-office" implies management and thus may have a positive connotation. On the other hand, if one belongs to the "rank and file" and has a grudge against the "front office," the connotations could be reversed. If a student tells me she "expects" my approval of her proposal, I am irritated. However, if she "requests" my approval, I am pleased.

Avoid Sexist Language

During the last twenty years, Americans have become sensitive to the subtle ways in which language can influence our attitudes toward persons of the opposite sex. Substantial changes have occurred in common usage in the following areas:

• job titles. We now hear

flight attendant	*instead of*	stewardess
firefighter	*instead of*	fireman
mail carrier	*instead of*	mailman
cleaner	*instead of*	cleaning woman

• letter salutations and attention lines. Common now are

| Dear Supervisor: | *instead of* | Dear Sir: |
| Attn: Research Dept. | *instead of* | Gentlemen: |

Still, writers have not yet found acceptable substitutes for the use of "he" and "his" to refer to both men and women. Most writers choose one of the following solutions:

1. Make the sentence plural.

NOT: Each technician *is* responsible for *his* own equipment.
BUT: Technicians *are* responsible for *their* own equipment.

2. Eliminate the pronoun or replace it with "the."

Each technician is responsible for *the* issued equipment.

3. Change the third-person pronouns to first-person or second-person pronouns.

We are responsible for *our* own equipment.
You are responsible for *your* own equipment.

4. Use "he and she" or "his and her."

Each technician is responsible for *his or her* (or *his/her*) equipment.

5. Alternate pronouns from chapter to chapter or paragraph to paragraph, using male pronouns in one and female in the next.

The five solutions in this list appear in order from the most graceful and inoffensive to the least graceful and inoffensive, so you should keep the ranking in mind when you choose alternatives. You can also see that each solution changes the meaning slightly. You need to think about the words you use and the effect you want to create. But above all, you need to keep your purpose and your readers clearly in mind.

EXERCISE 8-5   Finding Examples of Connotative and Sexist Language

Directions: To sharpen your sensitivity to word choice, find five examples of connotative or sexist language (or, conversely, five examples of denotative and nonsexist language). Look in newspapers, magazines, books, or technical reports. Clip out or copy the examples and bring them to class for discussion.

Clarify the Jargon of Your Specific Occupation

Every occupation develops its own terminology—abbreviations, symbols, words, and phrases that have specific meanings to the people in that field. "Shop talk" can be useful shorthand for communicating with coworkers; it saves time and limits meaning. Airplane pilots communicate with other pilots using terms like pitch, yaw, and sink rate. Automobile mechanics talk about universal joints, shackle bushings, and the front planetary gear set. Crab fishers discuss peelers, pots, and progging.

Each of these workers is using shop talk or jargon, and using it effectively and properly. However, the word "jargon" has two very different meanings, and technical writers need to understand both.

> JARGON #1: the highly specialized vocabulary of a particular group, trade, or profession. Call this *good* jargon.
>
> JARGON #2: meaningless or pretentious terminology; doubletalk. Call this *bad* jargon, pompous language, or *gobbledygook*.

Ironically, the same words or phrases can be either jargon or gobbledygook depending on the context and the reader.

Because good jargon is specialized language, you must know the terms of the field in which you work. Those terms can range all the way from mathematical symbols and acronyms to phrases and clauses. The key to using good jargon effectively in your writing is constant awareness of your intended reader or readers. You can use jargon *only* if you are writing to a technical audience (an audience of specialists and technicians) and *only* if you are positive that all of them know the terms you use. Otherwise, you should use common terms whenever possible, and carefully explain or define each jargon term the first time you use it.

EXERCISE 8-6   Compiling a List of the Jargon in Your Field

Directions: Write down as many of the jargon terms from your major field as you can think of. Then meet with other students in your class who are in the same major, and work together to define each term for a generalist reader. Add to the list when you encounter more jargon terms.

EXERCISE 8-7   Rewriting Jargon for the General Reader

Directions: The following sentences are taken from a variety of specialized fields and are written for experts in those fields. Using a general or technical dictionary to find common equivalents, rewrite each sentence for a general reader. Be ready to discuss what you can and cannot translate.

1. Psychology (from a textbook):

The counselor and client should engage in direct, mutual communication whenever it is useful, especially in times of stress and when the relationship seems aimless. Both underuse and overuse of immediacy in interpersonal relationships, including counseling, are stultifying.

2. Personal computers (from a salesperson):

Word processing on this system is really a snap. You just boot up your DOS, put your diskette in your drive, wait for the prompt, then check your defaults and your formatting, and you're ready to go.

3. Medicine (from a journal for paramedics):

Jugular venous distension (JVD) is absent, as is hepatojugular reflux (HJR). Auscultation of the lungs reveals good air exchange with normal vascular breath sounds. Rhonchi and rales are absent.

4. Law (from a promissory note):

In the event the herein described business property or any part thereof, or any interest therein is sold, agreed to be sold, conveyed or alienated by any of the promisors, or by operation of law or otherwise, all obligations secured by this instrument, irrespective of the maturity date expressed herein, at the option of the holder thereof and without demand or notice shall immediately become due and payable.

5. Ski Manufacture (from a technical report):

Aluminum and fiberglass laminates are commonly used in the load carrying layers of Alpine skis due to their great elastic properties relative to their weight. Despite the high elasticity of aluminum, permanent deformation will occur in an aluminum structure prior to occurrence of permanent deformation in an equivalent fiberglass structure.

Eliminate Gobbledygook

The word "jargon" comes from the same root that gives us the word "gargle," or "a noise made in the throat." *Bad* jargon is just that—gargling noise that doesn't mean anything. Other names for bad jargon are pomposity and gobbledygook. Texas Congressman Maury Maverick coined the word "gobbledygook" in the 1940s to describe bureaucratic writing that he said made as much sense as the gobbling of a Texas turkey. "Gobbledygook" is now a standard dictionary term; unfortunately, there's so much bad writing in government and industry that we need a term to describe it.

Gobbledygook has two primary ingredients: circumlocution and jargon. Circumlocution is a roundabout way of writing that uses many unnecessary words, and jargon is language that is specialized for a particular occupation. Add these two together, and you're writing gobbledygook.

Many writers think that to sound "technical" and "professional" they must use long, fancy words. They add verb endings to nouns or adjectives to get words like "incentivize," "prioritize," and "finalize." They take the word "solution," add "ing" to it, and say "solutioning a problem." Or even worse, they take an adjective like "ambiguous," add a prefix to make it negative— "disambiguous"—and then add a verb ending to the whole mess to turn it into a verb. The result (I didn't make this up!): "disambiguate."

Another way gobbledygook writers work is to inject fancy words into an ordinary sentence. Here are the directions for collecting student ratings at a university.

> To avert any procedural irregularities which can adversely affect your results, it is essential that you appoint a student who can reliably conduct the collection process described on each packet.

What does this sentence say? Something like this:

> Ask a reliable student to administer and collect the rating forms. If they're not filled out correctly, your rating might go down.

How can you avoid gobbledygook? Use the simplest and most accurate word you can, and don't try to impress your reader with the big words you know.

EXERCISE 8-8 Replacing Pompous Words

Directions: For each of the inflated words below, write a simple equivalent. Keep this list near your desk and use the simpler words whenever you can.

1. circumvent	11. peruse
2. utilize	12. initial
3. attempt	13. implement
4. unequivocal	14. minimal
5. initiate	15. effectuate
6. unavailability	16. endeavor
7. facilitate	17. increment
8. subsequently	18. rescind
9. sufficient	19. prioritize
10. ascertain	20. terminate

EXERCISE 8-9 Rewriting for Clarity

Directions: Rewrite each of the sentences below in simple, clear language. Be prepared to discuss problems of interpretation in class.

1. From an ad for a technical writer/editor:

> The position will require a degree in English, Journalism or equivalent; 3–5 years experience writing in an engineering/technical environment, fa-

miliarity with document preparation and production processes, and experience in expediting commitments from those providing inputs to documents.

2. From a government memo to employees:

... for those bonus goods or benefits that accrue to the Government, an agency may not set up internal procedures for the return of such goods for which it has been determined that the Government is unable to use such benefits or goods.

3. From an ad for ski boots:

This revolutionary design incorporates a traditional fitting system, which closes the shell around the foot, with high-tech fit control systems. The C series is a modern, rear-entry boot and features a single TD constant pressure buckle system which has two fulcrums for ease of adjustment.

4. From a government request for proposal:

If, in the view of the Contractor, due to specified Training System architectural, functional, and/or operational characteristics, it is technically reasonable and prudent to address the subject Software as comprising some number of essentially independent functional or operational units, each having clearly defined interactions with all other interfacing units, if any, each such unit shall be identified, assigned a specific nomenclature, and briefly described in terms of its functional characteristics and interfaces within the specified Training System.

5. From a course description:

This course will provide practitioners with guidelines and specific materials to plan and implement activities for the older adult in community and institutional settings.

8.3 CONSTRUCTING EFFECTIVE SENTENCES: SYNTAX

Choosing clear words, as you learned in the diction section, is essential to good technical writing. Equally important is arranging those words in an order that will communicate your intended message to your reader. Look at the following list of words:

boy	had	little	the
wagon	red	a	black

By themselves, these words do not convey a message. You can arrange them into an order that will say something meaningful, but you cannot be sure what the order should be. How many sentences can you write with these words? Spend a few minutes writing down three or four possibilities.

Did you notice that you could combine the words in several ways that changed the meaning radically? You could also, of course, combine those words in ways that made no sense at all. In putting the random words together, you completed an exercise in *syntax*, which deals with word order, or what can be called sentence patterns. To understand the importance of syntax to clear technical writing, look at the following sentence:

> This is the *only* tape rated "acceptable."

Suppose we move the word "only":

> This is the tape rated *only* "acceptable."

Do you see how important the position of a word can be in an English sentence? By the simple act of moving one word, you have completely changed the meaning of the sentence. And remember, your job as technical writer is to communicate clearly what you mean and only what you mean to your reader.

This chapter assumes a basic understanding of English grammar. Consult the Glossary in the Appendix if you need definitions of the grammatical terms used in the following pages.

Expectations of Readers

Technical writing often deals with complex ideas, and when the ideas are complicated, the writing must be crystal clear. A good technical writer will be like a meticulous groundskeeper who moves over the golf green to clip the grass and smooth lumps and bumps. When the green is free of impediments, the players can concentrate on their strokes. Just so, when the writing is free of distractions, the reader can concentrate on content.

One easy way to help your readers is to meet their subconscious expectations. Because readers come to any piece of writing with much practice in reading, they subconsciously expect sentences to be built in certain ways, following specific sentence patterns. In English the most common patterns are the simple

> subject-verb-object (SVO)
> subject-verb (SV)
> subject-verb-complement (SVC)

For example,

> The unit [S] can dump [V] the module's local memory [O] in 20 seconds.
> The report [S] will be rewritten [V].
> The Model 620 [S] is [V] the fastest digitizer [C].

How do we know these patterns are common? In 1967, Professor Francis Christensen studied samples of 200 sentences from each of 20 well-known American writers—10 fiction writers and 10 nonfiction writers. He analyzed the kinds of sentences they wrote and counted the different sentence patterns. He found that 75.5 percent of the time these writers constructed subject-verb-object, subject-verb, or subject-verb-complement sentences.

Another 23 percent of the time the sentences opened with some kind of adverbial modifier—a single word, a phrase, or a clause—and the rest of the sentence was in S-V-O order. Most of the adverbial openers served as transitions from a previous idea to a new idea. For example:

> Otherwise, designers should stay with industry-standard parts.

Adding these two figures together, you can see that respected professional writers use nearly the same sentence pattern over 98 percent of the time. The very few exceptions are sentences in inverted order (with a delayed subject) and sentences that open with verbals. For example:

> Included also is a projected cost analysis. (delayed subject)
> Elevating the beaker, he examined the precipitate. (verbal opening—present participle)

What can you learn from this as a writer or editor?

1. You communicate best when you meet your reader's expectations.
2. Those expectations are for sentences that follow the standard S-V-O, S-V, or S-V-C order most of the time.

That doesn't mean you have to write short, choppy sentences of the "Run, Spot, run" variety. It does mean that you don't have to write complicated sentences in order to communicate complex subject matter.

EXERCISE 8-10 **Analyzing Your Own Sentence Patterns**

Directions: Choose a 200-word sample (about a page of typed, double-spaced material) from a major paper for your technical writing class. Count the total number of sentences. Analyze each sentence by marking subjects, verbs, direct objects, and complements to determine the type. Count the number of each type and record them like this:

    Total number of sentences in sample: _____
    Number of sentences of S-V-O, S-V, or S-V-C type: _____
    Number of sentences beginning with an adverbial modifier: _____
    Number of sentences beginning with a verbal: _____
    Number of sentences with a delayed subject: _____

How does your writing compare with Christensen's results?

Problems with Syntax

When you look out your window in the morning, what prompts you to say, "It's really clear today"? Where I live, that means I can see all the way across my valley and the bay to hills that are clearly sculptured by sun and shadow. Beyond those hills I can even make out the peaks of distant mountains. In contrast, what is it that blocks a clear view? It might be fog, or smog, or rain, or smoke—perhaps snow. Whatever it is, the something preventing clarity is made up of many small particles. By itself, each particle causes no problem, but when all the small particles are present together, they obscure the clear view.

You probably never look at a sentence and say, "What a clear sentence!" In fact, you don't usually notice such a sentence at all, because its clarity leads you to focus directly on the content. A sentence like that— as a sentence—is invisible; it's doing its job well by clearly communicating its message. Sometimes you *do* look at a sentence and frown or even groan. You pick your way through it again, or you back up to the previous sentences seeking help in understanding this one. Where is your focus now? On content? On meaning? No. Your focus as reader has shifted to the sentence itself because the pieces of the sentence have obscured the clear view.

Just as many snowflakes in the air at one time can obscure the view, so too many words in a sentence can obscure the meaning. You might think that the answer is to write short sentences. In fact, some people give just that advice. "KISSS," they say: "Keep it short and simple, stupid!" But short words and short sentences will not, by themselves, guarantee clear communication. To communicate clearly, you must know the material, understand the purpose, and meet your reader's needs. If you *never* use three-syllable words (even if they could accurately say what you want), and if you *never* write a complex sentence, your writing may be short and simple, but it may also be lifeless, and you may force the reader to figure out the relationships between ideas that you should be making clear through syntactical constructions. For example, the very short sentences in the following description make the reader work too hard to relate one idea to another.

> The kilns to be compared in this report are electric kilns. They have a 7-cubic-foot capacity. The firing chamber is where the pieces to be heated are placed. It is a 23 3/8-inch decagon. The sides are 27 inches deep. The outside dimensions are 30 × 36 × 41 1/4 inches. The inside of the kiln is insulated. The insulation is refractory firebrick. This brick is made from a fine grade of clay. The clay will withstand temperatures of 2500 degrees and above. The firebricks are similar to the common red bricks used in construction. But they are much lighter. The firebrick has grooves cut into it. These grooves are to accept the resistance elements. These resistance elements encircle the interior of the kiln.

Syntax problems are not caused by length itself, but by these culprits:

- repetitious words
- long modifier strings

- packed sentences
- inappropriate expletives

Repetitious words   One cause of unnecessarily long sentences is redundancy, or saying the same thing in more than one way. This sentence is from a technical description of a typewriter.

> The main process of the typewriter is its printing mechanisms procedure.

This writer uses extra words to "sound technical." The two related words— "process" and "procedure"—both describe the main function of the typewriter. But what is a "printing mechanisms procedure"? The key word seems to be "printing." This sentence could mean

> A typewriter's main function is to print.

If so, the new sentence uses only half as many words to say the same thing.

Here are two more examples of wordiness caused by repetition. These sentences were finalists in a "Classic Clunker" contest IBM held among its employees in 1983 to call attention to bad writing within the company.

- You can include a page that also contains an Include instruction. The page including the Include instruction is included when you paginate the document but the included text referred to in its include instruction is not included.
- For a priority system based upon fixed numerical ordering of requesters, following the servicing of any given requester, a request from a lower priority requester, if deferred due to a conflict with the just-serviced requester, shall be honored prior to honoring a second request from the just-serviced requester.

Long modifier strings   Another kind of wordiness is created when the writer tries to pack too many technical words into one phrase—particularly in a long modifier string. Such packing yields a sentence like this:

> *The elevator primary power servo electrohydraulic servo* valve accepts series inputs from the AFCS autopilot control function to compensate for engine thrust pitch moments, using the elevator, during manual approach and landing operations.

Even an expert will grow pale reading seven modifiers for the word "valve"—the subject of that sentence. Here's an example from a different subject area:

> In my view, because I am a *biological aspects of management financial* analyst, diversification by management produces both more growth and safety than diversification by industry.

In this sentence the subject complement "analyst" is modified by six words. How can you fix sentences like these? You need to eliminate or move some of the modifiers and break up the sentence. You could move some modifiers to a prepositional phrase in the first sentence, saying

> The electrohydraulic servo valve is located in the elevator primary power servo. This valve accepts . . .

For sentence two, you might move some modifiers to a clause:

> I am a financial analyst who believes in biological aspects of management. Thus, I believe that diversification . . .

In general, you can follow these principles for improving sentences with long modifier strings:

1. Put key nouns first in the sentence.
2. Use possessives and hyphens.
3. Turn nominals into verbals ("management" into "managing").
4. Keep related ideas together.
5. Put prepositional phrases after important nouns.

Note: Technical writing also uses hyphens in modifier strings, so you need to understand when and how to use hyphens correctly. The punctuation section in the Appendix gives both explanations and examples of common hyphen uses.

Packed sentences   Some writers have so much information to present that they overload a sentence, stuffing in detail after detail until the sentence is bulging. Packed sentences overload the reader's short-term memory because there is simply too much information to process at one time. For example, here is a sentence explaining a rating system:

> Two points are given if all office software is available and if CAD software is available—because they are most important; one point is given if the software is UNIX based, if CAD software requires at least 512kb RAM, and the computer has the RAM, if CAD software needs 1Mb RAM or more and the system has the RAM, and if the computer has a co-processor; and a half point is given if the co-processor is available.

You can repair an overloaded sentence like this one in at least three different ways. You can:

1. Make individual sentences out of independent clauses. Periods provide stronger resting places for the reader than do semicolons.

> . . . important. One point is given if the . . .

2. Rewrite the sentence as a list, assigning numbers or bullets to subpoints of equal value. (See Chapter 6 for more on parallelism.)

- Two points are given if . . .
- One point is given if . . .
- A half point is given if . . .

3. Reorganize the information into a table or figure that will graphically present the data. (See Chapter 12 for more about tables and figures.)

EXERCISE 8-11   Unpacking a Wordy Sentence

Directions: Restructure each of the following sentences to present the data more clearly, choosing one of the three methods described above. Use each method at least once.

1. Until recently, the typical VCR offered an audio frequency response at its fastest speed of about 70 Hz to 10,000 Hz, dynamic range of 40 to 50 dB, wow and flutter of 0.15 percent, signal-to-noise ratio of around 40 dB, and harmonic distortion of 3 percent.
2. The *normal program type* will be shooting at 1/60 seconds and at f/4; to minimize camera shake, *program tele* will shoot at 1/250 seconds at f/2; and the *program wide* will shoot 1/30 seconds at f/5.6: the slow shutter speed is to have more depth of field.
3. The T70's shutter is an electronically governed, vertically traveling focal-plane shutter with stepless speeds on automatic from 2 to 1/1000 seconds and manual speeds from 2 to 1/1000 seconds plus B (Bulb).
4. The increased understanding of the printing mechanisms, the editing features including memory, the portability features, etc., made me realize the futility of the typebar printer, the versatility of the electronic typewriter, the great progress made in the last few years with the dot matrix, and the possibilities for increased productivity in the future.
5. Keeping in mind that many reading this report would be doing so to make the best use of their finances, to escape from the stressful life of the progressive, technological atmosphere, to determine whether they could build a log cabin themselves, to take the lowest risk possible in buying their own home, and to investigate the possible return on their investment, the following criteria were chosen: cost efficiency, esthetics, convenience, company reliability, and durability.

Inappropriate expletives   In English grammar, expletives are not swear words but the introductory words "it" and "there" appearing in the usual position of a subject. In a way, expletives are throat clearers; they prepare the way for what the writer wants to say. However, because they usually appear where the reader expects the subject of the sentence to be, the real subject is delayed while the writer clears his or her throat.

It is apparent that the shipment will be delayed for 15 days.
There was a manager assigned to every weekend work shift.

Expletives often add wordiness because (1) the introductory expletives themselves provide no content in the sentence—they're simply extra words, (2) when the usual subject-verb-object order is reversed, the subject must often appear in a clause or phrase, a process that adds extra words, and (3) expletives often combine with the passive voice, which also adds extra words.

Why not say:

The shipment will be delayed for 15 days. (still passive)

or

Apparently, the shipment will be delayed for 15 days. (still passive but with an adverbial modifier)

or

The wholesaler will delay the shipment for 15 days. (active)

or

Apparently, the wholesaler will delay the shipment for 15 days. (active with an adverbial modifier)

Why not say:

A manager was assigned to every weekend work shift. (still passive)

or

Every weekend work shift has an assigned manager. (active)

Note 1: Often you can eliminate expletives, but occasionally you will need an expletive for emphasis, to avoid an awkward construction, to move a long subject to the end of the sentence, or to lead the reader to a new topic. You need to assess each sentence individually and in the context of the paragraph to see if the expletive functions effectively or is superfluous. In the following sentences, the expletives work well:

- to emphasize:

  There are 14 subassemblies included in the working drawings.

- to avoid an awkward construction:

  There was no attempt to interrupt the speaker with questions.

- to move a long subject to the end of the sentence:

  It is clear that the project will be delayed six months.

- to lead the reader to a new topic:

  There are four possible reasons for the equipment breakdown: poor installation, inadequate maintenance, power surges, and defective parts.

Note 2: The pronoun "it" can also begin a sentence. Then the sentence follows standard word order because the "it" refers to a specific noun in the previous sentence.

It calls for a steel-alloy jacket.

EXERCISE 8-12 Evaluating Expletives

Directions: Analyze the following sentences to see if the expletive can be eliminated. Rewrite if appropriate, and underline subjects and verbs in the new sentence. Be prepared to discuss.

1. It should be pointed out that there are other things to consider in attempting to purchase an office communication system, which an analysis of this sort does not cover.
2. It is his awareness of the inherent difficulties raised by each of these positions that makes it difficult to classify him as either a materialist or a formalist.
3. There are many different ways to compare software.
4. It should be determined whether or not a tether will be required. Tethers must be bolted to the frame of the car and are very difficult to install. Lastly, it should be determined whether the safety seat will require an especially long car seat belt.
5. There are at least two basic tuning controls on the front panel of all oscillators: a switch to set the basic range and a vernier control knob to fine-tune the pitch.
6. It is mandatory that all employees who manually fill out their own time cards view this film and receive a copy of the appropriate time card preparation brochure. It is asked that supervisors divide their groups into two sessions for each shift to minimize impact on production operations.

Enhancements to Syntax

Writing well involves more than avoiding errors and problems. Fortunately, you can also take positive actions in choosing the order of words in your sentences. You can, for example,

- choose voice for effect
- verbalize the action
- emphasize by position

Choosing voice for effect   "Objectivity" and "detached abstraction" are often mentioned as the primary goals in technical and scientific writing. Writers used to think that to achieve objectivity they had to remove any mention of people. Thirty years ago, for example, some engineering schools *required* their students to write weekly lab reports in the impersonal third person (one, it, he, they), the past tense, and the passive voice. As a result of such teaching, much technical writing removed itself from the reality of the work place. The writing became detached and hard to read because the life had been sucked from it.

Today, the emphasis is on communication. In the introduction to this book I said that technical writing succeeds only if readers understand it. That's quite a different goal from objectivity or detached abstraction. To communicate effectively, you need to adapt your style to your goals. One way to change your style is to choose either active or passive voice for effect.

First some definitions:

- *voice*: the verb form that shows whether the subject is acting or acted upon
- *active voice:* the sentence construction in which the subject of the sentence *performs* the action of the verb

  T. Smith crated the machine.
  The manager moved the project to Dallas.

- *passive voice:* the sentence construction in which the subject of the sentence *receives* the action of the verb

  The machine was crated by T. Smith.
  The project was moved to Dallas by the manager.

Advantages of the active voice. Today most writing books and writing teachers advise using the active voice whenever possible, even in scientific and technical writing. Here are its advantages.

1. You can write shorter sentences. An active verb is generally only one word, and the doer of the action, as the subject, can also be one word. In the passive voice, you need a verb phrase; you also need a prepositional phrase if you intend to include the doer.

  Oil companies lease offshore oilfields from the federal government. (active voice, 9 words)
  Offshore oilfields are leased from the federal government by oil companies. (passive voice, 11 words)

Admittedly, this is not a large difference, but it's one that becomes large when passive voice sentences mount up in a paragraph or passage.

2. You can be more forceful. An active doer at the beginning of a sentence gets things going. The action then moves through the verb to the object, just as the batter connects with the ball and sends it flying. This makes the active voice especially useful for instructions.

  Stir the paint thoroughly. (You stir.)
  NOT: The paint should be stirred thoroughly.

3. You can make the reader's task easier. Active sentences link doer, action, and recipient in a logical progression like a chain link fence. Passive sentences either omit the doer (a crucial link) or force the reader (once that doer is produced) to reread the sentence and insert that information.

> You should address a job application letter to a specific, named person. (active)
> NOT: A job application letter should be addressed [by you?] to a specific, named person. (passive)

4. You can be more personal. Putting the doer "up front" in a key position lets you stress individuals instead of things. Most of us are more concerned with (and more interested in) other people than we are in objects. The active voice is especially useful in letters and memos, which are directed to specific persons. You can communicate better with those readers in the active voice.

> You ordered 14 heat shields on September 20, and we are shipping your order today.
> NOT: Fourteen heat shields were ordered on September 20, and your order is being shipped today.

5. You can assign responsibility.

> We will pay all costs if the unit fails.
> NOT: All costs will be paid if the unit fails.
> I designed the overload circuit.
> NOT: The overload circuit was designed by me.

Advantages of the passive voice. The passive voice also has its uses. Sometimes it's worth adding the extra words, being less forceful and personal. You need always to ask where you want the *focus* to be. Here are the advantages of the passive voice.

1. You can emphasize the receiver, events, or results of an action.

> Price-determination analyses were performed at a 0-percent DCFROR.

Who did the analyses is not important here, but the analyses themselves are, so they become the focus of the sentence. Emphasis on the receiver or results may make the passive voice useful in writing procedures or physical descriptions.

2. You can avoid assigning responsibility or soften the responsibility by removing the doer.

> The T-14 air scrubber was badly designed.

This sentence avoids the finger pointing that you'd get in an active voice sentence like this:

> You designed the T-14 air scrubber badly.

When you must convey bad news to a customer or a boss, you may find the passive voice useful.

> NOT: We are reducing your discount to 8 percent.
> BUT: Although the discount has been reduced to 8 percent, you will notice that by paying within 30 days your actual cost is less.

3. You can avoid the first person. Using the first person in technical writing is no longer considered a bad thing, but when you want to remove yourself as the doer, you can shift to the passive.

> NOT: I took samplings of the precipitate at 60-second intervals.
> BUT: Samplings of the precipitate were taken at 60-second intervals.

In this example, *who* took the samplings is not important, so the emphasis rightly shifts to the samplings.

4. You can avoid a long phrase as the subject, changing it to a prepositional phrase and moving it to the end.

> NOT: MINSIM, a computer-based comprehensive economic evaluation simulator, performed the DCFROR analyses.
> BUT: The DCFROR analyses were performed by MINSIM, a computer-based comprehensive economic evaluation simulator.

I have listed almost equal numbers of advantages for active and passive voice, so you might think that one is as good as the other. Remember, though, that you want to communicate as crisply and clearly as possible. When you want to be crisp and clear, you will use the active voice most of the time—for its shorter and more forceful sentences. Only in special circumstances will you use passive constructions. Wrongly used, passive constructions can confuse the reader, and overused, they can numb the brain. Look at this sample from a government publication (Davidoff and Hurdelbrink 1983, 4).

> Each hypothetical mineral operation was evaluated at three different predetermined profitability levels, referred to in this study as economic, marginal, and subeconomic. These three profitability levels are defined by the ore feed grades of the primary commodity that would yield, in constant terms, DCFROR's of approximately 18, 10, and 5 percent, respectively, under the Montana tax structure. The relative effects of taxation in each State were measured for each property type at each profitability level by comparing the derived tax payments and rates of return. Profitability levels are also discussed in chapter 2.

How could you make it easier to read?

EXERCISE 8-13   Evaluating Passive Sentences

Directions: Analyze each sentence to determine if the passive is used to advantage or disadvantage. Rewrite if necessary. Explain why you made the changes you did.

1. Analysis and evaluation of various machines that will be cost effective in the long run will be undertaken.
2. Local models were expected to be visited by September 22. None of the models is close enough to have been personally examined within that time period. They require weekend trips and are expected to be visited by November 11.
3. A study of the problem was conducted early in May, but the results were not released until the end of June.
4. The reliability of the sewing machines being studied will certainly not be ignored by me.
5. Detailed analysis of macroeconomic impacts is omitted because of the character of the results obtained in this part of the research.
6. Oil price controls were in part intended to prevent increased inflation that, in the absence of controls, it was thought would result from the increase in the world price of oil.
7. The area of specialty assigned me was to document an entire department and its interaction at every level inside and outside the company.
8. The flow of fluid in the airbrush and the amount of fluid released are regulated by two separate actions.
9. It was indicated that the Civiche is a good quality product, even though the organization of the company is poor, which could account for the lack of replacement parts.
10. The SLR (single lens reflex) 35 mm camera was designed to combine viewing and picture-taking functions into one lens, so the image seen in the viewfinder is, in most respects, the image that the film will record.

"Verbalizing the action" Most companies and individual employees want to be "action oriented"; they see themelves as aggressive and innovative— always moving forward. It's also good business to project that image to customers and competitors. Companies that succeed in fostering an action image succeed partly because of the care they take with all aspects of their public and internal relations. Writers need to take the same care when editing, studying the "internal relations" of each sentence to see if the sentence moves the communication forward. To produce effective action sentences, you should "verbalize the action."

To *verbalize the action* means to put the action of a sentence in the verb itself. I usually avoid words ending in "-ize" because they can easily lead to pompous diction. In this case, though, "verbalize the action" says very directly what you want to accomplish. First you want to find the action in a sentence and then lodge that action in the action word—the verb. If you don't put the action in the verb, you will often have a sentence problem called "nominalization."

Sometimes you must hunt for the action; it can be well hidden by writers who think they are sounding technical by using long words that end with "-ment," "-tion," "-ance," and "-ence." Here are the four most common places writers can hide the action:

1. in the *subject*

The supervisor's *intention* is to consolidate all rework requests.

Notice that action in the subject weakens the verb. If you move the action into the verb, the sentence becomes lean and purposeful.

The supervisor *intends* to consolidate all rework requests.

2. in the *direct object*

C. Rasch made the *motion* that funds be approved for the pollution study.

Again notice the colorless verb, "made." Put the action into the verb, and you can shed excess words.

C. Rasch *moved* that funds be approved for the pollution study.

or

C. Rasch *moved* that the pollution study be funded.

3. in a *verbal* or *prepositional phrase*

Downstream water users must take *into consideration* the sewage treatment facilities of the cities upstream.

If you move the action to the verb "consider," you can cut two dull words from the sentence.

Downstream water users must *consider* the sewage treatment facilities of the cities upstream.

4. in a *noun that follows an expletive*

It is a *requirement* of the federal granting agency that resumés be submitted with the proposal.

Turn the noun into a verb by dropping the "ment," and you can erase the ineffective expletive as well.

The federal granting agency *requires* that resumés be submitted with the proposal.

Notice in each case that by verbalizing the action, you have also tightened the sentence by eliminating extra words.

Here is a three-step sequence you can follow to put the action in the verb:

1. Search through the sentence, asking "Where is the action?" Be especially suspicious if the sentence includes words that end in "-ment," "-sion," "-tion," "-ance," or "-ence."

2. Turn the word that contains the action into a verb, or find a verb to express that action.

3. Rewrite the sentence, eliminating the excess words. If the sentence has no "doer," see if you can determine the doer from context. If you can, you can also write the sentence in the active voice. If you can't, you may have to retain the passive.

EXERCISE 8-14   Verbalizing the Action in a Sentence

Directions: Analyze each sentence to find the hidden action. Mark the word that contains the action. Then rewrite, putting the action into the verb and eliminating the excess words.

1. Since quality of work is a direct reflection of employee morale, this memo proposes that an investigation should be made of installing vending machines.
2. In the translation I was forced to choose just one of those words and convey the other implication by the context.
3. Your resumé has had a careful review by interested members of our staff; it is unfortunate that no openings exist at the present time.
4. An example of the sophistication level of the system is the ability of the instruction builder to discard corrupted memory information that appears in the cache without deviating from the execution stream in progress.
5. Blame for the slippage of the Civil Defense Program is given by R. Black to the reduced amount of funding from the county government.
6. There is a possibility of the reception by students of monetary awards from other agencies such as Vocational Rehabilitation or Social Security.
7. The occasion arises to manipulate the cursor when there is a need to branch to any point in a text to perform an editing operation.

Emphasizing by position   Why is it that in an argument, most people try to have the last word? They do it because they expect that's what their listener will remember. Why do lawyers save their most persuasive points for the end of their speech? They do it because they know the jury will remember best what it heard last. These examples illustrate an important principle for technical writers: where a word is placed in a sentence often determines the amount of emphasis it receives. Technical writers should know that:

- The most important position in a sentence is the end.
- The next most important position is the beginning.
- Words in the middle of sentences tend to be skipped over.

These statements may seem arbitrary, but if you think about your readers (and even yourself as you read), you will understand why position is so important. Reading forces us to keep a number of ideas in our short-term memory while we process the words that supply new ideas. Those new ideas must then be related to the old ones for understanding to occur. Because short-term memory can efficiently hold only up to nine items, those items need to be rearranged and reemphasized constantly as new information is introduced. A sensitive writer or editor will revise a sentence to ensure that most of the time the beginning refers back to previous information, while the end introduces new information.

However, sentence revising is never quite that simple. As a writer, you must balance the principle of emphasis with other demands of the sentence. You need to think about

- the advantages of subject-verb-object order
- the choice of a passive or active construction
- parallelism
- the location of the action in the sentence
- emphasis by position
- clarity

And you must balance all these enhancements at the same time! Thus, in order to emphasize certain words, you may have to disregard other principles of good syntax. You might, for example, put a sentence in the passive voice in order to move the receiver to the front of the sentence. You might begin a sentence with an expletive so that the new information you want to stress will come at the end. Whenever possible, you should play with sentence arrangement at the editing stage, shifting words around for various effects until you find an order that pleases you and fulfills your purpose.

After weighing the other things you want to accomplish, you should try to use the end and the beginning for your key words. This may mean:

1. moving a dependent clause

   We are providing a steadily increasing number of jobs, although we have an unemployment problem.

   Where does the emphasis seem to fall in that sentence? On the jobs? No, it seems to fall on the problem. But if you move the clause to the beginning, the emphasis shifts.

   Although we have an unemployment problem, we are providing a steadily increasing number of jobs.

2. moving modifier phrases or words

   We have seen several years of increased per capita income after a decade of no growth.

In this sentence, not only are the emphasized words negative, but the word "we", now in an emphatic position, is probably not important to the message. If you shift the prepositional phrases, however, the sentence becomes more positive, and "we" is moved to a less important position.

> After a decade of no growth, we have seen several years of increased per capita income.

3. making the sentence passive, using an expletive, or both
   (Be sure this doesn't distort other things you are trying to do in the sentence. See earlier sections on passives and expletives.)

> In Japan the private sector has carried the burden of research and development, even in basic research areas.

What are the key words in this sentence? Probably not "even in basic research areas," although these words appear at the end. More likely, the key words are "research and development" and "private sector," and those key words should not be buried in the middle of the sentence. If you rewrite the sentence in the passive voice, you might achieve the emphasis you are seeking.

> The research and development burden in Japan, even in basic research areas, has been carried by the private sector.

Note now how the two emphatic positions are filled by the key words "research and development" and "private sector."

EXERCISE 8-15   **Revising Sentences to Emphasize by Position**

Directions: Determine the key words in each of the following sentences. Then rewrite if necessary, moving those key words into positions of emphasis. Be prepared to justify any changes you have made.

1. Where the stays join on a 10-speed bicycle is where the wheels and rear gear mechanism are hooked up.
2. Because the list price of the D200 is relatively high, two different distributors are cited as sources of the D200 instead of the manufacturer.
3. Some of the older, traditional wood rackets have virtues for beginner or intermediate players who are not strong-wristed yet.
4. I must also mention that the Head Professional is an excellent aluminum racket with a reasonable price and is the most popular in history.
5. It has been determined by Michelin test sites that synthetic rubber is the key to longer tread wear.
6. Representing the tire's ability to stop on wet pavement are the traction grades A, B, and C, which are measured under controlled

conditions on surfaces of asphalt and concrete specified by the government.

7. We feel that the Cromalin process, which is also a screened process, is too costly and requires too much labor to get desired results.

EXERCISE 8-16   Putting It All Together: Editing for Clarity and Economy

Directions:

1. Bring to class a sample of your own writing, preferably something technical, of at least three pages.
2. Exchange writing samples with a classmate.
3. Choose at least five sentences from the sample and rewrite them for clarity and economy.
4. Return the sample and rewritten sentences to the writer.
5. When you receive your own edited sentences, read them carefully and consider their effectiveness as rewritten.
6. Write a response to the editor, commenting on each rewritten sentence and then on the editor's work overall.
7. Turn in all the papers to the instructor.

REFERENCES

California Senate. 1982. Bill no. 2051, section 1, chap 3.3.

Christensen, Francis. 1967. *Notes toward a new rhetoric*. New York: Harper and Row, 41–51.

Davidoff, Robert, and Ronald J. Hurdelbrink. 1983. Taxation and the profitability of mineral operations in seven mountain states and Wisconsin: a hypothetical study. *Mineral Issues*. Bureau of Mines. May, 4.

IBM. 1983. Classic clunkers. *Think*, March/April, 43.

PART TWO

TOOLS AND TECHNIQUES IN TECHNICAL WRITING

Tools are implements or instruments that enable people to do work. In construction, they might be hammers, saws, or mason's trowels that are held in the hand, or they might be machines like jackhammers or welding torches. Writers, like construction workers, use tools. You often take the simplest writing tools for granted: the pencil, the ballpoint pen, even the typewriter. Today, though, many writers have access to sophisticated tools like computers and word processors: tools that are so valuable in writing you need to learn how to use them.

A technique, on the other hand, is a method or way of completing a job. The construction of a large building requires many different techniques—ranging from plastering and plumbing to welding and cement finishing. Together, these techniques are used to finish the building, but they are often interdependent and complementary. In the same way, any writing assignment may call for a combination of several techniques, meshed in whatever way is necessary to fulfill the purpose and meet the reader's needs. For example, even a short proposal may call for definition, description, graphics, and documentation—all within the bounds of that form of writing we call a proposal.

Part 1 of this book traced a series of steps that you can follow in the process of writing any kind of technical document. In the five chapters of Part 2 you will learn how to apply the techniques of that process to specific tasks you face in communicating technical information to readers. You may have learned some of these techniques in your freshman composition classes, but the needs of technical writing will force you to use them more precisely.

During this term your instructor may ask you to complete separate exercises or writing assignments in order to practice such techniques as defining, describing, or illustrating with graphics—because concentrating on one technique at a time is a good way to master it. But you should be aware that you often need to use many techniques in order to construct a single writing assignment.

CHAPTER 9

Using a Word Processor

"Word processing" and "desktop publishing" are two popular terms that make it seem as though computers and printers are making technical writers and editors obsolete. As a student you may already have used computers extensively to research information, to crunch numbers, to manage accounts, or to write papers. If you're lucky, your writing class may even include the use of word processors. But if you have little computer experience, you will find it advantageous to take a short course in one of the popular word processing systems or get some hands-on experience at a local computer store. Chances are that whatever your career path, computers will play an important part—and when you must write, you will find word processing a valuable tool.

This chapter looks briefly at word processing for writing, for editing, and for publishing. Since the technology changes so rapidly, I will not discuss specific programs nor will I give advice about choosing a word processing system. I will, however, look at advantages and disadvantages of word processing and discuss the interaction between you as a writer and that writing tool called a computer.

9.1 DEFINING WORD PROCESSING TERMS

In this book a word processor is a computer (the hardware) into which one can type words that will be displayed on a screen and that can be saved for later printing. Some word processors are "dedicated"—that is, they are computers whose primary function is to manage words. Others are personal or mainframe computers that can be used for a variety of purposes, of which word processing is one. The term "word processor" can also be applied to the person who enters data into a word processing system.

A whole word processing system includes a central processing unit (CPU) and internal memory (the computer itself), a cathode-ray tube for display of words (the CRT or terminal), a keyboard, disk drives and disks for storage, and a printer for producing the "hard copy." Figure 9-1 shows how these parts interact to form a word processing system.

A word processing program (the software) is a computer program that accepts words, displays them, stores them, and provides a series of instructions to the user for manipulating those words in a variety of ways.

9.2 WRITING

Word processing has several real advantages for you as a writer because, in addition to simply accepting typed-in words, most word processors will

- insert, delete, and write over copy
- number pages automatically and insert a title line on each page (either "headers" or "footers" or both)
- exchange one word for another throughout a manuscript, or query the writer each time it locates a questioned word
- copy or move a block of text from one location to another
- format a page, inserting bold and large type for titles, creating even margins, centering titles, and word wrapping (moving words automatically to the next line and page)
- show on the screen exactly how the page will look when printed (sometimes called WYSIWYG or "what you see is what you get")
- print an entire document or just a piece of one
- single, double, or triple space and combine different spacing in one document

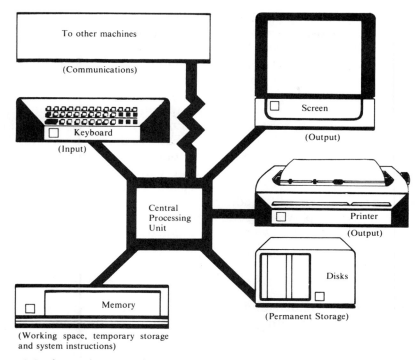

Figure 9-1   A word processing system

In addition, you can change part of a document and print out a new copy each time it's needed without retyping the whole thing, a process that encourages you to revise and improve your writing.

Many word processing programs will do fancier things, such as

- creating a table of contents and an index automatically
- creating a bibliography
- creating a glossary
- inserting and automatically numbering footnotes
- "unerasing" to restore the last thing you erased
- splitting the screen to let you read one document while working on another
- merging two documents or transferring information from one document to another
- providing superscripts, subscripts, and easy ways to construct formulas or mathematical equations
- calculating numbers
- storing text automatically to avoid losing material
- outlining by identifying headings with subheads below them and hiding or displaying subordinate information

No word processing package will carry out all of the functions listed here, nor will you need them all for any particular job. In choosing word processing software, you should first list what you need and then go looking for the software that will (1) work with whatever hardware you have and (2) fulfill the greatest number of your needs. Check to see what is available to you on your campus.

You should also consider graphics capabilities. While graphics production is not strictly a word processing function, computers increasingly give writers access to graphics programs that will design tables and draw diagrams and detailed graphs. A few examples of such computer-designed graphics appear in Chapter 12. Graphics programs are usually separate from word processing programs, though some word processing programs allow for many varieties of heading styles and page designs. Companies that need both word processing and graphics have "electronic publishing" systems that combine the two through sophisticated software.

9.3 EDITING

While most students understand that a computer will not actually do their writing for them, it's tempting to think that the editing functions in a word processing program will ensure well-written and error-free copy. Without question, word processors can be very helpful for editing tasks. Some programs contain

- a dictionary that will check your spelling against a stored vocabulary and highlight any questionable words
- a thesaurus that will suggest substitutes for words that aren't quite right in a sentence
- a set of rules about sentence length, constructions like passive voice, and diction problems like unspecific words, sexist vocabulary, and clichés
- a feature that will print out the first and last sentences of all paragraphs to let you look at continuity and coherence in a document
- automatic correction for "widows and orphans" (single letters or words left by themselves at the bottom or top of a page)
- a punctuation check for missing parts of pairs, like quotation marks and parentheses, and for the position of periods, semicolons, and colons in relation to double quotation marks

However, you can't sit back and assume that your own editing responsibilities are over. For example, spell-checking programs will not recognize real words that are incorrectly used in a sentence (such as principle for principal or two, too, and to). Punctuation editing functions on a computer are bound by strict rules and can't make stylistic exceptions—like deliberately not using a comma in a short compound sentence in order to stress the continuity of two equivalent ideas. Therefore, in order to use a word processor's editing functions most effectively, you yourself need to understand grammar and punctuation.

9.4 PUBLISHING

Your publishing capabilities will vary enormously depending on the type of printer available to you and the capabilities of the program you are using. In addition to your word processing program, you might have access to

- a dot-matrix, daisy-wheel, or laser printer. The quality of the printout will vary with each kind of printer and even within kinds, depending on the manufacturer.
- a variety of print fonts (types, sizes, and styles of letters), including capabilites for color, boldface, italics, and underlining.

Most students are concerned with producing a clean, legible document quickly, not with publishing their writing in order to make a profit. But so-called "desktop publishing" is much in the news, and you need to know what it is and what it involves because someday you may have to use it.

The term "desktop publishing" implies that you can go from idea to polished, printed, "published" work with equipment that fits on the top of your desk: usually a word processor, a page layout program, and a laser printer. But as one expert in desktop publishing points out (Strehlo 1987, sec. F, 1,2):

> ... giving people a hammer and saw does not equip them to build a house. Although a computer makes it possible for anyone with an aesthetic flair to acquire the necessary skills to design attractive pages, many business people may not have the time or the inclination. ... Just how tough is it to produce attractive publications with desktop publishing software? Running the software itself is no tougher than running a capable word processing program—but learning how to lay out attractive pages takes a bit of talent and a bit of practice. ...

Because of the time and talent involved, some companies hire trained graphics designers to run their desktop publishing systems.

9.5 ADVANTAGES AND DISADVANTAGES OF A WORD PROCESSOR

Word processing has many advantages for you as a writer—both as a student and as a working professional. Its primary advantage is the ease of correcting, changing, and reorganizing copy without having to retype. A second major advantage is page formatting of headings, margins, and spacing. And finally, most students especially appreciate a feature like automatic checking of spelling. As a writer myself, I enthusiastically endorse word processing. This book, in fact, has been written and edited on a word processor, and I would not have had the energy or the courage to write it without that help.

But as a writing teacher, let me caution you that word processing also has some limitations. The responsibility for planning, organizing, writing, and reviewing is still yours, and you need to learn and practice all those skills. Clarity and coherence will not magically appear in a document, nor will a word processing program pay attention to purpose and the needs of the reader. A beautifully presented document that is shallow, wordy, or full of errors will be judged more by its content than by its presentation. In addition, you may do yourself a real disservice if you rely on computerized style programs instead of developing your own sense of what constitutes good writing style.

Some companies use computer programs that assess readability by counting the number of words in each sentence and the number of short versus long words. These programs imply that shorter sentences and words make a document easier to read, but research in adult reading does not support that implication (Redish and Selzer 1985). In fact, short sentences may be harder to read because they do not indicate the relationships among ideas, and short words are no better than long words if your reader does not understand them. The problem with readability formulas is that they focus attention on words and sentences at the expense of larger concerns like organization and coherence.

Other programs assess style by noting matters like awkward and passive sentence construction and clichés. These programs may be useful in alerting a writer to potential problems at the editing stage so long as the writer then makes an informed decision about whether or not to change the sentence. For example, as you learned in Chapter 8, sometimes a sentence in the passive voice is better than its active counterpart. If a style program flags the passive, and you automatically change it, you may be making the reader's job harder instead of easier.

One writer (Catalano quoted in Andrews 1986) notes that such programs could "blanderize writing into a homogenized computerspeak." And syndicated columnist James Kilpatrick says (Andrews 1986, sec. F, 13), "To the extent that these mechanical devices remind us of things we already know, maybe they serve a useful purpose. But I think they can become a crutch and just wreck any effort in style. If you have to worry about putting a short sentence after a long sentence, all of sudden you're not the master of your prose, you're the slave of someone's advice on word count and average sentence length." Kilpatrick says he would rather see a program connected to a "bucket of slop that would pour over the writer's head whenever he used [a cliché like] 'it remains to be seen.' That would clean up a lot of prose in a hurry!"

Also, you must remember that because a computer is a machine, it is subject to breakdown, which can mean lost files, scrambled data, and printer malfunction. You'll need to learn a word processing program well before you have to produce any major paper on it so you are comfortable with the way it works.

As you can see, the advantages of having a word processor and all its related equipment are obvious, but you can't delude yourself into thinking that the work of planning, organizing, writing, and reviewing will disappear. The writing and much of the editing still must be done by you with your best concentration and all the skills you can learn.

CHECKLIST FOR WORD PROCESSING

1. What functions do I need in a word processor?
2. What hardware and software are available to me?
3. How can I best combine my skills with those offered by the computer?

EXERCISE 9-1   Analyzing Word Processing Software

Directions: Read three articles in current magazines or journals about at least two different kinds of word processing software. Write a short informal report discussing the advantages and disadvantages of each system and recommending one for use by college students in a technical writing course. Add a List of References to your report, and annotate each reference. See Chapter 13 for information on documentation.

EXERCISE 9-2   Learning How to Use a Word Processor

Directions: Well before your term project is due, spend some time learning a word processing system that you can use for completing your report. Practice with short assignments.

EXERCISE 9-3   Choosing Publications Software: Panel Discussions

Directions: Form groups of three to five students, with each group studying a different kind of word processing, graphics, or page layout software. Present the advantages and disadvantages in a 20-minute panel discussion before the rest of the class. See Chapter 23 for help in planning presentations.

REFERENCES

Andrews, Paul. 1986. Help or hindrance? Writers are undecided whether electronic editors are their type. *San Jose Mercury*, 9 March, F 13.

Redish, Janice C., and Jack Selzer. 1985. The place of readability formulas in technical communication. *Technical Communication*, 4th quarter, 46-52.

Strehlo, Kevin. 1987. IBM PCs are writing a new chapter. *San Jose Mercury*, 15 February, F 1, 2.

CHAPTER 10

Defining and Comparing

One of your major tasks in technical writing is to explain what you did or what you know to your readers; in other words, to teach them something new. When you teach, you have to understand where those readers start from, meet them at that point, and lead them to greater understanding. That's why it's so important to identify your potential readers at the beginning of a writing project. (See section 1.2 for information on reader identification.) Once you've identified the readers, you need to ask "Does my reader understand what's going on here?" This chapter discusses two ways you can explain new material: by defining terms and by using analogies. It also helps you decide where to place these explanations in a document.

10.1 DEFINING WORDS AND PHRASES

You can often explain what you mean by defining words and phrases, but in technical writing there are three reasons why you can't simply send your readers to a dictionary: First of all, technical readers are busy, and they often won't take the time to go to another source. Second, standard dictionaries seldom adequately define technical terms. Third, defining the terms yourself lets you stipulate the meaning—that is, specify what you want a word to mean in a particular context.

To decide which terms need definition, you must put yourself in the primary reader's place. Here are some questions you can ask:

1. Does the reader need this term, or can I avoid definition by using a simpler word? For example,

> From the air, the field appeared to be striated (striped).

In this case, the simpler word would say what you mean, so you could eliminate the definition.

2. How many words do I need to define? If you are writing to your peers (fellow architects, scientists, or technicians, for example), you might not have to define many or any words. When you do, you can use jargon of the field in your definition. More often, though, you'll be writing for generalists or you'll have more than one type of reader. For general and multiple readers, you need to define more often, and you must use simple language in the definitions. A definition like this one is appropriate for generalists because, even though the word is formidable, the explanation is simple:

> Epikeratophakia is a surgical procedure that involves sewing a machine-tooled donor lens onto the patient's cornea.

But the next definition (Syva n.d.) would be acceptable only to medical specialists and technicians, who could understand the terms used to define the test.

> A MicroTrak™ test is a monoclonal antibody-based immunofluorescent culture confirmation test that enables rapid detection of all known human seriological variants of *C. trachomatis.*

3. If I'm new to this subject, am I defining too much? Generally, it's better to define too much than too little. Sometimes, though, when you first approach a subject, you are both overwhelmed and fascinated by its vocabulary. In an attempt to share your new knowledge, you may overload the reader with detail. Keep in mind your reader's needs as well as your own purpose, and ask yourself "What does the reader want out of this?" Do *not* write a definition for a triangle like this:

A triangle is a closed plane figure having three sides and three angles. A plane in mathematics is a surface generated by a line moving at a constant velocity with respect to a fixed point. A figure is a combination of geometric elements dispersed in a particular form or shape.

Unless you are teaching geometry, or these terms form the basis of your discussion, this much detail is smothering. Most adult readers will already have in mind the basic concept of a plane and a figure.

When you seem to need more than about five definitions in a single document, consider consolidating them in a glossary and highlighting the words in the text to send the reader to that glossary. You will find examples of a glossary in the student report in Chapter 21 and in the Appendix at the back of this book. Depending on your context and your reader's needs, you can define terms in three different ways:

1. informally, with a word or phrase
2. formally, with a sentence
3. by expansion

Informal Definitions

An informal, or parenthetical, definition is useful when a close synonym exists for the word you are defining, or when you want to stipulate (specify) the particular meaning you want to use. (Notice that I have just used an informal definition.) The following informal definitions provide adequate explanations of the unfamiliar words.

- Nearly all species have clocks that regulate their metabolism in circadian (daily) rhythm.
- Psittacosis, so-called parrot fever, is transmissible to humans, causing high fever and infecting the lungs.
- Geologists today almost universally accept the theory of plate tectonics, which evolved from the phenomenon formerly called continental drift.
- Form in this book means the way a document looks on the page based on organization and layout. [a stipulated definition]

To write an informal definition, you can simply enclose the new word in parentheses, commas, or dashes, or you can add a sentence or phrase of explanation.

Formal Definitions

A formal definition is so named because it follows a set form and is stated in a sentence. A formal definition follows the method of organization called classification/division; you first put the word into a larger class, and then you add the details that distinguish your word from the other words in that class. The order is *word to be defined + verb + class + distinguishing characteristics.* Table 10-1 shows the pattern to follow in writing formal definitions.

Table 10-1 FORMAL DEFINITIONS

Word	Verb	Class	Distinguishing characteristics
Gravity	is	a fundamental force	that operates between any two objects having mass, causing them to attract one another (Rensberger 1986, 176).
An enzyme	is	a molecule	that causes other molecules to undergo specific chemical reactions, breaking them into smaller pieces or making them combine with other substances (Rensberger 1986, 140).
A cave	is	a natural opening in the ground	that extends beyond the zone of light and is large enough to permit human entry (Davies and Morgan 1980).
Asphalt	is	a mixture of hydrocarbons	that is used, when mixed with gravel or crushed rock, for roadbuilding.

You have to think carefully about a word in order to write a formal definition because a dictionary definition by itself usually won't give you the precision you need. Putting the word in a larger class helps the reader see relationships; pointing out distinguishing characteristics narrows the focus. If you also keep the class reasonably narrow, you can limit the number of comparisons the reader must make. For the word "harrow," for example, "implement" is too wide a class; "agricultural implement" narrows the field considerably.

For good formal definitions you also need to remember:

- not to repeat the term you are defining in the class or among the distinguishing characteristics

  NOT: A harrow is an agricultural implement used to harrow the ground.

  BUT: A harrow is an agricultural implement that uses spiked teeth or upright disks to level and break up clods in plowed land.

- to maintain the form of *word + verb + class + distinguishing characteristics.* If you use either "when" or "where" in your definition, you are telling time, place, or circumstances, not what something is.

  NOT: Hatching is when you draw a series of parallel lines to shade or to differentiate one part from another.

  BUT: Hatching is a series of parallel lines used in a drawing to shade or to differentiate one part from another.

- to use simpler language for the definition than the word itself. If you define with words the reader doesn't know, you have defeated your purpose, which is to explain or teach. Readers are seldom insulted by simple language, but they are easily angered by long, pretentious words.

Descriptions and introductions often begin with formal definitions; that way the reader knows at once what the key terms mean.

EXERCISE 10-1   Writing Informal and Formal Definitions

Directions: From the list below, choose two terms that would be suitable for an informal definition and define each one in a sentence. Then choose two terms that require a formal definition. Think about the terms first and then consult regular and specialized dictionaries to find a suitable class and write a list of clear distinguishing characteristics. Be prepared to discuss your choices and your definitions.

musical synthesizer	trachea
cirrhosis	interstice
gratuity	pepsin
crankshaft	exhaust orifice
emulsion	optical fiber
tornado	radial tire
a pond	tear gas
toxin	

EXERCISE 10-2   Writing Definitions for a Short Report

Directions: Select two words that need to be defined for a short report you are writing. Write a clear formal definition for each one and submit them for review.

Expanded Definitions

Useful as a formal definition is, sometimes you need to explain a word even further. You can do this by choosing one or several of these six methods of expanding a definition: description, examples, comparison, classification/division, etymology, illustration. In all these methods you begin with a formal or informal definition and then add to it.

1. **Description**   Since technical writing usually deals with either physical objects or processes, one of the best ways to expand a definition is to describe how the object looks or how the process works. Description can include weight, size, color, composition, and component parts. Such de-

scriptions can be simply a sentence or two added to a formal definition. For example, you might add this description to the formal definition of asphalt given in Table 10-1:

> Asphalt cement is a brownish-black substance that can range from (1) a liquid during mixing, transporting, and placing, to (2) a plastic during compaction and service under traffic, to (3) a solid under low temperatures.

Description can also be the major purpose of a document; description of objects and places is covered in detail in Chapter 11, and description of procedures is covered in Chapter 18.

2. Examples   Giving one or more examples helps the reader understand a word by making it more concrete and easily visualized.

> Two of the most famous caves in the United States are Mammoth Cave in Kentucky and Carlsbad Caverns in New Mexico.

> Examples of nitrogen fixers in the grasslands are lupines and clovers, both in symbiosis with bacteria of the genus *Rhizobium.*

3. Comparison   Like examples, simple comparisons help the reader visualize something unknown by relating it to something that may be known. Comparisons can point out likenesses or differences (contrasts).

> Asphalt cement is like concrete in being a mixture of bonding matter with sand or gravel. However, asphalt and concrete differ in their type of bonding material: in asphalt the bonding material is bituminous, and in concrete the bonding material is a calcined mixture of clay and limestone.

Synonyms, or words that mean almost the same thing, are a kind of comparison.

> Asphalt is also called blacktop or tarmacadam.

Comparison is also a method of organizing material; you can group items by the ways in which they are alike or different. See Section 3.2 on organization.

4. Classification/division   You can group related items into a single class or divide a larger class into related items to explain them to your reader.

> Caves can be classified into four main types: solution caves, lava caves, sea caves, and glacier caves.

Like comparison, classification/division is also a method of organization. You could, for example, structure a portion of a report on caves based on the four types listed above. See Section 3.2 for more on classification.

5. Etymology   Etymology tells where a word comes from and what its history is. Etymology is particularly useful to explain a word combined from two or more source words, such as edentate—meaning toothless and referring to mammals such as armadillos, sloths, and anteaters. The word comes from the Latin prefix "e-" (or "ex-") meaning "out of," and the Latin root "denti" meaning "tooth."

6. Illustration You can illustrate meaning either with words or with graphics. Usually when you illustrate with words, you give examples. But drawings, diagrams, and photographs can effectively expand the meaning of a word. The following student example (Roulin 1985) combines a drawing with a definition:

> Torsion is the twisting that occurs sideways in a ski when a force acts on only one of the bottom edges of the ski. In Figure 10-1, the front of the ski encounters an irregularity in the terrain, but the ski is able to twist slightly without disturbing the skier's balance. The middle section under the skier's foot remains stable.

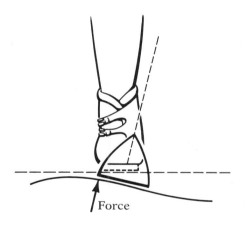

Force

Figure 10-1 Torsion in a ski

To show how definition works in the context of a larger report, the following passage gives a student's definition of surgical lasers (Leonard 1985). This definition forms part of the introductory section of the comparative report discussed in Chapter 21. Notice that the writer begins with a formal definition giving three functions of surgical lasers: to vaporize, to coagulate, or to photoradiate tissue. Then he expands the definition by the division method—devoting each of the next three paragraphs to one of these functions. Within the paragraphs he uses comparisons: "identical to water boiling," "like a scalpel," and "when an egg is cooked."

Surgical Lasers Defined
    Surgical lasers are medical instruments which use radiant energy to vaporize, coagulate, or photoradiate tissue. Consequently, the general term laser surgery involves any of these three laser operations.
    Using a laser to vaporize tissue is specifically called laser surgery. In this application, a very fine, dense beam of radiant energy is converted to intense heat at the target tissue. The response is identical to water boiling:

intracellular liquid vaporizes. Directing the beam along a path produces a fine line of vaporized cells: an incision. In this case, the surgical laser is like a scalpel except the blade is made of light instead of steel.

Using a laser to coagulate tissue is called photocoagulation. The process is similar to vaporization except that the thermal response is not as intense in photocoagulation. Before liquids can vaporize, protein rearrangements cause clotting. The same process occurs on a larger scale when an egg is cooked.

Using a laser in photoradiation is quite different from either vaporization or photocoagulation. Photoradiation therapy is a cancer treatment in which a nontoxic chemical (a hematoporphyrin derivative) acts selectively on malignant cells. Laser radiation reaching the chemical stimulates a toxic product (singlet oxygen). Thus, the laser treats a specific area with a local toxin.

Surgical lasers, then, are the tools that enable these valuable medical procedures.

Other terms needed to understand this report are defined in a glossary, but the term "laser surgery" is central to understanding the report itself, so it is defined in the introduction.

EXERCISE 10-3   Writing an Expanded Definition

Directions: Choose one of the terms from Exercise 10-1 and expand the definition in two of the following ways:

description	classification/division
example	etymology
comparison	illustration

EXERCISE 10-4   Analyzing Expanded Definitions

Directions: Study the following student definitions and decide which expansion devices have been employed. Write a paragraph discussing what the devices are and whether others are needed. Be prepared to defend your conclusions in class.

Example 1. Definition of a Mountain Bicycle (Cording 1985)

A mountain bike is a bicycle designed to withstand the rigor of off-road riding. The mountain bicycle is a successful meshing of three bicycle designs:

1. The Beach Cruiser. The mountain bike borrows the wide 1.75 by 26 inch tires of the beach cruiser, as well as the upright frame design. The tires and the frame design increase the bike's stability and handling in dirt and bumpy trails.
2. The Moto-cross Bicycle. The mountain bike adopts the pedals, brake levers, and handlebars from the rugged little moto-cross bicycle. The pedals,

with their sawtoothed edges and wide platforms, offer a good grid and stability for the foot. The handlebars and brake levers are very similar to those of the off-road motorcycle. They offer greater control and stability.

3. The Standard 10- to 15-Speed Bicycle. The mountain bike uses a 15-speed gear range to allow for easy pedaling up steep trails and hills.

Example 2. Definition of a Piano (Boruk 1985)

A piano is a stringed instrument with steel wire strings that sound when struck by felt-covered hammers operated from a keyboard. Specifically, a grand piano, in the correct sense of the term, is a piano that is constructed on a horizontal plane, with strings stretched across the piano plate and framework. The four essential elements that work together to produce the sound are:

1. strings
2. action
3. soundboard
4. framework

The steel strings generate sound when struck by padded hammers, flat felt-covered pieces of wood. This process of the hammers striking the strings and emanating sound is performed by the action of the piano. The action of the piano is defined as the principal mechanism involved in producing the sound of the piano. The soundboard, a flat piece of finely grained wood upon which the strings are strung, amplifies the sound made by the vibrating strings. The framework includes the inner and outer rim assembly and serves to support the stress of the piano, as well as to enhance the sound.

10.2 USING ANALOGIES

Analogy is a kind of comparison that is especially useful when you want to explain something to generalists. In fact, all types of readers appreciate an analogy if you are explaining a complex phenomenon that can't itself be visualized. In an analogy you compare what is unknown to something that is known, usually something belonging to a different class. For example, the laminated layers in a ski can be compared to a sandwich. Analogies can help readers understand objects, processes, or abstract ideas.

To Explain Objects

The following analogies explain new or unfamiliar objects by comparing them to familiar objects.

- Carbon filter elements look like solid or broken charcoal briquets, but seen under a microscope, the carbon granules look like sponges, with little canals and pockets that trap suspended solids (Cochran 1985).
- A simpler definition would be to compare a varistor to the volume con-

trol knob on a radio. A person manually adjusts this knob (to raise and lower volume) regulating the amount of voltage passing through. A varistor automatically adjusts to the amount of voltage entering it. As a surge enters its circuitry, the "knob" is turned down to compensate for the increased voltage, and the output remains constant (Cook 1985).

- No expression describes the typical comet better than "dirty snowball." Basically, that's what comets are—chunks of ice and frozen gases mixed with rocky debris. The snowballs range in size from 100 yards in diameter to 50 or 60 miles (Rensberger 1986, 96).

To Explain Processes

Terms for unfamiliar processes can be clarified with simple analogies, as the following student examples show.

- A disk duplication center is much like a book printing company. Just as an author might employ a printing company to print a required number of books, so would a programmer employ a disk duplication center to duplicate a number of his programs. He would send the duplication company a copy of his program on a floppy disk, and the duplication company would send him the required number of exact duplicates (Chandler 1985).
- Lamination is the process of making something by bonding different materials together. Snow skis are laminated. A ski cut in half would look like a sandwich that has been cut in half. The base is the bottom piece of bread, the wood core of the ski is the meat of the sandwich, and the top skin is the top piece of bread (Freeman 1985).

To Explain Abstract Ideas

Eminent scientists have often used analogies to explain difficult or abstract concepts. Albert Einstein was a master at helping others visualize his abstract theories through simple comparisons, as you saw in section 1.2. More recently, biochemist Harold J. Morowitz (1985, 27-8), in an essay titled "Mayonnaise and the Origin of Life," used an analogy with homemade mayonnaise to explain the components of a living cell. Here are the first two paragraphs of that essay.

The volume in my kitchen on the art of French cooking has a basic mayonnaise recipe that lists only three constituents: vegetable oil, egg yolk, and vinegar. A recipe for mixing up a living cell would be vastly more complicated, but the microstructures of both cell and salad dressing depend on a class of molecules of the greatest importance in all living processes. These ubiquitous chemicals are designated the amphiphiles. Since that noun occurs neither in the Oxford English Dictionary nor in the Webster's New Collegiate Dictionary, I feel safe in assuming it is not a household word. Actually, it was necessary to go to Chemical Abstracts to confirm the usage. What is surprising about the relative obscurity of the amphiphiles is the fact that as requisites for life they are just as important as proteins or DNA.

The word itself is a compound of *amphi*, meaning "both kinds," and *philos*, "to love." The two kinds referred to here are oil and water, for these molecular structures have one end that is attracted to oil and another end that is attracted to water. The presence of amphiphiles in the egg yolk is what makes the smooth, homogeneous mayonnaise sitting in my refrigerator belie the old adage that oil and water don't mix. The egg material contains lecithin, a phospholipid molecule with a water-soluble phosphate moiety and a lipid portion that dissolves in oil. Lecithin therefore occupies the interface between the oil droplets and the surrounding vinegar and thereby produces a smooth emulsion—pleasant, tasty, and fattening.

Notice that within the analogy is an expanded definition: the etymology of the word "amphiphiles."

You must use your imagination to develop useful analogies. Whenever you want to explain something for which no synonyms or easy comparisons exist, see if you can write an analogy. I often use brief analogies in this book to help you understand the writing process; watch for them as you read and see if they clarify the explanations.

10.3 PLACING DEFINITIONS AND ANALOGIES IN A LONGER DOCUMENT

Depending on the type of document you are writing, you can place definitions and analogies in any of the following four places:

1. *In the body*. Informal definitions and analogies are most likely to go in the body of a report where they help explain key terms in the discussion. If the term is one that underlies the whole paper, however, the definition might go in the introduction. Analogies may also work well in the introduction, especially for generalist readers.

2. *In a footnote or endnote*. If you are writing to a mixed group of readers (specialists and managers, for example), too many definitions or analogies will slow the pace of your explanation. In this case, you should put a definition or analogy in a footnote or in an endnote (at the end of a chapter or section), making it available to the group of readers that may want it but unobtrusive to the other.

3. *In a glossary*. If you need to define more than four or five terms in any one document, you should include a glossary. In addition, you should define the word within the text the first time it's used; the glossary makes the definition available the next time the word appears in case the reader has forgotten the meaning. Terms in glossaries are alphabetized for easy access. In a formal report, the glossary is usually in an appendix. (See Chapter 21 for the form of a report.) Occasionally, you will include a glossary as part of an introduction, especially if all the terms defined are essential to understanding the report.

4. *In an appendix.* Long analogies or extended definitions (etymologies, for example) may simply be of added interest—not central to the content of the paper. In this case, you should put the information in an appendix, even though you may not have a formal glossary.

CHECKLIST FOR WRITING DEFINITIONS AND COMPARISONS

1. For this document, should my definitions be
_____ informal?
_____ formal?
2. What method or methods shall I use to expand the definition?
_____ description?
_____ examples?
_____ comparison?
_____ classification/division?
_____ etymology?
_____ illustration?
3. Could I effectively use analogy for
_____ explaining objects?
_____ explaining processes?
_____ explaining abstract ideas?
4. Is the most effective place for my definition or analogy in
_____ the introduction?
_____ the body?
_____ a footnote or endnote?
_____ a glossary?
_____ an appendix?

EXERCISE 10-5 Writing Definitions or Analogies for a Major Paper

Directions: Choose two or more terms that need to be defined for your proposal, interview, research, or product study. Consult specialized dictionaries and other resources to help you write formal definitions. If necessary, expand the definitions or include analogies. If you are working on a project as part of a team, be sure that one person is responsible for defining terms. Review those terms in the group so you agree on the definitions.

EXERCISE 10-6 Writing Analogies for Generalists

Directions: Choose a technical term or object from your major field that would probably be unfamiliar to a generalist reader. Define it by using an analogy, keeping the explanation simple.

REFERENCES

Boruk, Charlie Sue. 1985. *The grand piano: A competitive brand analysis.* Formal report, San Jose State University.

Chandler, Richard. 1985. *Definitions of disk duplication terms.* Informal report, San Jose State University.

Cochran, Marian. 1985. *Evaluation and recommendation of drinking water sources for residents of Spartan City.* Formal report, San Jose State University.

Cook, Ronald. 1985. *A definition of surge suppressor terms.* Informal report, San Jose State University.

Cording, Richard. 1985. *Off-road mountain bicycles: A technical consumer report.* Formal report, San Jose State University.

Davies, W. E., and J. M. Morgan. 1980. *The geology of caves.* U. S. Geological Survey, no. 311-348/34. Washington: Government Printing Office.

Freeman, Greg. 1985. *Definitions of ski terms.* Informal report, San Jose State University.

Leonard, David. 1985. *An evaluation of Nd:YAG surgical lasers for Allied Hospital.* Formal report, San Jose State University.

Morowitz, Harold J. 1985. *Mayonnaise and the origin of life.* New York: Scribners, 27–28.

Rensberger, Boyce. 1986. *How the world works.* New York: William Morrow, 96, 140, 176.

Roulin, Kim. 1985. *Physical description of a ski.* Informal report, San Jose State University.

Syva. n.d. Microtrak™ test data sheet STD011 10M. Palo Alto, CA: Syva Company.

CHAPTER 11

Describing Objects and Places

What are you doing when you describe a favorite car or a lake to a friend? You are sketching that car or lake, using words instead of pencil or paint. In fact, if your description is good enough, your friend should be able to turn your words into a drawing of that car or lake.

You will often use description's picture-making ability in technical writing because it is a good way to communicate what you know to your readers. Description can be of objects, of places, or of processes. Procedure description has a distinct form and is closely related to instructions, so it will be discussed in Chapter 18.

220

11.1 DEFINING DESCRIPTION

A *description* is a verbal picture of an object or place. Often that word picture can be combined with an illustration like a drawing or photograph, giving the reader even more help.

Technical description can be either generic or specific; for example, you can describe a "typical," or generic, compact disk player or a specific one, such as Sony's model CDP-610. Both kinds of description are useful depending on the specific purpose, the type of reader, and the reader's needs. If your purpose is to evaluate four potential locations for a helicopter landing pad, you might need to begin by describing the "ideal" (generic) landing pad before going on to describe each location specifically. Then your evaluation can compare each site to the generic one in order to draw conclusions. On the other hand, if you are writing an operator's manual for a vacuum cryopump (which pumps gases at very low temperatures), the operator will want a specific description of that model and make of cryopump.

You must also consider types of readers and their needs. Generalists and managers may need generic descriptions to help them understand the thing or place itself, while specialists and technicians may want specific descriptions to help them visualize a particular variation. Each time you write, you need to analyze your specific readers and their needs.

In technical writing, as in most writing, a description is usually not a separate document but forms part of a longer whole, whether report, manual, or proposal. Description can also be part of a letter, memo, or set of instructions or procedures. In addition, as noted in Chapter 10, description is sometimes used to expand a definition. As a student, you may write a separate description paper as a way to learn the skill, but most of the time on the job you will incorporate description into some larger document. Frequently you will combine description with graphic (visual) representation. Pictures and words together give a reader solid information about a thing or place.

11.2 SOURCES OF INFORMATION

Before you can describe an object or a place, you must have detailed and accurate information about it. You can obtain that information yourself by examining, weighing, and measuring the object, or you can find the details in written material. Obviously, the best source is the object or place itself, and related explanatory material like advertisements, catalogs, manuals, and specification sheets. If you use sales literature as source material, though, be careful to strip away biased or persuasive sales language from the factual data. Technical description ought to be objective and unemotional, so your facts must be accurate and your language carefully chosen.

Look at the following paragraph of description from a 1986 Volkswagen sales brochure:

> The spacious interior accommodates five comfortably. Then pampers you with standard convenience features like fully reclining bucket seats, tinted glass, rear window defroster and intermittent wipers. The huge cargo area offers up to 39.6 cubic feet of useable space, thanks to the fold-and-tumble feature of the rear seat.

As advertising, this description is primarily emotional rather than objective, aiming to convince you of the superiority of this interior. If you want to describe this car's interior objectively, you can use the facts (reclining bucket seats, tinted glass, 39.6 cubic feet of cargo space), but you must eliminate emotion-laden words like "spacious," "comfortably," "pampers," and "huge."

Besides sales literature, sources of descriptive information include technical dictionaries and encyclopedias, textbooks, handbooks, and catalogs. You can find most of these in your library; ask a reference librarian for help. In addition to written sources, you can interview specialists, technicians, and operators. You should, if possible, get hands-on experience with whatever you want to describe—seeing it and using it, measuring and weighing it if necessary.

11.3 INFORMAL DESCRIPTIONS

Informal descriptions are generally short and may appear almost anyplace in a long document. Frequently, though, you will find them in the introduction or in that portion of the body dealing with definitions, materials, or the explanation of a problem. An informal description may be only a paragraph, yet its purpose is the same as a more formally designed long description: to enable the reader to picture the object or place mentally and understand its function.

For example, here is an informal place description written for pilots:

> Thomasville Municipal Airport (on the Jacksonville Sectional) is located 7 mi. NE of the city of Thomasville. It has two 5,000 ft. asphalt runways, a rotating beacon, runway lights (available on request until 10 p.m.), hangars, tiedowns, fuel, and Unicom (AOPA 1965, 73).

If this description were more complete, it would give the altitude of the airport and the compass orientation of the runways; with those added facts, a pilot would have a good mental picture of how the Thomasville Airport looked from the air and how it would function.

The following student paragraph (Bernucci 1981) informally describes the feed system of an Elna SU68 sewing machine. Coupled with the illustration in Figure 11-1, the description gives the reader a good

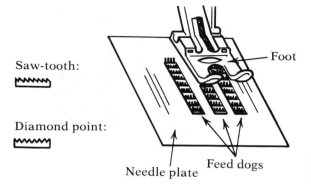

Saw-tooth:

Diamond point:

Foot

Needle plate

Feed dogs

Figure 11-1 Diamond point feed dogs

picture of the feed mechanism and leads to an understanding of how it works.

One of the features that distinguishes the Elna from the Riccar, Pfaff, and Singer is its feed system. The feed system passes the fabric under the needle with feed dogs, a series of moving teeth that come up through tracks in the needle plate. The feed dogs must pull the material just enough so that it does not bunch up, but not so much that it catches and snags on the feed dogs' teeth. Elna's feed dogs have diamond-pointed teeth instead of the traditional slanted saw-tooth (Figure 11-1).

The diamond points do not snag sheer fabric such as chiffon. Slanted saw-tooth feed dogs have a tendency to "eat" sheer fabric and the lining on bonded wool by pulling it down into the needle hole in the needle plate. Because of their gentle pull, diamond-pointed feed dogs also keep stitches from stretching out of shape on bathing suits and tee-shirts.

11.4 FORMAL DESCRIPTIONS

Formal descriptions are generally organized in three parts: (1) an introductory overview, (2) a breakdown into parts with part descriptions, and (3) a concluding discussion of how the object works or is used.

Overview

In a formal description the introductory overview should begin with a definition, as does this student example (Davis 1985).

A catalytic combustor is a device mounted in the firebox or flue of a wood-burning stove that causes fuel to be burned more completely than in a conventional stove.

Following the definition, you should tell the purpose of the object.

Its purpose is to increase the energy efficiency of the wood-burning stove by enabling smoke to burn at a lower temperature and to decrease the amount of creosote-forming gases and pollutants released into the air.

Next, you should describe its general appearance, including here, if possible, a cutaway or exploded drawing. (These drawings are explained in Chapter 12.) In this section you should also discuss the object's material composition and number and list its major components. This list of components will serve as a transition to the breakdown of parts in section two of the description.

A catalytic combustor, as shown in Figure 11-2, is typically a ceramic cylinder 6–8 inches in diameter and 2 inches high. It has a porous interior that is coated with catalytic metals such as platinum, palladium, or rhodium. These metals cause a chemical reaction to take place with the incoming fuel molecules.

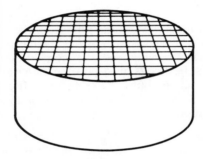

Figure 11-2   Catalytic combustor

In this example, the catalytic combustor does not itself have component parts; instead, it works with other parts of the wood stove to produce efficient combustion. Those parts might be named next.

Breakdown and Description of Parts

The parts description of any object or place can be organized in several ways:

- *spatially*. You can organize the description from left to right, top to bottom, or outside to inside. Spatial organization is especially good if there are no moving parts.
- *chronologically by order of operation*. This organization is related to the function or process of the object or place. If you were writing a description of a golf course, for example, the parts (each hole) would be covered from Hole 1 to Hole 18—the way the course would be played.
- *chronologically by order of assembly*. This organization is more ap-

propriate for an operator's manual or in a description intended for a technician who must build or reassemble a mechanism made up of number of parts. If you were writing a description of a sink trap, for example, the parts could logically be discussed by the order in which they would be assembled: sink drain, hexagonal nut, washer, trap, hexagonal nut, washer, drain pipe. You could also include an exploded drawing showing that relationship, as does Figure 11-3.

• *from general to specific or complex to simple*. This order works best when some parts of a mechanism or place are clearly more important or more complex than others.

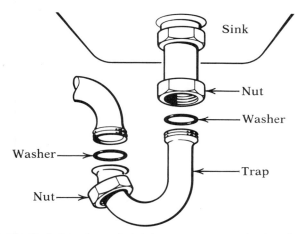

Figure 11-3 **Exploded drawing of sink trap components showing the order of assembly.**

As you describe each of the parts, you should cover the same points as in the overview: definition of the part, its purpose, its general appearance (perhaps with a graphic illustration), its composition, and if necessary, its parts (and those parts may need to be described as well!). You continually move from more complex parts to simpler ones so the reader can see the larger picture first and then the details.

Conclusion

Some descriptions do not require a conclusion; once you have described the individual parts, you can stop. In other cases you might add a short section—called a process description or procedure—that describes how the object works or how it is used. Notice how easily one kind of description shades into the next. Process descriptions are discussed in Chapter 18.

In the following student example (Davis 1985) the writer describes a specific brand of wood-burning stove for generalist readers. Basic terms have already been defined, and the overall method of organization is spatial—first the exterior is described and then the interior. Notice how the writer first gives a general description and then moves to a description of parts. The description concludes with a brief process description.

Technical Description: Hearthstone I

Exterior

The Hearthstone I is a wood-burning stove whose soapstone exterior makes it unique. Soapstone is a type of mineral known for its ability to store heat, later releasing it slowly and gently. Available in buff-grey, green, or brown, the stove comes either in the natural textured finish or the highly polished marble-like finish. Black porcelainized enamel on cast iron trims the door and frames the slabs of soapstone on the top and sides of the stove. Because of the soapstone, the overall weight of the stove is 700 lbs.

The Hearthstone's body dimensions are 32 in. high by 39 in. wide by 28 in. deep, and its short legs give it a rather squat appearance, as shown in Figure 11-4. The depth of the firebox allows it to accept firewood 24–26 in. in length. The double doors are 13 3/4 in. high by 17 1/2 in. wide, and each holds a rectangular insert of thermal glass, which allows a view of the fire inside. An ash lip extends like a narrow shelf below the doors, making loading and cleaning easier. The flue, with an 8-in. stovepipe diameter, can be made to extend through the rear as well as through the top of the stove. Air inlet control knobs are located on the lower left side and on the front of the stove below the left door.

Interior

Looking inside the stove (Figure 11-5), one can see the firebox with its floor of firebrick. Toward the top of the stove, a heavy cast iron baffle arches toward the front of the stove. Just above the baffle in the pathway to the flue is the catalytic combustor, a 2-in. by 6-in. cylinder with a porous interior. A temperature-sensing device called a catalytic monitor probe extends from the catalytic combustor to the back of the stove, where a thermostat displays the temperature of the combustor. Above the catalytic combustor and just beneath the top surface of the stove, an array of heat transfer fins lies ready to radiate heat through the top of the stove. The Hearthstone I is available with a water heating unit inside the firebox.

Burning Process

When a fire is kindled in the firebox, air flows through the air inlets and into the burn area. The smoke and gases mix with the incoming air and circulate through the firebox giving off heat. When the temperature reaches 500 degrees F, the catalytic combustor begins to burn the leftover smoke and gases, releasing more heat and decreasing the pollutants that would otherwise foul the chimney and the outside air.

Figure 11-4   Exterior view of the Hearthstone I wood stove

1. Dual adjustable spin drafts and window air ports
2. Venturi pressure system
3. Catalytic combustor
4. Catalytic monitor probe
5. Directional smoke baffle
6. Heat transfer fins
7. Catalytic monitor probe

Figure 11-5   Interior view of the Hearthstone I wood stove

11.5 HINTS FOR WRITING GOOD DESCRIPTION

As in all types of technical writing, you must shape description to fit the particular needs of each assignment, remembering your purpose and reader. The wood stove description in this chapter provides good information for generalists, but for technicians or specialists it might be inadequate. These readers might want detailed parts descriptions of elements in the stove's interior, like the baffle, the catalytic monitor probe, the thermostat, and the heat transfer fins. For them, the description might need to be more like a specification sheet, with detailed measurements as well as descriptions of each part.

Your description assignments will vary depending on your field, your purpose, and your reader, but these general hints will help you write good descriptions:

- Record measurements, dimensions, and weight accurately. A small discrepancy in a measurement might mean that the object won't fit where your reader needs it, and weight might be a factor in your reader's decisions.
- Be specific and concrete. Use terms at the bottom of the abstraction ladder (see Chapter 8); choose words that can be visualized easily. Give parts a name (even if your source doesn't name them) so they can be identified more readily. For example:

the rubber plug
the C-shaped insert
the 6-inch cork gasket

- Use simple language and analogies to help generalist readers visualize (see Chapter 10 for information on analogies). We visualize in the following five ways, and you can use many of them in a description.

1. size. In addition to measurements, compare to common objects: a football field, a paper clip, a nickel.
2. shape. Use terms like "circular," "rectangular," "concave," and "square." Compare to common objects: U-shaped, sawtoothed, threadlike.
3. texture. Describe the texture as accurately as you can: mirrorlike, honeycombed, spongy.
4. color. Describe both hue (red, blue, green) and tint (amount of white added).
5. position. Fix the position of your object or place relative to other objects or places. Use terms like "above," "below," "parallel to"; use directions like northeast, west southwest.

To see how a writer focuses on technical description within the context of a long report, study the following student description of the Nd:YAG surgical laser (Leonard 1985). This description is part of the introduction to the comparative report discussed in Chapter 21. Notice how the writer lists eight components in the overview and then organizes the rest of the description by detailing each of those components. The components are arranged chronologically by order of operation, thus also providing information about the process of producing and using the laser beam in surgery. Terms that appear in capital letters are defined in the glossary of the report.

Surgical Lasers Described

Nd:YAG surgical lasers represent a specific class of surgical lasers which use a solid Nd:YAG crystal LASING MEDIUM. To a large extent, this medium determines the applications and the structure of the laser system.

The applications of the Nd:YAG laser and its relation to other surgical lasers are discussed in section 2.1.3.2. The structure of Nd:YAG systems is described here. The Nd:YAG system has eight basic components:

1. optical energy source
2. optical resonator
3. optical wave guide
4. surgical handpiece
5. lens system
6. electronic shutter system
7. water cooling system
8. control panel

These components are typically housed in a rectangular cabinet approximately 2 feet wide, 1.5 feet deep, and about waist height. The laser head may be separate and is often situated above the main unit. Different models vary somewhat in shape and size. The units range in weight from about 400 pounds to 800 pounds. See Figure 11-6.

The OPTICAL ENERGY SOURCE consists of one or two high power xenon or krypton FLASHLAMPS. The lamps supply the optical energy required by the next component, the OPTICAL RESONATOR.

The OPTICAL RESONATOR is contained in the laser head. It is the critical component in laser production. The resonator consists of a cylindrical cavity with plane parallel mirrors at each end. The Nd:YAG crystal is contained within the cavity. When optical energy reaches the crystal, laser light starts resonating between the mirrors producing the beam. This process is described with more detail in section 2.1.2.

The laser beam leaves the resonator and enters the OPTICAL WAVE GUIDE. The wave guide is usually a flexible OPTICAL FIBER, but it can be rigid for special applications. The wave guide transports the beam from the resonator to the SURGICAL HANDPIECE.

The SURGICAL HANDPIECE directs the beam onto the target tissue. A Nd:YAG system typically has interchangeable handpieces; each is designed for a different application.

Figure 11-6   The LaserSonics Model 8000: A representative Nd:YAG surgical laser

The lens and shutter systems regulate the effect the beam has on the target tissue. The LENS SYSTEM concentrates the energy of the large beam produced in the resonator into a minute spot directed at the target. FOCUSing occurs once as the beam enters the wave guide and again as it exits the handpiece. The ELECTRONIC SHUTTER SYSTEM controls the duration and frequency of laser exposure by repetitively opening and closing an aperture to the wave guide. Most systems allow CONTINUOUS WAVE or PULSED MODE with PULSE LENGTH adjustable from 0.1 to 10.0 seconds.

The WATER COOLING SYSTEM prevents the resonator from overheating. Ordinary tap water is circulated around the resonator removing excess heat. The flow rate ranges from less than 2 gallons per minute to more than 6 gallons per minute among different systems.

The CONTROL PANEL is generally positioned near the top of the cabinet and angled for easy hand operation. It provides the interface between the operator and the other components in the system. Controls allow adjustments in power and pulse characteristics, while meters indicate the status of system components.

When the Nd:YAG laser is operated, optical energy flows into the resonator producing a laser beam concentrated and pulsed according to the operator's specification. The system is generally self-contained; the operator can devote his or her attention to the surgical procedure.

CHECKLIST FOR WRITING DESCRIPTION

1. What is my purpose?
2. Who is my primary reader and what are his or her needs?
3. Do I need a generalized description as an overview or a very specific description?
4. Where can I find the information needed for this description? If I use sales literature, have I deleted any biased language?
5. Should my description be informal or formal?
6. If I choose to write a formal description, does my overview contain
   _____ a definition?
   _____ a purpose statement?
   _____ general appearance?
   _____ illustration?
   _____ a list of components?
7. How is the breakdown and description of parts organized? Is this a logical organization? Have I thoroughly described each part and broken it into components if necessary?
8. Do I need a conclusion? Should I add a process description?

EXERCISE 11-1 Evaluating Descriptions in Advertising

Directions: Choose an ad from a trade magazine in your field and make a copy of it. In one or two paragraphs, write an analysis of the ad discussing what parts of it are biased and what parts are objective. Give examples from the ad to back up your points. Be prepared to discuss in class.

EXERCISE 11-2 Writing Formal Descriptions

Directions: Choose one of the following objects or places and write a generic description of two to three pages. Follow the three-part structure explained in section 11.4. Specify the purpose and the reader.

lawn sprinkler	rolling pin	putty knife
auto cigarette	electric razor	folding lawn chair
lighter	steam iron	smoke detector
air popcorn popper	ski pole	Bunsen burner
roller skate	thermometer	the human ear
baseball glove	your room	the football stadium
the human eye	the fire house	the nearest park
local golf course		

EXERCISE 11-3 Writing Informal Descriptions

Directions: Choose one object or place from the list in Exercise 11-2 and write a paragraph-long informal description.

EXERCISE 11-4   Writing Description for Your Project Report

Directions: Study the outline for your term project to see where you have included (or should include) description of an object or a place. Write a draft of a formal or informal description (or both). Be sure to consider your purpose and reader. After editing, incorporate the description smoothly into the appropriate section of your report. If you are working on a collaborative project, be sure that one person is responsible for writing any needed descriptions.

EXERCISE 11-5   Analyzing a Description

Directions: Study the following description (Bejar 1985), analyzing it for language (is it specific and unbiased), for organization (does it have an overview, analysis by parts, process description), and for completeness of content. Be prepared to discuss in class.

> The installation of a burglar alarm in a home is now easy enough for one to do without professional help. A self-installed home alarm system, usually referred to as a local alarm, simply sounds a siren on the premises whenever an intruder is detected by the system.
>
> A typical alarm system works similar to an automotive dome lamp. When all openings are closed, the system is off and armed: ready for detection. When any door or window is opened, the circuit is broken. Then the siren sounds, just as the dome light in a car lights up. Another type of alarm system uses an energy field to detect intruders. It works similar to the electric eye used in many stores. When the beam of light is crossed, the alarm sounds. In home applications, the field takes on a bubble shape, protecting a large portion of the interior. The first system is usually referred to as perimeter protection, while the second is called specific-area protection. Both have advantages and limitations which will be discussed in the main text.
>
> A typical alarm system consists of four main parts: sender, circuit path, control module (receiver), and siren. The sender is the component that initially detects an intruder. It notifies the control module via the circuit path.
>
> The circuit path can be either in wire or wireless form. The wire form uses small gauge wire, similar to speaker wire, for connecting the components. Wireless systems use a noise-emitting box for a sender. It is similar to a cigarette box in size and shape. Instead of wires, the circuit is made complete by sound or radio waves inaudible to the human ear. The circuit works similar to a TV remote control.
>
> The control module resembles a small, wood-grain clock radio with indicator lights on the front and terminals at the rear for connecting the system's components. The control module receives the signal from the sender when an intruder is present and activates the siren, usually through a wired circuit.
>
> The siren is usually in loudspeaker form and resembles the hand-

held speakers used by police. When an intruder is detected, the siren is set off, emitting an extremely loud noise throughout the exterior perimeter of the premises.

REFERENCES

AOPA (Aircraft Owners and Pilots Assn.). 1965. *Places to fly.* Washington: AOPA, 73.

Bejar, John. 1985. *Description of a burglar alarm system.* Informal report, San Jose State University.

Bernucci, Julie. 1981. *Comparative study of top-of-the-line sewing machines: Elna SU68, Riccar 808E, Pfaff 1222E, Singer Touch-Tronic 2010.* Formal report, San Jose State University.

Davis, Mary. 1985. *An evaluation and comparison of four wood-burning stoves.* Formal report, San Jose State University.

Leonard, David. 1985. *An evaluation of Nd:YAG surgical lasers for Allied Hospital.* Formal report, San Jose State University.

Volkswagen of America. 1986. Introducing the 1986 Volkswagen, form no. W64-006-6001.

CHAPTER 12

Illustrating with Graphics

As a student you have looked at—sometimes even studied—hundreds of graphics like bar graphs, cutaway drawings, maps, and tables. Because you studied them for the information they contained, you probably accepted uncritically the construction and effectiveness of the graphics themselves. Now, though, as a writer seeking to present information clearly and efficiently, you should start looking at graphics from another viewpoint. Ask yourself "What kind of graphics will best achieve my purpose in this piece of writing?" and "How can I ensure that the graphics I choose do their job?"

Just as different forms of technical writing are suitable for specific tasks, so different forms of graphics are suitable for specific purposes. This

234

chapter will look at graphics from a task-oriented perspective, presenting each graphic aid in light of what it can do for you as a writer. In addition, you'll learn how to present each type of graphics most effectively.

12.1 DEFINITIONS OF GRAPHICS TERMS

In this book the word *graphics* means any visual form of presenting information: that is, pictures or arranged numbers as opposed to sentences. Graphics is thus a general term, as are the related words *visuals* and *illustrations*. Graphics increase understanding by appealing to the eye.

Graphics are divided into two large categories: tables and figures. *Tables* present data (often numbers) in columns for easy understanding or comparison; the presentation is visual, but the data is set in type. All graphics that are not tables are called *Figures*. The main divisions of figures are graphs, diagrams, photographs, drawings, maps, and printouts. In a formal report or proposal, tables and figures are numbered separately in the text—there can be both a Table 3 and a Figure 3. Following the Table of Contents, you place a List of Tables and a List of Figures.

For a quick overview of your choices among graphics, look at Figure 12-1, which diagrams the main types of graphics and shows their relationships.

12.2 ADVANTAGES OF USING GRAPHICS

Graphics are widely used in technical literature to enhance and supplement the written text. While graphics can seldom totally replace sentences, they offer several advantages over printed words:

1. Graphics increase reader interest and understanding. They appeal to the reader's visual sense and pattern-making capabilities, and they usually are easier to understand than sentences saying the same thing. Tables and figures also help break up a page of solid words, providing rest and focus for the reader's eyes. (See Chapter 6 for more on document design.)

2. Graphics display and emphasize important information. In his book *How to Write and Publish Engineering Papers and Reports*, Herbert B. Michaelson (1986, 79) even suggests that at the organizing step writers should (1) decide on the key points, (2) sketch a figure for each key point, and (3) then plan the manuscript around those figures. Certainly it makes sense to focus the reader's attention on key information through graphics.

3. Graphics summarize and condense information, pulling together many isolated bits and displaying them in one place. This capability makes them especially good in conclusion and recommendation sections.

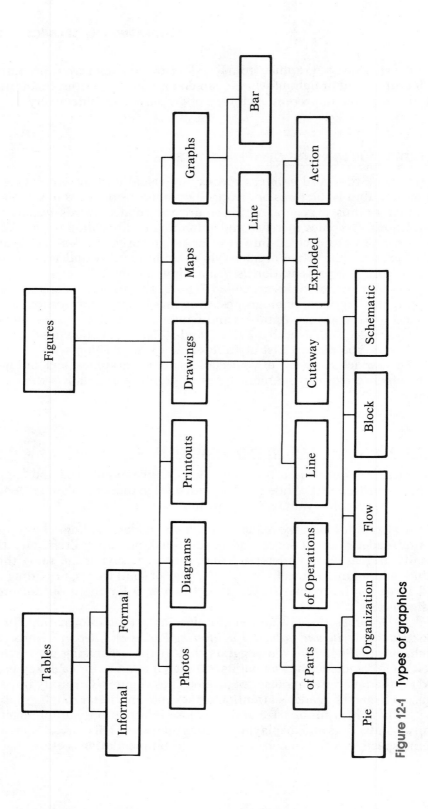

Figure 12-1 Types of graphics

4. Because of their visual presentation, graphics convey quantitative relationships—likenesses, differences, percentile rankings, trends—better than words.

5. Increasingly, graphics can be designed using a computer. In many industries, 80 to 90 percent of graphics are designed by computer. As more low-cost computer-aided design (CAD) systems become available for microcomputers, students will be able to produce tables, diagrams, and drawings of professional quality. Figures 12-2 and 12-3 show examples of computer-designed graphics produced by students to illustrate key concepts in their reports.

Figure 12-4 shows a more sophisticated line drawing generated by a computer-aided design (CAD) system in industry. When used in a document, this drawing would be labeled and referenced.

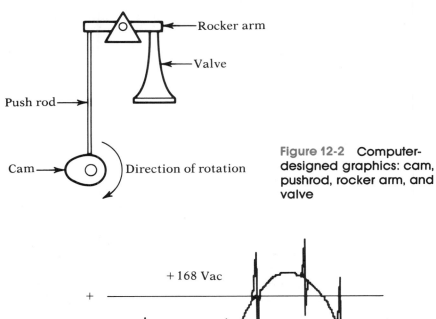

Figure 12-2 Computer-designed graphics: cam, pushrod, rocker arm, and valve

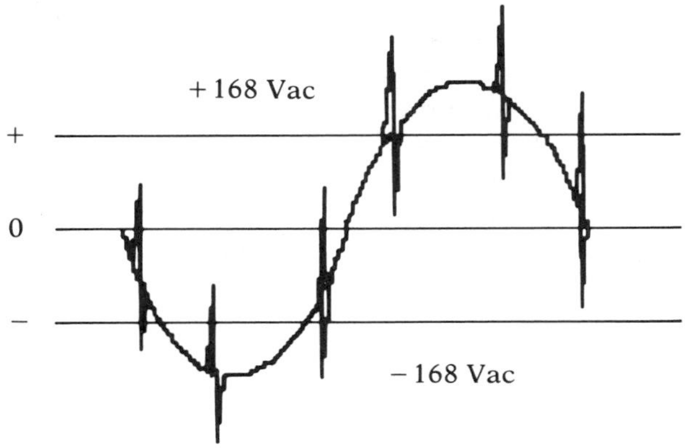

Figure 12-3   Computer-designed graphics: sine wave with noise interference

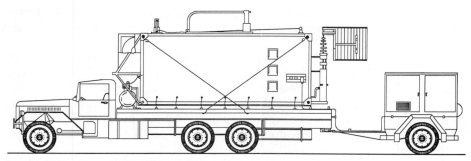

Figure 12-4 Line drawing by computer

6. Graphics can be borrowed from other sources and incorporated into your own reports, as long as you *identify* and *give credit* to your sources. (You will, of course, have to secure written permission from copyright holders if you are going to publish or sell this borrowed material.) You can copy graphics, reduce or enlarge them on a copy machine if necessary, paste them on your typed page at appropriate locations, and then copy the whole page. This process will give you a professional-looking page. But be careful, because color photographs do not copy well and reduction may destroy details.

7. Many graphics can be traced or hand-drawn with a ruler and pen; you need not be artistic. You can type captions or use rub-on letters.

12.3 CHOOSING TYPES OF GRAPHICS BY PURPOSE

Your central question about graphics at the draft-writing stage is "What kind of graphics will best accomplish my purpose?" To answer that question, this section presents six general purpose statements and suggests appropriate graphics choices for each purpose. Remember that you must also evaluate the type and background of your reader: specialists often want different graphics than do generalists. Although sometimes only one type of graphics will work, at other times you will have a choice. Remember too that you are not limited to these suggestions. Just be sure that the graphics you choose will accomplish your purpose. Table 12-1 summarizes these six purposes and appropriate graphics, but for details about how to use each one, read the explanations that follow.

Purpose 1. To Condense or Summarize Information; to Present Reference Data; to Give Detail

Tables   Tables are the best choice for condensing and referencing data or for giving exact numbers and measurements. A table is an arrangement of numerical or factual data by rows and columns; it can be either informal or formal.

Table 12-1   CHOICES OF GRAPHICS FOR SPECIFIC PURPOSES

Purpose	Tables	Photos	Diagrams	Printouts	Drawings	Maps	Graphs
1. Condense and summarize information	X			X	X	X	
2. Compare amounts, show relationships and trends							X
3. Show parts of a whole and relationships of parts to the whole and each other		X	X		X	X	X
4. Show steps in a procedure or the function of each element in a system			X		X	X	
5. Show how something looks externally or internally		X			X	X	
6. Give examples or reproduce what the reader might see		X		X	X		

Informal tables are lists that are part of the text itself but have been pulled out of the paragraph and put into column format. For example, I can break out a list of reader types and appropriate graphics for that type into column format, like this, as long as I explain within the text what I mean by the list.

Reader type	*Especially effective graphics*
specialists	tables, line graphs, schematic diagrams
technicians	line graphs, flow diagrams, block diagrams, schematic diagrams
operators	bar graphs, flow diagrams, block diagrams, schematic diagrams, maps
managers	bar graphs, block diagrams, organizational diagrams
generalists	bar graphs, block diagrams, organizational diagrams

| all types | pie diagrams, maps, photo-graphs, line drawings, examples, organizational and hierarchical diagrams, exploded drawings |

This list gives general guidelines for choosing graphics by type of reader; however, the categories overlap considerably, and many graphics are suitable for all readers. Informal tables like this are not titled or numbered because they require additional text in order to be understandable.

By contrast, a formal table must be both understandable *without* additional text and referenced or explained *within* the text. A formal table has a table number and a title, which appear on a line above the table, either flush to the left margin or centered. Table 12-2 is an example of a formal table from a student report.

To keep the table free of clutter, use a minimum of lines; use white space for separation whenever possible. Place one horizontal line above and one below the column titles to form what's called a "boxhead," and one line at the bottom to separate the table from the following text. The first column on the left is called the "stub column"; it contains the list of items, and it may or may not have a title. If you need footnotes for explanation, use lowercase superscript (above the line) letters; put footnotes at the bottom of the table.

Don't load a table with too much information. Sometimes it's better to design several tables, each of which contains a manageable information load.

The advantage of a table is its ability to present a large amount of specific information in a small space, thus making comparisons and summaries easy. The disadvantage is that it does not point out or stress relationships; either the reader must make comparisons from the data presented, or you must discuss the relationships in your comments on the table.

Because the information in any table can usually be arranged in two or three different ways, you may need to test several options to find the most effective one. Use these suggestions to help you design an effective table:

- Confine the table to one page if possible and design it to fit vertically on the page.
- Place items that use names in the stub column (on the left) and items that use numbers under appropriate column titles in the rest of the table.
- Write column titles that clearly identify the items and the units of measure used.
- Align decimal points and limit decimals to two places if possible.
- Add notes to any items that need further explanation. Use lowercase letters to label the notes.

table number and title   column head   footnote designation

stub heading

Table 12-2   SPECIFICATIONS OF 10-INCH TABLE SAWS

boxhead

Make and Model	Working size L W H in inches			Consumable square feet	Base material	Net weight in lbs.	Rating[a]
De Walt 7756	27	49	35	9.2	cast aluminum	135	3
Powermatic 66	28	38	24	7.4	steel	450	1
Rockwell 34-710	22	40	34.5	7.5	cast iron	130	2
Shopsmith Mark Five	19	71	41.5	9.4	cast aluminum	198	2

rows

[a]Rating: 1 = Above Average   2 = Average   3 = Fair   4 = Poor

rules to set off table from text

footnote

columns

Tables of summary are best placed in the conclusions and recommendations section or in the appendix of a report. Tables of detail appear in the body or appendix. Specialists appreciate tables; they like the specific numbers and don't mind making the comparisons themselves.

Drawings, maps, and printouts   At a more visual level, cutaway and exploded drawings (especially if they include dimensions and measurements) will provide specific detail of objects and machines. Maps that display many locations simultaneously can also condense large amounts of information effectively, and printouts of computer screens, EKG monitors, and other devices work well to summarize information. (See the following sections for descriptions of drawings, maps, and printouts.)

Purpose 2. To Compare Amounts; to Show Relationships and Relative Quantities; to Show Trends, Distributions, Cycles, and Changes Over Time

The best way to compare amounts and show relationships and trends is with a graph. A graph is a pictorial presentation of numerical data. In other words, graphs can give the same information as tables, but they *picture* comparisons and trends, making the reader's job easier.

Line graphs   Use line graphs to show trends, distributions, and cycles. Standard usage calls for plotting variables of distance, time, load, stress, or voltage on the horizontal axis (called the abscissa) and variables such as money, temperature, current, or strain on the vertical axis (called the ordinate). Figure 12-5 shows an example.

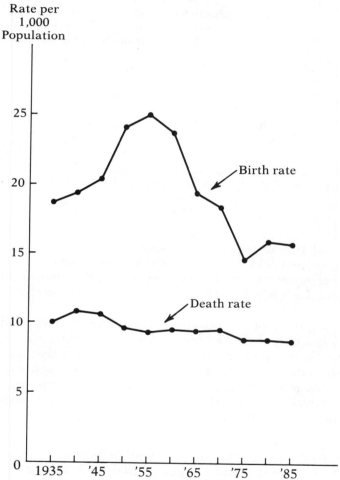

Figure 12-5   Line graph showing birth and death rates: 1935–85

The figure number and title usually appear below the graph, and the source line appears at the lower left. In a line graph, start your scales at zero, and show a break in the scale if necessary. Take care not to distort increments by making the scales too wide or too narrow. Percentages should always go from 0 to 100. Include a key if needed, and make the lettering horizontal, not vertical.

Bar graphs Use bar graphs to compare amounts and show proportional relationships. The bars should be of the same width but can be either vertical or horizontal. Unless they are tied to time, bars usually are arranged in increasing or decreasing length. Figure 12-6 is a typical bar graph showing how computing time in a laboratory is reduced by multiprocessing.

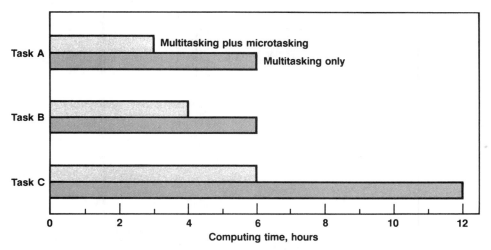

Figure 12-6   Bar graph showing how computing time is reduced by multiprocessing

The bar graph has labeled vertical and horizontal axes, and as with other figures, the title and figure number are usually located below the graph. For more precision, bar graphs can include exact numbers—a good location for numbers is at the top of each bar. All graphs should be plotted on graph paper with a horizontal and a vertical axis. Since you won't want the grid lines to clutter your final figure, use blue graph paper for plotting and then copy the graph on a copy machine to get only the lines or bars of your graph. Bar graphs can be truncated to show a break in the scale, as in Figure 12-7, which shows the enrollment trend at a university over 10 years.

Two special kinds of bar graphs are *segmented bar graphs* and *pictographs.* In a segmented bar graph, a single bar represents 100 percent. It is divided into parts to show proportional relationships, as shown in Figure 12-8. The pictograph uses a single simple drawing to represent one unit, and then repeats those drawings to assemble the bars, as shown in Figure 12-9. Pictographs are not as accurate as standard bar graphs, but they add interest and are therefore good for generalists. Be careful in using pictographs not to increase the size of the drawing to show a greater quantity, because that will distort the relationship.

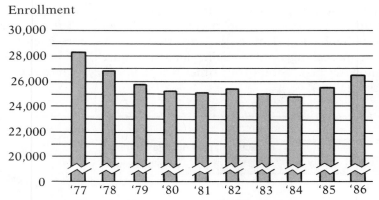

Figure 12-7   **Truncated bar graph showing Smith University enrollment patterns 1977–86**

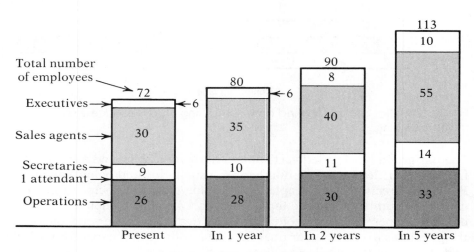

Projected Growth of Allied Insurance
Over the Next Five Years

Figure 12-8   **Segmented bar graph**

The advantages of graphs lie in their visual appeal; they are more interesting than tables, and they show comparisons better. Generalists and managers appreciate bar graphs for the quick and easy view they give of relationships. Specialists and technicians like line graphs for their detail. However, unless graphs also give specific numbers, they may be less accurate than tables, and they can easily distort information if the scales are not properly proportioned.

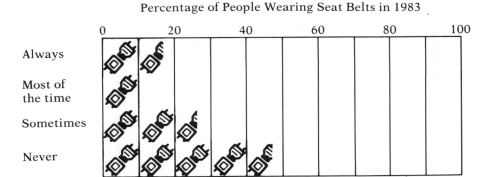

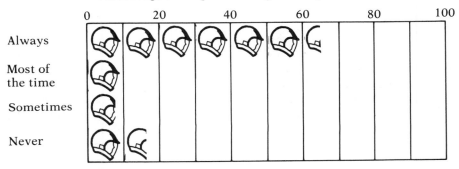

Figure 12-9   Pictographs

Purpose 3. To Show Parts of a Whole and/or Comparisons of One Part to Another; to Show the Chain of Command in an Organization; to Give an Overview of a Whole

For presenting wholes and the parts that make up the whole, you can use pie diagrams, segmented bar graphs, cutaway or exploded drawings, and organizational and hierarchical diagrams.

Pie diagrams   A circle or pie diagram shows both the parts of a whole and the relative size of each part. It uses a circle to represent 100 percent, so that each percentage point equals 3.6 degrees. The first radius should be drawn at 12:00, and the segments should be drawn clockwise around the circle starting with the largest segment, as in Figure 12-10. Pie diagrams are most effective if they contain no more than seven segments, and if each segment is large enough to be seen easily. When you draw a pie diagram, double-check that your percentages total 100 percent. If you shade segments, start with the lightest shade at 12:00 and work clockwise around the circle to the darkest shade.

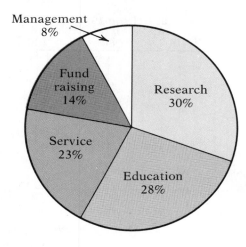

Figure 12-10 Pie diagram showing allocation of American Cancer Society Funds, 1986

Segmented bar graph   A segmented bar graph shows the same relationships as a pie diagram. Both appeal especially to generalists and managers and are good additions to conclusions and recommendations sections or to the analysis sections of a report. Pie diagrams and segmented bars work well to represent large sums like a whole budget or one year's tax revenues. Figure 12-10 shows the relationship of parts to the whole in a budget allocation; Figure 12-11 shows the same relationship in a segmented bar graph.

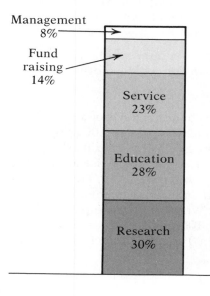

Figure 12-11 Segmented bar graph showing allocation of American Cancer Society Funds, 1986

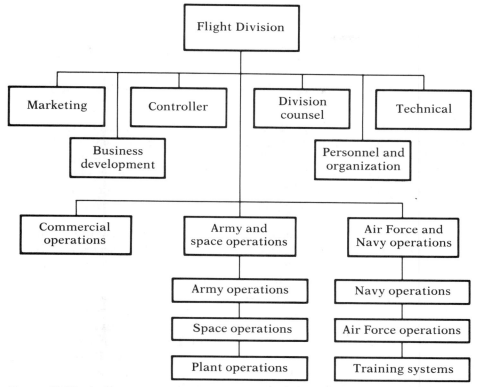

Figure 12-12 Organizational diagram: organization of flight division

Drawings A cutaway or exploded drawing can be used to show the parts of a whole object like a geological formation or a generator. Examples of cutaway and exploded drawings appear under Purpose 5 in this chapter.

Organizational and hierarchical diagrams An organizational diagram is designed to show areas of control and responsibility; a hierarchical diagram shows the relationships among the parts of a complex subject. As in a pie diagram, a whole is divided, but this time what is divided is a department, a company, or a complex idea with many subpoints. These diagrams consist of a series of connected boxes and are read from the top down. The title and figure number usually appear below the diagram. Diagrams are unequalled for showing areas of responsibility and the chain of command when they refer to people and positions. When they refer to ideas, diagrams effectively outline hierarchical organization. By placing items of equal importance along the same horizontal, they show the equality of main points and the relationships among subpoints. Figure 12-12 shows a typical organizational diagram for a company division; Figure 12-1 at the beginning of this chapter is a hierarchical diagram showing the relationships among the types of graphics.

Both organizational and hierarchical diagrams meet the needs of all types of readers because they provide an overview of relationships. They can be placed in the introduction of a report as an overview, in the conclusion as a summary, or in the appendix as added information.

Purpose 4. To Show Steps in a Procedure; to Show the Function of Each Element in a System

A flow diagram, sometimes called a flow chart, can greatly simplify the description of a procedure because the reader simultaneously sees an overview (the whole process) and the individual steps making up that process. Block, schematic, and wiring diagrams all show how each element in a larger system functions and how it relates to other elements. Maps also show relationships—for example, between cities within a state.

Flow diagrams and related drawings   Flow diagrams visually represent the steps of a procedure. They can be constructed with a series of labeled blocks, drawings, standardized symbols, or photographs. The steps are connected by arrows showing the direction of flow or movement from one step to another. Usually flow diagrams are read from top to bottom or left to right; if you want to show a repetitive process, you might design a clockwise flow. For clarity, give the flow diagram a descriptive title, placing it and the figure number below the diagram. Label the parts clearly and spend some time planning the most effective presentation. Figure 12-13 is a simple flow diagram; Figure 18-5 in Chapter 18 is more elaborate.

An *action drawing* is very similar to a flow diagram, but the parts themselves may be isolated, with the movement of those parts indicated by arrows. For assembly of parts, an *exploded* drawing shows each part in the order in which it should be used, making the steps in the procedure self-explanatory. (Exploded drawings are explained in Purpose 5). All these diagrams are useful for all types of readers, but especially for operators and technicians who must follow the procedures or carry out the instructions.

Block diagrams   Block diagrams are like flow diagrams, but they show relationship rather than movement. When you must present very complex circuitry to inexperienced readers, it often pays to use a block diagram first to give them a simplified overview. Block diagrams are well placed in overview sections of manuals and also at the beginning of (for example) a maintenance section. They can also be used for managers who need a basic understanding without the detail of a schematic. See Figure 12-14.

Schematic and wiring diagrams   Schematics are detailed maps of circuits showing the *logical* connections and current and signal flow. Wiring diagrams show the *actual* point-to-point connections. The American National

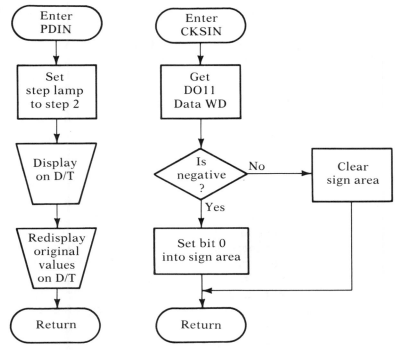

Figure 12-13   Flow diagram: typical subroutines of a computer program

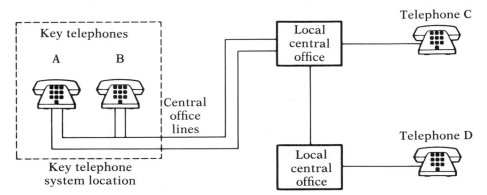

Key telephones A and B contain the same central office
lines. A can call C or D by dialing directly. C and
D can call either of the key telephones and both will
ring.

Figure 12-14   Block diagram of key telephone system

Standards Institute (ANSI) issues guidelines and lists of the standardized symbols for schematics and wiring diagrams. Be aware, though, that the symbols and layout may vary from industry to industry. Figure 12-15 shows a simple schematic. Figure 12-16 shows a wiring diagram.

Schematic and wiring diagrams are mostly used by specialists, technicians, and operators, while block diagrams are useful for managers and generalists. *Maps* can also show functions and relationships, but in this case the "systems" are geographical. Maps are explained in Purpose 5.

Purpose 5. To Show How Something Looks Externally or Internally; to Show Details and Locate Parts

Photographs Photographs can show external suface better than other graphics because of their absolute reproduction; they can show damage, record phases of a phenomenon such as growth, identify or locate parts, or show how equipment functions. However, they can also overwhelm with detail or introduce distracting background. Sometimes, therefore, a line drawing, which can be simplified to show only important features, is a better choice. Photographs and external line drawings are often used on the cover of a manual or report or with introductory material. They can be used for all types of readers.

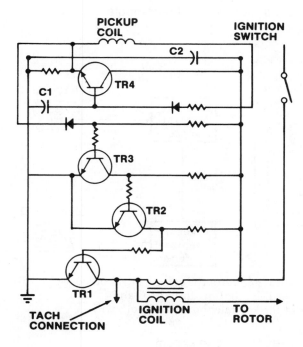

Figure 12-15 Schematic diagram

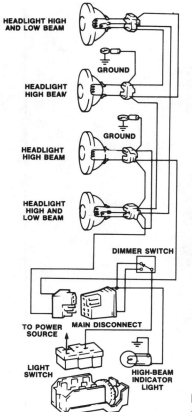

HEADLIGHT HIGH
AND LOW BEAM

GROUND

HEADLIGHT
HIGH BEAM

GROUND

HEADLIGHT
HIGH BEAM

HEADLIGHT
HIGH AND
LOW BEAM

DIMMER SWITCH

TO POWER   MAIN DISCONNECT
SOURCE

LIGHT
SWITCH

HIGH-BEAM
INDICATOR
LIGHT

Figure 12-16   Wiring
diagram

Internal views are most commonly shown by photographs and three kinds of drawings: cutaway, line, and exploded. Photos can show internal views with disassembled parts or through a "blowout": a telescoped view of a small part or assembly "blown out" of the interior of the machine and displayed. Blowouts are also common with line drawings. Cutaway drawings show things or places in cross section.

Whether showing exterior or interior views, photographs must be clear enough to reproduce well, and they should be cropped to show only the desired portion. Parts should be labeled carefully, and the title and figure number should be placed below the photograph. To show scale, you can include in the photograph a ruler or familiar object such as a coin. Photographs are especially good for introductory sections because they provide an overview. The disadvantages are that surrounding areas may clutter the photograph, or you may be unable to highlight the part you wish. See Figure 12-17 for an example of a photograph used in industry.

Figure 12-17   Photograph for industry

Line drawings   Drawings are useful in technical writing because they can emphasize a single part and show as much or as little detail as desired. Sometimes called "overall" drawings, line drawings show the outside of something. They are often shaded perspective drawings that give a three-dimensional impression. Line drawings are useful in overviews and introductions or anyplace an exterior view is needed. Sometimes line drawings are simply titled without explanations, the title and figure number going below the drawing. At other times, parts are labeled on lines or arrows extending from the parts, as in Figure 12-18.

Cutaway drawings   In a cutaway drawing, the view is either of a cross-section or a part from which the outside covering has been removed. "Phantom" drawings are similar except that the outside covering is drawn in but appears to be transparent. Cutaway drawings are excellent for showing how interior parts are arranged and how they relate to the exterior. Figure 12-19 shows a cross-sectional view of an airbrush.

Exploded drawings   An exploded drawing shows the correct order of assembly of a complex mechanism, and does it better, more easily, and in less space than a written explanation could. In an exploded drawing, the

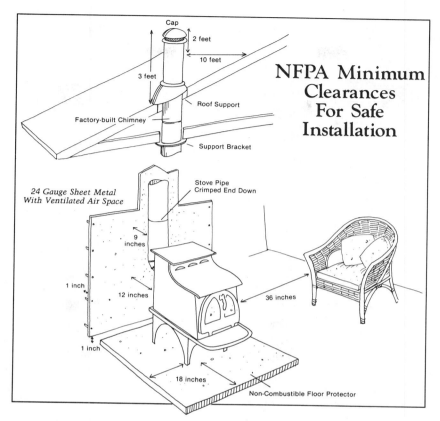

Figure 12-18   Line drawing

parts are spread out along an imaginary line that indicates the axis. Each part is drawn in perspective where it would fit in the assembly. Parts are either labeled for identification or numbered, with the numbers identified in a key. The figure number and title appear at the bottom; source references usually appear in the lower left corner. Figure 12-20 shows an exploded drawing of an injector cover assembly drawn by computer.

Action drawings   Action drawings are related to flow diagrams in the way they help explain what happens. However, flow diagrams indicate sequence, while action drawings may simply show the part in motion. Action drawings can be three-dimensional or stylized. See Figure 12-21.

Maps   Maps are designed to identify and locate places within a larger geographical area. Like diagrams, maps show the relationships among parts and how the parts make up a whole, but the whole represented by a map is located on land, sea, or in the sky. All types of readers use maps

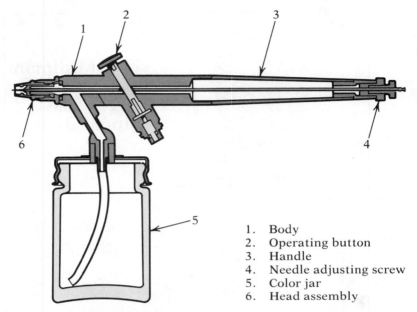

1. Body
2. Operating button
3. Handle
4. Needle adjusting screw
5. Color jar
6. Head assembly

Figure 12-19 Cutaway drawing

at one time or another, but operators—whether pilots, hikers, or traveling sales representatives—are the most frequent users. Aeronautical and nautical maps are usually called charts. A map will be most effective if it presents only information relating directly to the reader's purpose and needs. Thus, maps concentrating on rivers may not include cities, and road maps do not usually show changes in elevation. Label those items you want emphasized, and below the map give a title that clearly explains what the map shows. Figure 12-22 shows a simple map.

Purpose 6. To Give Examples; to Reproduce What the Reader Might See in an Operation; to Enhance Speeches or Oral Presentations

Printouts   Readers who operate equipment or conduct experiments often want references to what they might expect to see on a computer screen, oscilloscope display, or through a microscope. You can reproduce screens, computer printouts, or photographic slides for this purpose. These are graphic presentations through examples, and they can be helpful to all types of readers. See Figure 12-23.

Transparencies, slides, and flip charts   In speeches and oral presentations, you will communicate most effectively if you appeal to both sight and

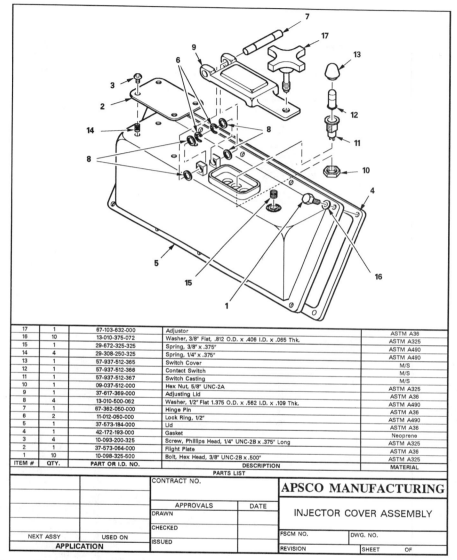

17	1	67-103-632-000	Adjustor	
16	10	13-010-375-072	Washer, 3/8" Flat, .812 O.D. x .406 I.D. x .065 Thk.	ASTM A36
15	1	29-672-325-325	Spring, 3/8" x .375"	ASTM A325
14	4	29-308-250-325	Spring, 1/4" x .375"	ASTM A490
13	1	57-937-512-365	Switch Cover	ASTM A490
12	1	57-937-512-366	Contact Switch	M/S
11	1	57-937-512-367	Switch Casting	M/S
10	1	09-037-512-000	Hex Nut, 5/8" UNC-2A	M/S
9	1	37-617-369-000	Adjusting Lid	ASTM A325
8	4	13-010-500-062	Washer, 1/2" Flat 1.375 O.D. x .562 I.D. x .109 Thk.	ASTM A36
7	1	67-362-050-000	Hinge Pin	ASTM A490
6	2	11-012-050-000	Lock Ring, 1/2"	ASTM A36
5	1	37-573-184-000	Lid	ASTM A490
4	1	42-172-193-000	Gasket	ASTM A36
3	4	10-093-200-325	Screw, Phillips Head, 1/4" UNC-2B x .375" Long	Neoprene
2	1	37-573-064-000	Flight Plate	ASTM A325
1	10	10-098-325-500	Bolt, Hex Head, 3/8" UNC-2B x.500"	ASTM A36
ITEM #	QTY.	PART OR I.D. NO.	DESCRIPTION	ASTM A325
			PARTS LIST	MATERIAL

		CONTRACT NO.		**APSCO MANUFACTURING**		
		APPROVALS	DATE	INJECTOR COVER ASSEMBLY		
		DRAWN				
		CHECKED		FSCM NO.	DWG. NO.	
NEXT ASSY	USED ON	ISSUED				
APPLICATION				REVISION	SHEET	OF

Figure 12-20   Exploded drawing

sound—emphasizing the main points listeners *hear* with graphics or with projected key words that they can also *see*. All the graphics mentioned already can be used in speeches, but most of them need to be simplified for projection. Many professionals use transparencies (also called *viewfoils* or *foils*) to project simple tables, line or bar graphs, diagrams, and the like. Others use flip charts or slides on which they write or draw appropriate

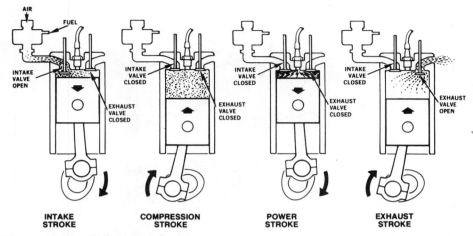

Figure 12-21   Action drawing

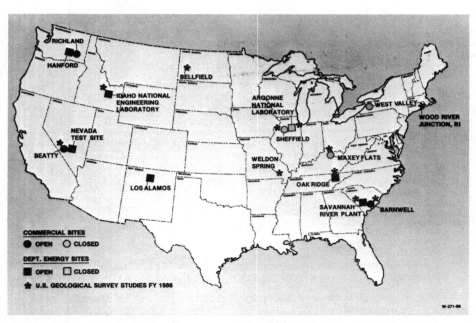

Figure 12-22   Simple map: Low-Level Radioactive Waste Sites and U.S. Geological Survey Studies

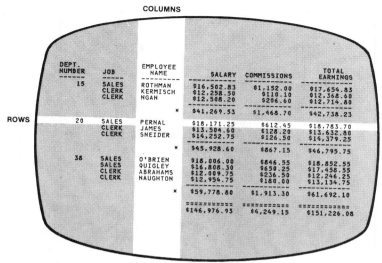

COLUMNS

ROWS

DEPT. NUMBER	JOB	EMPLOYEE NAME	SALARY	COMMISSIONS	TOTAL EARNINGS
15	SALES CLERK	ROTHMAN	$16,502.83	$1,152.00	$17,654.83
	CLERK	KERMISCH	$12,258.50	$110.10	$12,368.60
	CLERK	NGAN	$12,508.20	$206.60	$12,714.80
		*	$41,269.53	$1,468.70	$42,738.23
20	SALES	PERNAL	$18,171.25	$612.45	$18,783.70
	CLERK	JAMES	$13,504.60	$128.20	$13,632.80
	CLERK	SNEIDER	$14,252.75	$126.50	$14,379.25
		*	$45,928.60	$867.15	$46,795.75
38	SALES	O'BRIEN	$18,006.00	$846.55	$18,852.55
	SALES	QUIGLEY	$16,808.30	$650.25	$17,458.55
	CLERK	ABRAHAMS	$12,009.75	$236.50	$12,246.25
	CLERK	NAUGHTON	$12,954.75	$180.00	$13,134.75
		*	$59,778.80	$1,913.30	$61,692.10
			$146,976.93	$4,249.15	$151,226.08

Figure 12-23   Reproduction of a computer screen

graphics. All audiences respond well to graphics, but you can seldom include as much detail as in a printed document. Chapter 23 gives more detail on using graphics in speeches.

12.4 GENERAL RULES FOR USING GRAPHICS

Follow the general rules listed below to present all your graphics clearly and effectively:

1. Keep graphics simple and uncluttered; keep lines and wording to a minimum. Design the table or figure so that it is large enough to be read easily. If you reduce it for better fit on the page, be sure that the reduced numbers and letters are readable.

2. Insert graphics immediately after their first mention in the text. Referring to them in the text specifically by number (Table 6, Figure 1), tells the reader when to look at them and why, especially if conclusions are involved. Keep your terminology consistent from text to table or figure and from one graphic aid to another. A major fault in technical writing is using graphics without referencing them.

3. Make each graphic aid self-sufficient: that is, intelligible without the explanation in the text. The keys to self-sufficiency are (1) a title that has both key content words and function words, like "How the Four-Stroke Cycle Works," (2) labels or column heads that clearly identify all parts, and (3) simplicity in detail so that the reader is not bogged down by extraneous information.

4. Unless your company style guide has a different system, place the table number and title *above* the table. Tables are usually read from top to bottom, so this arrangement is sensible. Place the figure number and title *below* the figure. Label the parts clearly with lines or arrows to the part (these are called "callouts"). Provide a key either on or below the figure.

5. Identify the source of the table or figure or the source of information in small print at the bottom of the graphic—for example: "Source: U.S. Dept. of Labor." Often sources are put at the lower left corner, but they may be placed elsewhere as long as you are consistent within a document. *Always* identify your sources; otherwise, you could be accused of plagiarism.

6. Separate the graphics from the text either with thin lines or with a frame of white space. Keep graphics on one page if possible, and try to avoid printing them sideways and making the reader turn the document. Make all wording read horizontally (left to right, not top to bottom).

7. Use graphics only if they are functional. If your table or figure won't directly help your reader understand your point, don't use it.

8. Include a List of Illustrations in the front of a long report or proposal—especially if you use more than three graphics within the document. If you have both tables and figures, make two separate lists, since tables and figures are usually numbered separately. Some readers are more interested in the graphics than in the text itself, so make it easy for them to find graphic material.

CHECKLIST FOR CHOOSING AND USING GRAPHICS

1. As I consider my document, do I have places where graphics can
_____ increase reader interest and understanding?
_____ display and emphasize important information?
_____ summarize and condense information?
_____ convey quantitative or other relationships?
_____ be designed by computer?
_____ be borrowed from other sources or hand-drawn?

2. What purpose do I want to accomplish with any specific graphic choice?
_____ to condense or summarize information?
_____ to compare and show relationships or trends?
_____ to show parts of wholes and relationships of parts to the whole and to each other?
_____ to show steps in a procedure or the function of each element in a system?
_____ to show external or internal views?
_____ to give examples or reproduce what the reader might see?

3. Who is my primary reader?
   _____ generalist
   _____ manager
   _____ operator
   _____ technician
   _____ specialist
4. What specific type of graphics should I use to meet that reader's needs?
   _____ tables (formal or informal)
   _____ graphs
   _____ maps
   _____ drawings
   _____ printouts
   _____ diagrams
   _____ photographs

EXERCISE 12-1   **Finding Examples of Graphics**

Directions: Search in textbooks, magazines, and newspapers for examples of at least four different kinds of graphics. Notice the purpose and group of readers for which each is used and analyze how effectively each one is presented. Clip out or photocopy the graphics and bring them to class. Be prepared to discuss their good and bad points.

EXERCISE 12-2   **Choosing Appropriate Graphics**

Directions: For each of the following situations, choose the most appropriate graphic aid. If more than one would work, note that. List the reasons for your choices and the type of primary reader. Be prepared to discuss your choices in class.

1. Your state has (or is considering) a lottery to raise money for education. The graphic aid should show the breakdown of revenues, with the percentage to go to each educational level.
2. A company manufacturing burglar alarms needs to explain the operation of its burglar alarm system.
3. You wish to compare current stock prices for three companies that make integrated circuits.
4. You must verify storm-caused roof damage for an insurance claim.
5. For a consumer report, you need a comparison of three top-of-the-line washing machines according to cost, capacity, warranty, and special features.
6. A major publisher wants to show the sales coverage in the United States of its 125 sales representatives.

7. Your college or university needs to show a comparison of the enrollment in technical writing classes over the last five years.

EXERCISE 12-3 Using Graphics in a Report or Proposal

Directions: Prepare a minimum of one table and one figure to include in your formal report or proposal. Be sure that titles, labels, and keys are included correctly. Indicate the source of your graphic or your information. Place the graphics immediately following their first mention in the text. If you are working on a project as part of a team, decide as a group what graphics to include, and assign one person to create the graphics and write the text references.

EXERCISE 12-4 Evaluating Graphics Effectiveness

Directions: Using factual or statistical data from your report, construct a table. Then use the same data to construct a figure. Exchange with a fellow student. Analyze that person's figure and table to see which is more ap-

Table 12-3   U.S. POSTAL SERVICE RATES FOR LETTERS AND POST CARDS: 1958 TO 1986

| Date of rate change | Surface mail | | | Postal and post cards | Domestic airmail letters[1] | Express mail[2] |
| | Letters | | | | | |
	Each ounce	First ounce	Each added ounce			
1958 (Aug. 1)	4¢	(x)	(x)	3¢	7¢	
1963 (Jan. 7)	5¢	(x)	(x)	4¢	8¢	
1968 (Jan. 7)	6¢	(x)	(x)	5¢	10¢	
1971 (May 16)	8¢	(x)	(x)	6¢	11¢	
1974 (Mar. 2)	10¢	(x)	(x)	8¢	13¢	
1975 (Sept. 14)	(x)	10¢	9¢	7¢	(x)	
1975 (Dec. 31)	[3](x)	[3]13¢	[3]11¢	[3]9¢	[4]	(x)
1978 (May 29)	(x)	15¢	13¢	10¢	[5]	(x)
1981 (Mar. 22)	(x)	18¢	17¢	12¢	[5]	$ 9.35
1981 (Nov. 1)	(x)	20¢	17¢	13¢	[5]	$ 9.35
1985 (Feb. 17)	(x)	22¢	17¢	14¢	[5]	$10.75
in effect Oct. 1986	(x)	22¢	17¢	14¢	[5]	$10.75

X = Not applicable.   [1]Each ounce, except as noted.   [2]Post Office to addressee rates. Rates shown are for weights up to 2 pounds, all zones. Beginning Feb. 17, 1985, for weights between 2 and 5 lbs, $12.85 is charged. Prior to Nov. 1, 1981, rate varied by weight and distances. Over 5 pounds still varies by distance. [3]As of October 11, 1975, surface mail service upgraded to level of airmail. [4]17¢ for first ounce; 15¢ for each added ounce. [5]As of May 1, 1977, domestic airmail service, as a separate class of mail, was discontinued.

propriate for the writer's purpose and content. Write a paragraph of evaluation.

EXERCISE 12-5   Graphing Statistical Data

Directions: Use the information in Table 12-3 from the 1987 *Statistical Abstract of the U.S.* to create an effective graph for generalists (fellow college students). Consider which kind of graph will best meet their needs.

REFERENCE

Michaelson, Herbert B. 1986. *How to write and publish engineering papers and reports.* 2nd ed. Philadelphia: ISI Press.

CHAPTER 13

Documenting Your Sources

As a technical writer, you may be writing either to peers or to readers who know less about your subject than you do. In either case readers come to writing seeking information, instructions, or recommendations, and they rely on you to be accurate. They do not, however, expect you to have conjured up what you say from thin air. They expect that in researching the field, you have gained bits and pieces of information from a variety of books, periodicals, and experts, as well as from your own observation and experimentation. Your contribution as a writer is to synthesize those bits and pieces of information into something new. At the same time, you must tell readers the sources of your information—first, to establish your credibility (prove that you've done your homework), and second, to tell readers where they can go to learn more about a particular aspect of your subject. This chapter covers the four areas related to documentation listed above.

13.1 DEFINITIONS OF DOCUMENTATION TERMS

In this book the term *documentation* applies to the process of identifying your sources of information. A similar often-used term is "citation." To document is to furnish with evidence; the words "document" and "documentation" have as their root the Latin word meaning "to teach." Scientific and technical papers seldom quote another source directly, usually summarizing or paraphrasing the key ideas. The sources of these summaries or paraphrases are listed in a section called "Reference List," "Works Cited," or "Literature Cited."

Documentation is primarily found in journal articles, books, and formal technical reports. Even manuals, procedures, and descriptions can require documentation, however, if they contain material from some other source. It's important to acknowledge sources of information. For one thing, plagiarizing is a form of stealing, and the penalties for being caught can be severe, possibly destroying your academic or professional career and jeopardizing the organization for which you work. Second, you will be more believable if you admit that all your impressive information did not simply spring out of your own brain. In both science and technology, you need to convince readers that you know what's already been done and said on a subject; in fact, this increases the likelihood that the reader will be persuaded to accept your conclusions and recommendations.

Note: The term "documentation" is also used in industry to identify a series of publications that provide instruction for users of particular products. "Documentation" can also mean writing down what was done in a procedure or experiment. These specialized definitions of the term are different from the one used in this chapter.

13.2 WHAT TO DOCUMENT AND WHERE TO FIND THE INFORMATION

What to Document

As you write, it pays to keep track of what you should document, because it's easier to note the sources correctly in your first draft than it is to go back and find the information later. Either write the source on the page itself or keep a separate 3×5 index card for each source, keying it to the draft. All of the following kinds of information should be documented.

A direct quotation   Though technical and scientific writers seldom quote an authority directly, it's obvious that if you do, you must identify the source of the quotation.

A paraphrase or summary To paraphrase is to restate someone else's words in your own. Students often fail to document a paraphrase or summary since they are putting the information into their own words. Nevertheless, if the information or the idea is someone else's, you should document it. For example, suppose your information source (Shaw 1987, 96) says:

> It is now beyond question that even in rural areas acid deposition (both wet and dry) almost always stems from the activity of human beings: primarily the combustion of fuel for power, industry, and transportation.

You condense and rewrite the explanation this way:

> Scientists are now sure that air pollution both from acid rain and from acid particles (called acid deposition) nearly always comes from human activity, mostly the combustion of fuel.

Even though almost all the words are your own, the major point (as well as the authority for scientists being "sure") comes from your source, so the information should be documented.

Controversial statements or positions It's important to give the source of any position or statement that is arguable. Your readers may want to examine the sources themselves in order to understand the controversy in more detail or to check on a related point.

Where to Find the Information

The best place to find the information to include in your reference list is the original source—the title page of the book, journal, or manual. For a book, you always need the

- author's name
- title of the book
- publication information, including
  volume number and edition
  city of publication
  publisher
  date of publication

For an article, you need the

- author's name
- title of the article
- title of the journal, manual, or newspaper
- publication information, including
  volume number and issue number (if available)
  date
  inclusive pages of the article

If more than one city of publication is listed, copy only the first. If the city of publication is not well known, list the state or country also (Menlo Park, CA). If you list a company's manual, give its order number or document number and the date. For examples of how to treat material from nonprint sources such as interviews and personal letters, see section 13.4.

In all the systems of documentation, you indicate the source information in two ways: (1) *within* the text, usually at the first natural sentence break following mention of the information, and (2) at the *end* of the text in a "Works Cited" or "Reference List" or "References" section. This final list contains complete information about the source that will allow readers to find it on their own.

13.3 HOW TO CHOOSE A DOCUMENTATION SYSTEM

Unfortunately, no one documentation system is universally accepted by writers even in the physical sciences, let alone the social sciences and humanities. Thus, you must choose the documentation system most likely to be acceptable to the specific readers for whom you are writing. Here's how to make that choice:

- If you are working, ask to see either the company style guide or a previous paper that has an accepted documentation system, and follow the suggested system.
- If you are submitting an article to a journal, you can (1) look in the journal for the "Guidelines for Authors," (2) write and ask for such guidelines, or (3) study current issues of the journal to see what documentation system is used and then follow the examples.
- If you are a student, ask your writing teacher or professors in your major field what documentation system they recommend and where it is explained, and follow that system exactly.

You have choices among several systems, but once you've chosen a system, it's best to follow it closely. Don't try to choose a bit from here and a bit from there. As *The Chicago Manual of Style* points out (1982, 440), a consistent style must be followed "not just because the publisher says so but because inconsistency in bibliographical details confuses readers and suggests careless research methods."

If your major field follows the documentation system of a particular style guide, you may want to buy that style guide or borrow it from a library. Be sure to use the most recent edition. Some of the style guides used in science and technology are the following:

- American Chemical Society. *Handbook for Authors.*
- American Institute of Physics. AIP *Style Manual.*
- American Mathematical Society. *Manual for Authors of Mathematical Papers.*

- American Medical Association. *Style Book: Editorial Manual.*
- American Society of Civil Engineers. *Author's Guide to the Publications of the ASCE.*
- Council of Biology Editors. *CBE Style Manual.*
- Institute of Electrical and Electronics Engineers. *Information for IEEE Authors.*

Four other widely known style guides are

- American Psychological Association. *Publication Manual.*
- Modern Language Association. *MLA Handbook for Writers of Research Papers.*
- University of Chicago Press. *The Chicago Manual of Style.*
- United States Government Printing Office. *Style Manual.*

This book follows the author-date documentation method outlined in *The Chicago Manual of Style*, 13th edition.

13.4 COMMON DOCUMENTATION SYSTEMS

The three common documentation systems may be called the *author-date* system, the *author-page* system, and the *number* system. The author-date system identifies the source in the text by the author's last name followed by the year of publication (Davis 1986). The author-page system is similar but gives the author's name and the page number (Davis 342). The number system simply numbers the source and puts that number in parentheses within the text (6). For most systems, the reference list is alphabetized by the author's last name. In the number system, sometimes the numbers are assigned in order as the citations appear, and the citations are listed in that order.

The authoritative *Chicago Manual of Style* (1982, 400) calls the author-date system "most economical in space, in time (for author, editor, and typesetter), and in cost (to publisher and public)—in short, the most practical. . . ." The author-date system is widely used in the biological, physical, and social sciences and is the system followed in this textbook. The author-page system is used primarily in the humanities and is the system supported by the Modern Language Association (MLA). The number system is often used in the sciences because it minimizes distraction for the reader and keeps printing costs down in journals by eliminating extra words within the text.

Author-Date System

The author-date system is explained in the *Publication Manual* of the American Psychological Association (APA) and in *The Chicago Manual of Style*, though the details are slightly different in each source. In the author-date system, you indicate the source in the text of your document at a natural break or at the end of the sentence following the first mention of the information. For example:

> A red dye (Rose Bengal) is added to the preservative to facilitate the sorting and identification process (Saloman 1976).

The complete reference is then listed at the end of the book or article in a Reference List. The example would be listed in this way:

> Saloman, C. H. 1976. *The benthic fauna and sediments of the nearshore zone off Panama City, Florida*. Report no. 76-10. Fort Belvoir, VA: Coastal Engineering Research Center.

Because the references are alphabetized by author's last name, that is listed first, followed by first name or initials. The year of publication is given next because that helps identify the source. The report title is given in full, followed by the report number. In this case, the report was published in Fort Belvoir, and since that place is not well known, the state is also given. The publisher is the Coastal Engineering Research Center.

Several minor variations on this order can be used, but the important thing is to be consistent. Generally, *within the text* you include in parentheses the author's last name and the date with no punctuation. You can also add a comma followed by the page number, and if you are quoting directly, you must include the page number. In the Reference List, you include information in the following order—each major bit of information separated by a period.

For Books and Reports:

> Author's last name, author's first name or initials. Year. Title of book (underlined or in italics). City of publication: publisher.

For Journal, Magazine, and Newspaper Articles:

> Author's last name, author's first name or initials. Year. Title of article. Title of journal or magazine underlined or in italics volume number (or date): inclusive page numbers.

Notice that the second and third lines of the reference are indented two spaces. The date, if given, appears in parentheses. In titles of articles, books, and journals, only the first word and names are capitalized. Specific examples follow based on *The Chicago Manual*.

Sample Entries

A Book with a Single Author:

> Laws, E. A. 1981. *Aquatic pollution*. New York: Wiley.

A Book with Two or More Authors:

> Ehrlich, P. R., A. H. Ehrlich, & J. P. Holden. 1977. *Ecoscience: Population, resources, environment.* San Francisco: Freeman.

Two Books by the Same Author (arranged chronologically by date of publication):

> Calabrese, E. J. 1978. *Pollutants and high risk groups: The biological basis of increased human susceptibility to environmental and occupational pollutants.* New York: Wiley.
> ———. 1981. *Nutrition and environmental health: The influence of nutritional studies in pollutant toxicity and carcinogenicity.* New York: Wiley.

Note that for the second book, the author's name is replaced by a line three spaces long (the space of three letter *m*'s).

Two Works by the Same Authors in the Same Year (arranged alphabetically by the first major word of the title—lowercase letters *a* and *b* differentiate one work from another):

> Lindskov, K. L., & B. A. Kimball. 1984a. *Streamflow in the southeastern Uinta Basin, Utah and Colorado.* USGS water supply paper no. 2224. Salt Lake City, UT: Geological Survey.
> ———. 1984b. *Water resources and potential hydrologic effects of oil-shale development in the southeastern Unita Basin, Utah and Colorado.* USGS professional paper no. 1307. Salt Lake City, UT: Geological Survey.

Magazine or Journal Article:

> Demyan, J. P. 1983. Producing disinfectant on-site is effective at wastewater plant. *Public Works* 114 (August): 56.
> Shaw, Robert W. 1987. Air pollution by particles. *Scientific American*, Aug., 96–103.

Interview:

> Brown, Paul. Interview with author, Chicago, Illinois. 7 Aug. 1987.

Television Program:

> Struggle to survive: China's great panda. Host Sam Fowler, ABC. 23 March 1987.

Note: The *Chicago Manual* does not list television programs, but this form is consistent with other entries.

Personal Communication: References to personal communications are best given in the text itself. If they are included in the Reference List, they are listed this way:

> Bell, Karla. Letter to author, 10 September 1988.

To summarize, if you plan to use the author-date system, you should (1) identify the source within the text by putting the author's last name and the year in parentheses, and (2) compile the Reference List alphabetically, following the examples in your style guide *exactly*, including punctuation and spacing. For most documents you will write as a student, this summary of the author-date system will give you enough detail. However, if you need more specifics, consult the *Chicago Manual* or the APA *Publication Manual*, as mentioned above.

Author-Page System

The author-page system is also called the parenthetical reference or parenthetical documentation system. It is useful when referencing whole books and long articles because it sends the reader directly to the page on which the information is located. This system is explained in detail in the *MLA Handbook for Writers of Research Papers* (1984), and explanations from that handbook are summarized here.

Like the author-date system, the author-page system places references within the text, in this case the author's last name and the page number, in parentheses. Do not separate them with commas. For example:

> Glue words hold the working words together, but if the sentence has too many glue words, it is verbose (Wydick 7).

The reference is to page 7 of the book written by Wydick. In what is called the Works Cited section at the end, this source would be listed in full, alphabetized by the author's last name. For example:

> Wydick, Richard C. *Plain English for Lawyers*. Durham: Carolina Academic P, 1980.

Note that the general form is the same as in the author-date system, but in this case the publication date is moved to the end of the entry and key words in the title are capitalized. Here are the general forms to follow in the Works Cited section.

For Books:

> Author's last name, author's first name. Title of book (underlined or in italics). City of publication: publisher, date.

For Reports and Journal, Magazine, and Newspaper Articles:

> Author's last name, author's first name. Title of the article enclosed in quotation marks. Title of the journal or magazine underlined or in italics. Volume number (Year of publication in parentheses): inclusive page numbers.

Notice that each major entry is followed by a period. Lines after the first are indented five spaces.

If the author is not identified in your source, alphabetize by the first major word of the title. Below are typical sample entries.

Sample Entries

A Book with a Single Author:

> Laws, Edward A. *Aquatic Pollution.* New York: Wiley, 1981.

A Book with Two or More Authors: Notice that only the first author is listed last name first; the others are listed first name first.

> Ball, John E., Michael J. Humenick, and Richard E. Speece. *The Kinetics and Settleability of Activated Sludge Developed under Pure Oxygen Conditions.* Austin: Center for Research in Water Resources, 1972.

Two Books by the Same Author: Notice that a line made with three hyphens replaces the name for the second entry.

> Calabrese, Edward J. *Nutrition and Environmental Health: The Influence of Nutritional Status on Pollutant Toxicity and Carcinogenicity.* New York: Wiley, 1981.
> ---. *Pollutants and High Risk Groups: The Biological Basis of Increased Human Susceptibility to Environmental and Occupational Pollutants.* New York: Wiley, 1978.

Magazine or Journal Article:

> Huck, Chris. "Burning the Wilderness: Forest Service's Planned Ignitions." *American Forests.* Aug. 1985: 13.

Article or Book with No Listed Author: Alphabetize by the first major word of the title.

> *Valley of the Drums: Bullitt County, Kentucky.* Washington, DC: U.S. Environmental Protection Agency, 1980.

Television Program:

> Struggle to Survive: China's Great Panda. Host Sam Fowler. ABC, KGO, San Francisco. 23 March 1987.

Data from Computer Search: If the information is from a computer on-line data search, add the vendor's name and the number in the system.

> Sun, Qiangnan, ed. "Computer Aided Techniques in Manufacturing, Engineering, and Management." *Computers in Industry* vol. 8, nos. 2–3, April 1987:111–269. DIALOG file 8, item 1766278 EI 8706055851.

Personal Communication: If the information comes from a letter, list the writer's name first, then the writer's title and company (if appropriate), then "Letter to the author," then the date in military style.

> Green, Frank. Manager of Engineering Documentation, XYZ Company. Letter to the author. 14 April 1987.

Do the same thing for a personal interview, identifying it as such.

> Smith, May. Director of Marketing, RRR Company. Personal Interview. 3 March 1987.

In summary, if you plan to use the author-page system, you should (1) identify the source within the text by putting the author's last name and the page number in parentheses, and (2) compile the Works Cited section alphabetically, following the examples in your style guide *exactly*, including punctuation and spacing. For most documents you will write, the samples given here will provide enough information; however, if you need more detail, consult the *MLA Handbook for Writers of Research Papers*.

Number System

The number system of documentation is often used in the sciences. A simple reference number is put in parentheses after the reference in the text, like this (12). Sometimes the relevant page reference is also included, separated from the reference number by a comma, like this (12, 243).

The list of references (called References Cited) can be done in two ways: In the first method, the reference number refers to the position in a list alphabetized by the author's last name. This method is sometimes called the alphabet-number system. Here, the citation numbers will not be sequential in the text. In the second method, called the citation-order system, the reference numbers are given sequentially in the text, and the references themselves are listed in that order in the References Cited section. The number system is explained in detail in the Council of Biology Editors *CBE Style Manual* (1983), which is the basis for the following summary.

The general form for entries in the References Cited section is the same as for the other systems, with these exceptions.

1. Titles of books and articles are not capitalized (except for the first word), nor are book and journal titles underlined or italicized.
2. No quotation marks appear around an article title.
3. The date appears at the end of the entry preceded by a semicolon.
4. The system is generally streamlined, often using initials instead of an author's first name, omitting subtitles, and abbreviating journal titles and publishers' names.

Here are the general forms to follow.

For Books:

> Author's last name, author's initials. Title of book. City of publication: publisher; year.

For Reports and Journal, Magazine, and Newspaper Articles:

> Author's last name, author's initials. Title of article. Title of journal. Volume number inclusive pages of article; year.

Remember that this form will vary slightly from one journal to another, so be sure to consult a specific journal or style book in your field. Following are examples.

Sample Entries

A Book with a Single Author:

> **1.** Laws, E. Aquatic pollution. New York: Wiley; 1981.

A Book with Two or More Authors:

> **2.** Ehrlich, P.; Ehrlich, A.; Holden, J. Ecoscience. San Francisco: Freeman; 1977.

Article in Journal or Magazine:

> **3.** Keller, W. Inventory of toxics discharged into the Hudson. Environment 24; 1985.

Technical Report:

> **4.** U.S. Environmental Protection Agency. Valley of the drums: Bullitt County, Kentucky. Washington, DC: EPA-430/9-80-014; 1980.

Unpublished Letter:

> **5.** Smith, H. [Letter to F. Whitehouse, Lawrence Livermore Laboratory]. 1987 February 14.

To sum up, if you plan to use the number system, you should (1) identify the source within the text with a number in parentheses, and (2) compile the References Cited, either

- alphabetically with the numbers keyed to the source, or
- sequentially by the order of the citations

In either case, follow the samples in your style guide *exactly*, including punctuation. If you need more detail, consult the *CBE Style Manual* or the *Chicago Manual of Style*. The two style manuals differ slightly in details.

Annotation of Reference Lists

When you get ready to document your sources, consider the advantages of annotating each source. In an annotated reference list, you follow the bibliographic information with a very brief summary of the contents of the article or book. Annotations seldom run more than 200 words and are often shorter, but they provide valuable information for the reader because they go well beyond the scanty identification provided by a title. Figure 13-1 shows several entries from an annotated bibliography by a student. This student followed the author-date system of documentation.

REFERENCES

Dixon, J. 1983. Surgical applications of lasers. *AORN Journal* 38: 223–231. Dixon's article gives a justification for surgical lasers and a general description of types and uses.

Nathanson, M. 1984. Hospital preps for outpatient laser. *Modern Healthcare* January: 138. Nathanson presents the benefits of minimally invasive surgery, and he discusses the implementation of the Laserscope OMNIplus surgical laser at El Camino Hospital in Mountain View, CA.

Pfister, J., and J. Kneedler. 1983. *A guide to lasers in the O.R.* Aurora, CO: Education Design. This manual has a good section on how to evaluate a hospital's needs for a surgical laser.

Sliney, D., and M. Wolbarsht. 1980. *Safety with lasers and other optical devices.* New York: Plenum Press. This book has a chapter devoted to laser safety in the clinical facility. It explains what safety hazards to be aware of and what procedures can reduce risks of injury.

Tatton, L. Clinical Coordinator of Surgery, Allied Hospital. Interview with author, San Jose, CA. 13 February 1985. This discussion provided the valuable information about Allied Hospital's specialties, laser objectives, facilities, and budget that allowed me to shape my criteria to this facility.

Figure 13-1 **Annotated reference list**

CHECKLIST FOR DOCUMENTATION

1. What specific material in my manuscript should be documented? Do I have any
   _____ direct quotations?
   _____ paraphrases or ideas from my sources?
   _____ controversial statements or positions?
2. Which documentation system is commonly used in my field or recommended by my instructor?
   _____ author-date
   _____ author-page
   _____ number

Exercise 13-1 Analyzing Documentation

Directions: For practice in using the style of documentation preferred in your major field or one recommended by your instructor, organize and rewrite the following citations:

Berry, Richard, "All the Angles on Magnification," *Astronomy*, vol. 9, no. 4, (April 1981), pg. 47–49.

Berry, Richard, "The Telescope Revolution," *Astronomy*, vol. 10, no. 12 (Dec. 1982), pg. 75–77.

Pananides, Nicholas A. and Arny, Thomas T., *Introductory Astronomy*, Reading, Massachusetts, Addison-Wesley Publishing Company, Inc., 1979.

Swihart, Thomas L., *Journey Through the Universe*, Boston, Houghton Mifflin Co., 1978.

Exercise 13-2   Documenting a Major Paper

Directions: Consult your writing instructor or a professor in your major field to find out what documentation system to choose. Then follow that system exactly to document the sources in the paper for your term project. If you are working on a collaborative project, decide which documentation system to use before writing the draft. Make each writer responsible for providing full documentation of his or her own written material. Assign one person in the group to prepare the in-text references and the final reference list; assign another to review that documentation.

REFERENCES

CBE Style Manual Committee. 1983. *CBE style manual: A guide for authors, editors, and publishers in the biological sciences.* 5th ed. Bethesda, MD: Council of Biology Editors, Inc.

The Chicago manual of style. 1982. 13th ed. Chicago: University of Chicago Press.

Gibaldi, Joseph, and Walter S. Achtert. 1984. *MLA handbook for writers of research papers.* 2d ed. New York: Modern Language Assn.

Shaw, Robert W. 1987. Air pollution by particles. *Scientific American*, August, 96–103.

PART THREE

THE FORMS OF TECHNICAL WRITING

What is a form? On the construction site of a large building, you have probably seen workers building "forms"—those hollow wooden structures into which a fluid mix of concrete is poured to make foundations and walls, steps and sidewalks. When the concrete hardens, it takes on a shape dictated by the form into which it was poured. In other words, the form shapes the content.

In the same way, technical writing forms shape their content. For each of the forms in the next 10 chapters, you can follow the general process explained in Part 1, and you can incorporate many of the techniques discussed in Part 2. However, these 10 forms of technical writing have evolved to meet different practical needs, and each one is constructed differently and looks different on the page. In this book the forms are arranged from the simplest and most commonly used (memos and letters) to the most complex and demanding (recommendation and research reports). The last chapter takes up speeches and oral reports because they are a special case.

Each chapter takes up a separate form—first explaining what it is and how it is used and showing examples written by students and by professionals in various fields, then explaining how the general writing process can be applied to that form. You can use these chapters as a reference whenever you need to write a particular kind of technical document. The 10 common forms in which technical writing appears are shown on the preceding page.

CHAPTER 14

Memorandums and Informal Reports

Much of the day-to-day work of business and industry is carried on by means of memorandums and informal reports, so you will find yourselves writing—and reading—hundreds of them in a single year. Even with electronic mail, the basic forms persist; technology has simply changed the way they are transmitted. In many ways, a memorandum is the easiest kind of technical document to write because it follows a standard form, has specific identified readers, is short and direct, and follows a clear method of organization. Informal reports often follow the memorandum form. This chapter discusses both memos and informal reports.

14.1 DEFINITIONS OF MEMORANDUM AND INFORMAL REPORT

A *memorandum* is a short document directed to a reader or readers within the same organization as the writer. The plural form is either memorandums or memoranda, and the word is often shortened to memo. Memos generally run to fewer than four typed, single-spaced pages, and they carry on routine business, such as giving or requesting information.

An *informal report* is a document of fewer than 10 pages, which is often directed to a decision-maker. If intended for internal use, it may follow memo form or be attached to a transmittal memo. If directed outside the organization (to a customer for example), it may be in the form of a letter or attached to a transmittal letter. The form may vary depending on the purpose. In general, informal reports can

1. provide factual information about tests, events, meetings, experiments in the laboratory or the field, surveys, interviews, the progress or status of a project

2. analyze or evaluate information in order to recommend action. Such reports might evaluate courses of action or products or discuss the impact of an event or procedure.

Both types are shortened and simplified versions of the formal reports called feasibility, recommendation, and research reports that are discussed in detail in Chapters 21 and 22.

14.2 MEMORANDUM FORM

Memos always begin with four standard items: the date, the reader's name, the writer's name, and the subject. You can use any of the arrangements shown in Figure 14-1, but when you are working, you should learn the accepted form of your company and follow that. Headings can be as they appear here or in all capital letters.

It's important to provide the full calendar date (April 10, 1989) for later reference. How formal you are with the reader's name depends on four things:

1. who the reader is (whether a coworker or a supervisor, for example)
2. how well you know the reader
3. what kind of information is in the memo
4. how many people will read the memo

Because they are official pieces of business, most memos use either first and last names or first initials and last names. Include job titles as well as names in a memo that will become a permanent record (one that reports information or results) or a memo going to a person you do not know.

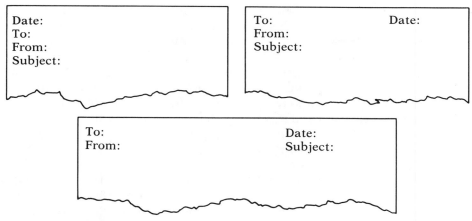

Figure 14-1   Possible arrangements for memo form

> To: J. Lockwood
>   Manager, Final Test
> From: S. Chu
>   G-15 Module Designer

Direct the memo to those people who must *do* something with the information. Others, who will simply read it to be informed, should be "copied." That is, their names should be listed at the end, preceded by cc for "carbon copy" or "courtesy copy."

> cc: C. Lorimar
>   H. Sanchez
>   T. Watt

As the writer, be sure to sign or initial your memo either immediately after your typed name or at the end of the document. Follow company style in this matter. Your signature or initials indicate that you have read and approved the memo; with that signature you bear responsibility for what it says.

The subject line is like a report title. It should specify the memo's purpose and indicate its content with key words. For example:

> Subject: Schedule of Meetings for the A2 Task Force
> Subject: Report on Design Agreements with the Madison Facility
> Subject: Biweekly Status Report on Analysis of Surge Protectors

Some memos also include a reference line that directs the reader to a previous memo, telephone conversation, or report. See Chapter 6 for more information on titles and reference lines.

The memo itself should be typed single-spaced with double spacing between paragraphs. Most memos follow the traditional organization of introduction, body, and conclusion, which is explained in Chapter 5. Keep memo paragraphs reasonably short, and deal with only one item of information per paragraph. In longer memos use headings to identify major topics. But keep a memo generally under four pages; if your document will be longer, consider making it an informal or formal report and attaching a transmittal memo to it. Whatever the length, be sure to include all relevant information and remember that once it leaves your hands, any memo can become a permanent record. Figure 14-2 shows a typical routine memo announcing a meeting. Like many organizations, MBL uses a printed memo form that reminds writers of the items to include. This memo is written to all members of Department D78; sometimes individual names will be listed instead. The writer signs or initials the memo. Courtesy copies are sent to others who need the information—in this case, members of another department. The manager's name appears first in the cc list, followed by an alphabetical list of other names.

14.3 INFORMAL REPORT FORM AND ORGANIZATION

The form and organization of short informal reports vary from company to company and from purpose to purpose. Four different kinds of informal

---

```
            To:  DEPARTMENT D78
          Date:  June 9, 1989
          From:  J. S. Ingle 335-5467    J. S.      MB
          Dept:  Cross Systems Support
     Mail Stop:  D78/C44 South Third Laboratory

       Subject:  WEEKLY DEPARTMENT D78 MEETING

     Reference:  Previous Department Meeting of June 4, 1989

                 This week's department meeting will be held in room
                 C466 (please note that the room has been changed) on
                 Thursday, June 11 at 10:00 A.M.

                 This will be a joint meeting with Gene Tolleff's
                 department to exchange ideas and discuss ways that
                 we can make the development/test process more
                 productive and effective from both sides.


                 cc:
                     E. W. Tolleff
                     J. T. Ferra
                     D. Ketchum
                     S. A. Tao
```

Figure 14-2 A routine memo

reports will be discussed here because these four typify the reports you might write and because each one follows a different method of organization. The four are

- narrative reports
- laboratory or field reports
- status or progress reports
- interview reports

Narrative Reports

A *narrative report* is an informal report, or segment of a report, that describes events in the order in which they occurred in the past. In some situations, the most objective and efficient way to present information for the record is sequentially, reporting what happened first, next, and so on. The excerpt from a police report in Figure 14-3 uses such a method of organization coupled with a printed form to identify the persons involved, date, time, and other facts. This organization has no clearly defined introduction, body, and conclusion; the events are simply presented in the order in which they occurred. Notice the deliberate attempt in this report to write objectively: the author speaks of himself or herself in the third person ("the reporting officer"). Most of the sentences are in the active voice because the narrative concerns the actions of the reporting officer and the suspect.

Narration concerns one-time actions that have already taken place. As a way of presenting information, narration can appear in any kind of memo or informal report. For example, safety inspectors and insurance adjusters write accident reports telling how an accident happened and what the resulting damage was. A geologist might report on the eruption of Mt. St. Helens—a narrative of a one-time occurrence. However, if the geologist wrote an explanation of how a volcano erupts—a recurring action—the document would be a procedure, which is explained in Chapter 18.

Laboratory and Field Reports

A second type of informal report, with which most students are familiar, is the *laboratory report*. These reports explain the procedure by which an operation was carried out and the results of that procedure. Such reports are not usually organized sequentially, but they do include required items, which are identified with subheadings. The introduction for a lab report will typically include the following:

- purpose
- problem or objective
- scope

ARREST REPORT—8 January 1989 8:00 P.M.

At the above date and time the reporting officer (R/O) observed the above vehicle southbound on Smith at Howard. The traffic signal at that intersection turned red, after which the suspect's vehicle continued through the intersection without stopping. The R/O turned around, activating the emergency equipment as the vehicle stopped for another light at Peninsula and Smith. The R/O advised via the PA system for the driver to turn off at the next side street to the right. On the green light, suspect's vehicle proceeded, almost passing the street, Barry, then abruptly turning right and crossing over to the north curb and stopping. R/O stopped and approached the vehicle as the driver, JONES, got out.

R/O observed JONES, noting his eyes to be bloodshot, his pupils to be dilated, and a strong odor of alcoholic beverage about his breath and person. After checking each of his pockets twice, JONES advised that he did not have his driver's license with him. R/O moved JONES to the sidewalk (flat concrete type) and administered a field sobriety examination. R/O noted that JONES'S words were slurred and his conversation confused.

The field sobriety examination was conducted as follows. JONES was asked by the R/O to

1. Stand with feet together, arms at sides, head back, and eyes closed. JONES swayed excessively.
2. Stand as in #1 with arms out and touch his index finger to the tip of his nose.
   a. Right hand. Touched left nostril, then moved to tip.
   b. Right hand. JONES relaxed and looked at R/O, asking: "Right?" Then he touched the left nostril again.
   c. Left hand. Touched upper lip, then moved to tip of nose.
   d. Right hand. Touched right nostril.
3. Stand on one foot. JONES stated he had nothing wrong with his legs or back. He was unable to keep either foot off the ground for more than 3–5 seconds before placing it back on the ground for balance.
4. Say the alphabet. JONES stated he knew the alphabet: "A B C D E F G H J J K L M N O P Q R S T H I J K—uh."
5. Repeat the alphabet. "A B C . . . U V M N O P W X Y Z."
6. Reverse count from 20–1. Began with 19 and ended with 0, left out number 11.
7. Walk a straight line, heel to toe, 10 paces and return. JONES did not maintain heel to toe contact/alignment, nor did he stay on the line. The line used was the expansion joint running the length of the sidewalk.

[The narrative continues, describing JONES'S vehicle; it concludes with the action taken by the reporting officer.]

Figure 14-3   A narrative report

- equipment or materials
- procedure or methods

The central portion, or body, reports on the findings or gives the data acquired. This section is often called "Results." It's tempting in this section to follow a sequential or narrative organization, reporting results in the order they were discovered, but in fact the narrative method probably wastes the reader's time. Instead, you should select significant data and present it in whatever hierarchical organization best fits the purpose. (Review Chapter 3 on types of organization.) You may also want to include a table or graph of some kind.

The conclusion of a lab report is usually called the "Discussion." In this section you explain

- what principles or generalizations the results support
- how these results agree (or disagree) with earlier published or unpublished results
- what the results mean and how those conclusions are supported

Field reports (in which you report on data collected from some location) follow a similar introduction-body-conclusion organization, but have subheadings relating directly to the material you are presenting. The introduction of a field report includes the purpose, the problem, and the methods used to collect information; the body details the findings; and the conclusion discusses what the findings mean. A field report might also include a recommendation based on that conclusion. In Figure 14-4, the writer reports on a field inspection of automatic swimming pool chlorinators. Notice the specific detail in the report and the support provided for the conclusions and recommendations. Because this report is short and internal, it is written in memo form.

Status or Progress Reports

Most people who work in technical occupations will regularly write either status or progress reports. A *status report* is a written account of your accomplishments on the job during a designated period of time. In business and industry, status reports are written at regular calendar intervals— usually weekly, biweekly, or monthly. A *progress report* is a written account of your accomplishments at stated intervals in a project—after completion of a test phase, for example. Since both types of reports are often for internal use, they follow memo form. Progress reports may also go to clients or customers and then may be in letter form or short report form attached to a cover letter. The form and organization for these reports can vary widely, and your company or instructor may have specific requirements. If not, the following form works well and can be adapted to a variety of occupations.

Date:
To:
From:
Subject:
Reporting Period: (From   to   )

   1.0 Status/Progress/Accomplishments (one of these)
    1.1 Tasks Completed

AUTO—CHEM, INC.

Date: September 15, 1989

To: F.C. Powers

     Quality Control Manager

From: S. Hioki   *S. Hioki*

     Customer Service Engineer

Subject: Inspection of C6432T Automatic Pool Chlorinators

INTRODUCTION
     On September 14, 1989, I inspected the C6432T Pool Chlorinators at four locations in Addison County to determine the reasons for customer complaints of residual chlorine levels below the legal minimum of 1.0 ppm. These four chlorinators were all less than one year old and installed in 20,000 to 30,000 gallon public pools in either apartment buildings or motels. The locations were

    1. Highlands Motel
       342 Green Avenue
       Centerville
    2. Conway Motel
       1442 Olney Drive
       Harkness
    3. Crestview Apartments
       836 River Parkway
       Centerville
    4. Olympia Apartments
       1135 Fourth Avenue
       Rockland

I collected information on these chlorinators in four ways:
1. Checked the chlorine level in the pool water
2. Visually inspected the chlorinators and the method of installation
3. Disassembled the chlorinator and checked the components
4. Interviewed the person responsible for pool maintenance

(continued)

Figure 14-4 **A field report**

Be very specific about what you have done, including the date. For example:

The clutch assembly test unit was completed on 4-17-88. Eight units completed destructive torque tests on 5-1-88. P. Smith wrote the report on the results.

  1.2 Tasks in Progress
Again be specific. Give completion dates if appropriate.

FINDINGS
1. Chlorine levels at all four locations were below 1.0 ppm.
2. The pump at location 3 was improperly installed, leading to a
   reduced feed of chemical crystals and a consequent chlorine level of
   0.80 ppm. In addition, chlorine was stored in a plastic garbage can
   instead of the 30 gallon crock with a pump mount we recommend.
3. At locations 1, 2, and 4, all the equipment was properly installed,
   but chlorine levels were 0.92, 0.85, and 0.87 ppm, which had led to
   warning citations by the county pool inspector.
4. Disassembly of the chlorinators at locations 1, 2, and 4 showed that
   the metering mechanism was clogged with chemical crystals in all
   three.
5. When I interviewed pool maintenance personnel, I found that
   a. The instructions for acid wash of the metering mechanism at
      location 1 were missing, and the pool cleaner had never performed
      the required acid washes.
   b. Pool cleaners at locations 2 and 4 said they performed acid
      washes approximately twice a month.

CONCLUSIONS AND RECOMMENDATIONS
        Clogging of the metering mechanism by chemical crystals is caused
primarily by improper or insufficient acid washing. Although our
instructions recommend weekly acid washes, maintenance personnel
performed this procedure only two times a month. Most maintenance
personnel are apartment or motel managers rather than trained pool
technicians, so we should either simplify the acid washing procedure or
reduce the number of acid washes needed.
        I recommend the following actions:
1. Rewrite the instructions to clearly specify times for acid washes
   based on hours of chlorinator operation.
2. Simplify the acid wash instructions by numbering the steps and
   adding an illustration of the metering mechanism.
3. Redesign the metering mechanism—perhaps with a squeeze–feed roller
   tube—to eliminate the need for acid washes.

Figure 14-4   Continued

Eight clutch assemblies are currently in "Normal Use Life Test." This
test is on schedule and targeted for completion on 5-30-88.

1.3 Tasks Scheduled

Include new assignments or tasks you have been assigned that
you have not yet started.

2.0 General Information

You can include here (or in a section 1.4) your log of time spent
on various tasks, such as writing and preparing letters for mailing. You can also include anticipated problems or potential
sources of information in this section.

Even though the sections of the status report are not labeled introduction, body, and conclusion, the form does contain those elements. Tasks Completed serves as an introduction or overview, while Tasks in Progress makes up the body, and Tasks Scheduled or a General Information section serves as a conclusion.

Sometimes you will have to include a Problem section in your status report. This can go in any section but is most likely to be included in 1.2. Divisions of the problem might be as follows:

—Old Problems: Resolved
—Old Problems: Unresolved
—New Problems

Another format is the Problem Report. Here is a possible outline:

—Statement of the problem
—Impact of the problem (cost, schedule, procurement, safety)
—Activity or action plan
—Person responsible

In Figure 14-5 a student reports the status of her comparative study. This report is one of a series written at two-week intervals. Notice how she adapts the suggested form to her specific requirements.

Interview Reports

You may want to use an informal report to communicate the information learned from an interview with a specialist or technician in some field (see section 2.3). In technical writing your reader is probably more interested in the information itself than in the personality of the informant, so the form of the report should reflect that. Two methods of organization that work well for interviews in technical fields are the question-and-answer format and the thesis-support format.

You are probably familiar with the question-and-answer format, which is often used in newspapers and magazines. In this method of organization, information is presented in the form of a direct question and its response:

Q. As a soil and plant testing laboratory, what kinds of analyses do you perform?
A. This lab does chemical analyses on either soil or plants. Depending on the client's needs, we can give either raw data on a sample or a recommendation of what to do to alleviate a problem. Our clients are nurseries, agricultural producers, and landscaping concerns.

The written order of questions and answers will probably *not* be the order in which they occurred during the interview. The writer needs to shape, cut, and rearrange the material so that it is coherent and clear, moves smoothly from one topic to the next, and presents only the information the reader needs.

To: Professor Rew
From: Pamela Nelson   *P.N.*
Date: October 11, 1988
Subject: Status of Research on Child Car Safety Seats
Reporting Period: September 13 to October 11, 1988

I. Tasks Completed/Results
1. I sent letters of inquiry to eight manufacturers on
September 21. To date, I have received one response, and
one letter was returned as undeliverable. So far I have
been unable to obtain another address for this
manufacturer.

2. On September 20 I completed library research on
California safety seat law and obtained a copy of the law.
This law is modeled on a previous federal law, and
standards for impact resistance at given speeds are not
stated in the California version. There is no list of
approved manufacturers.

3. I interviewed three people who own car seats on
September 20. Only one knew the brand he owned. Two owned
seats that can be fastened and unfastened with one hand;
the third regretted not buying such a seat. All commented
on the difficulty of obtaining seat belt clips.

4. On October 9 I sent follow-up letters to companies that
had not responded.

5. I drew up a formal consumer questionnaire on October 9.
The principal points are brand owned, reason for buying
that brand (if any), advantages, disadvantages.

II. Tasks in Progress
1. Library research for phone numbers of companies that
have not responded.

2. Interviews with people who own car seats (I project 20
interviews).

3. Research Federal Law 213 for specific standards on
crash protection and impact resistance.
                                        (continued)

Figure 14-5 **A status report**

III.   Tasks Scheduled
1.   Comparison shopping the week of October 14 at J.C.
Penney, Sears, Kiddie World, and Toys-R-Us. If possible, I
will also interview merchants to determine whether they
recommend a particular brand and/or which brands sell
best.

2.   Distribution of questionnaires during the weeks of
October 14 and 21.

IV.   Log of Time Spent
9-18   1 hour library research, 2 hours typing letters
9-20   1.5 hours library research, .5 hours interviewing
          consumers
10-4   1 hour library research on federal law and
          manufacturers' addresses
10-7   1 hour library research on federal law and
          manufacturers' phone numbers
10-9   1.5 hours composing and typing questionnaire
10-10 1 hour preparing status report

Figure 14-5   Continued

A more typical presentation of interview information is a straight-forward discussion of the facts and opinions gained from the source. In this case, the method of organization should reflect the major purpose of the document. Usually, the introduction should explain the problem in general terms, identify the informant, and describe his or her background and expertise. The body should describe the substance of the interview. The conclusion should either provide recommendations based on the body or summarize the main points.

Informal reports that are not in memo form can begin with a title, either with or without the writer's name. Follow your company's or instructor's requirements in this matter.

14.4 THE PROCESS OF WRITING MEMOS AND INFORMAL REPORTS

Planning

In planning a memo, first determine its purpose. Ask yourself why you intend to write this information instead of telephoning. Usually you'll choose to write if (1) you can't reach the person by phone, (2) the information must go to several people at the same time, or (3) you want to provide a permanent, written record. Here are common purposes for writing memos:

- to request information or action
- to respond to a question or request
- to give information: to announce, explain, describe, or recommend
- to formalize policies or confirm agreement reached in a phone conversation or meeting
- to boost morale or compliment people for work well done
- to document action taken. This might be a "memo to file": that is, written to yourself, for the record and to help you remember what was done.
- to propose new procedures or ideas. Short internal proposals are often written in memo form. See Chapter 20 for details on how to write proposals.

In planning an informal report, your first step is also to clarify the purpose. It could be any of those listed above or one more specific to your assignment. If your report will run more than 10 pages, consider using the formal report form instead. The additional front matter (such as Abstract, Table of Contents, and List of Figures) and the additional back matter (such as List of References and Appendix) will help clarify complex matters for the reader of a long report. Chapter 21 tells how to create front and back matter.

Your second consideration in planning is the reader. Most memos go to a specific reader, and you need to identify that reader by background and job. Then you'll know how much detail to include and whether you can use technical terms without definitions. (See section 1.2.) If you have only one reader (who is not your supervisor), you can write informally, but the more readers you have, the more formal the memo or report should be. Your relationship to the reader will also affect the tone, which you should keep in mind as you write. Even though you are writing a short document, it pays to think it through by using the Planning Sheet in section 1.6.

The memo shown in Figure 14-6 gives information briefly and directly. Notice that the reference line tells both the purpose ("clarification") and the topic ("air exchanger test"). Technical terms can be used because the reader is familiar with the tests and the equipment. Notice too that this memo has an attachment—another memo and a figure that expand on the topic.

Gathering Information

You might have the information for a memo before you begin to write, but if you are writing a short report, you might have to do additional research,

---

National Aeronautics and
Space Administration

Ames Research Center
Moffett Field, California 94035

NASA

Reply to Attn of: DF:213-12   January 26, 1984

TO: Frank Penaranda, Director, OAST Office of Facilities

FROM: Frank Nichols, Deputy Project Manager, 40x80x120 Wind Tunnel Project

SUBJECT: Clarification of Air Exchanger Test Results

Model tests of the new air exchanger configuration show significant flow
improvement into the fan drive system. The old air exchanger produced a
substantial velocity defect (a reverse flow, in fact) at the courtyard wall
which resulted in a poor velocity distribution coming into the fans. (One fan
on the courtyard side did show evidence of stall during the IST.) The new air
exchanger configuration energizes the courtyard wall boundary layer, removing
the velocity defect and giving a much more uniform velocity distribution into
the fans. This more uniform flow should result in more efficient operation of
the full-scale fan drive system.

The attached memo and figure will give you more detailed information.

Frank
Frank Nichols

Enclosure:
Memo with Figure

Figure 14-6   Informational memo

interview experts, or conduct additional experiments. Concentrate on accuracy, remembering that you are providing a record of your work.

Take special care in constructing documents like status reports, which report directly on the work you have been doing. Some employees are very casual and general in writing status reports, not realizing that at evaluation time managers frequently study a series of status reports to determine how well the employee has been performing. In effect, you are evaluating your own performance when you submit a status report, so it pays to have all the information at hand and to organize it well. The same is true of nearly all memos you write. Someone is likely to look at them when your performance is evaluated.

Organizing

Clear organization of a memo or report will help the reader to understand the content. As in all writing, you need to decide on your one major point, choose supporting subpoints, and arrange those subpoints in a method of organization that will work for that document. Most memos and informal reports have a three-part organization:

1. statement of the purpose and subject. This section may require background or a summary of previous actions.

2. details and discussion, organized in a way that fits the purpose and subject.

3. a conclusion that tells what you want the reader to do (action). Memos that report or provide information, like the one in Figure 14-6, do not require a conclusion.

If your document contains bad news, however, you may need to modify this three-part organization. Most readers will be offended if they are banged on the head by a terse opening statement containing bad news. In this case, you need an opening that begins positively or with the background of the problem. In the middle section, once the reader is prepared, you can present the bad news. In the conclusion, you indicate action but try to close on a positive note. See section 15.4 for more on this method of organization.

Reviewing, Writing, and Editing

Review your organization before you write; review and edit your memo or report before you send it out. Make sure that you have answered the pertinent questions posed in Chapter 7 for reviewing documents. Sign or initial the final copy of a memo, and remember that your initials indicate your approval and acceptance of responsibility for the content and form. Informal though a memo or short report might be, its contents can be of vital importance to you personally and to the company for which you work.

CHECKLIST FOR WRITING MEMOS AND INFORMAL REPORTS

1. In writing this memo, have I included the
_____ date?
_____ reader's name?
_____ my name and signature?
_____ subject line?
_____ subheadings for major topics?
_____ cc (carbon or courtesy copy) if appropriate?
2. In writing this short report, what form is appropriate?
_____ narrative (sequential)
_____ introduction, body, conclusion
_____ lab report
_____ field report
_____ status or progress report
_____ interview report
_____ other

EXERCISE 14-1   Writing a Preproposal Memo

Directions: Write your instructor a short memo explaining the topic you have chosen for your term report. Include the reason for your choice.

WRITING ASSIGNMENT #7   Writing a Narrative Report

Directions: Compose a list of five specific situations from your major field in which a narrative report might be needed. Choose one of these situations and write a two-page narration, following the guidelines in this chapter.

EXERCISE 14-2   Adding Narration to a Term Project Report

Directions: Study the raw material for your term project to see if you need a section of narrative. Write such material and submit it for review.

WRITING ASSIGNMENT #5   Status or Progress Reports

Directions: Using the memo form, write a status report to your instructor. Include as much information as is pertinent from the status report outline given in this chapter. If you are working as a team, assign one person to write any required status reports. Three days before each status report is due, submit to that person one or two paragraphs describing the status of your part of the project. The writer can then assemble this data into one report.

EXERCISE 14-3   Adding Interview Material to a Term Project Report

Directions: Write up the interview information you obtained and include it as a part of your term project report. Be sure to credit the source in your list of references.

WRITING ASSIGNMENT #8   Writing Up an Interview as an Informal Report

Directions: After completing a major interview (see Chapter 2), follow the suggestions in this chapter to write an informal report in either the question-and-answer or thesis-support format. Ask your instructor for specific guidelines.

REFERENCES

Nelson, Pamela. 1988. *Status report on child car safety seats.* Informal report. San Jose State University.

CHAPTER 15

Business Letters

Even though you work in a technical field, you will write many business letters during your professional career. Much day-to-day business in scientific and technical fields is conducted by telephone, but the letter remains important both because it can be read at a time convenient to the reader and because it provides a permanent record. Sometimes you will write a letter when you can't reach a person by telephone. Sometimes a letter is more appropriate than a phone call. On sensitive matters you can plan a letter carefully to project the tone, style, and information you want to convey. Letters, like internal company memos, are fairly easy to write because they follow a standard form, are directed to specific, identified readers, are short, and have a clear method of organization. In this chapter you will learn the four things about letter writing listed above.

15.1 THE DEFINITION OF A BUSINESS LETTER

A business *letter* is a short document addressed personally to a reader or readers who are usually *outside* the writer's organization or place of business. You will write letters both as a private individual and as a representative of your company. Generally, the letters you write for your company will be typed on letterhead stationary, and most letters should take only one or two typed, single-spaced pages. On very formal occasions, letters may be written to people within a company, but usually internal correspondence is by memo. A short informal report can be written in letter form if it goes outside the company. However, if the material will cover four or more pages, it should be written as a formal or informal report and accompanied by a one-page letter, called a cover letter or letter of transmittal.

15.2 LETTER FORM AND PARTS

Most business letters follow a standard form and include the same basic parts. These parts identify the reader and the writer, give the date, and open and close the message in a traditional way. With standard parts, the reader can focus on the letter's content rather than its form. Entire books have been written on proper letter form and organization, so the material in this chapter is necessarily condensed, but all the basic information you need to write successful letters is here.

Form

Follow your company's policy for letter form, but if there is no policy or you are writing on your own, you can choose one of the forms shown in Figures 15-1, 15-2, or 15-3. By following the form faithfully, including spacing and punctuation, you will write correct, acceptable-looking letters.

Figure 15-1 shows the block or flush left form preferred by most businesses and many students because it is easy to use: you simply line up all the parts of the letter at the left margin. Paragraphs are not indented but are set off by double-spacing.

Figure 15-2 shows the semiblock or modified block form, in which the heading (the writer's address and the date) and the complimentary close and signature are placed to the right of the center of the page. In this form, paragraphs can either begin at the left margin or be indented five spaces. If you choose the semiblock form, line up the heading and complimentary close along an imaginary line just to the right of the center. Always think of a letter as a picture to be framed on the page, and balance margins and white space to provide an attractive frame. All letters should have margins of at least 1 inch on the top and each side and 1.5 inches at

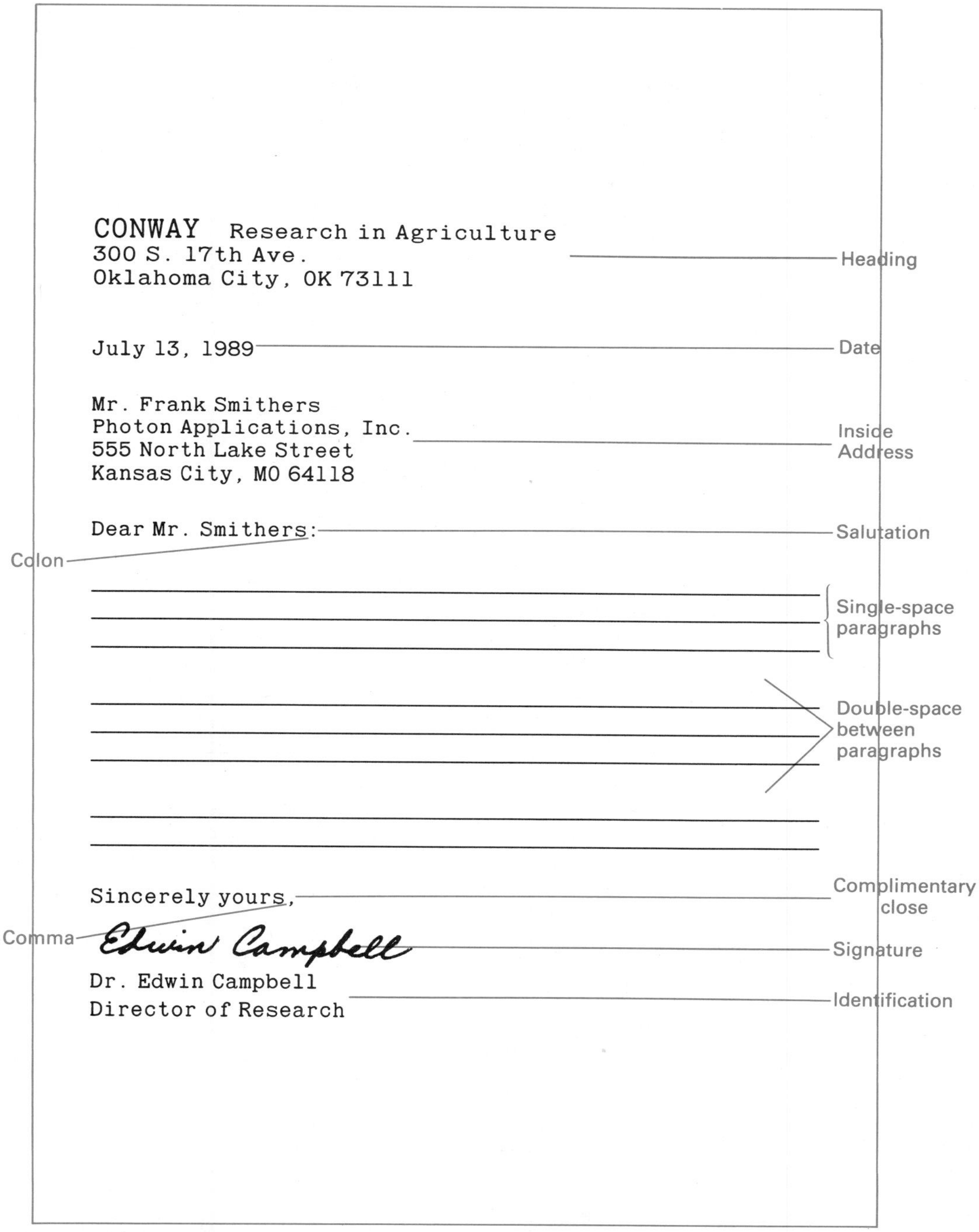

CONWAY   Research in Agriculture —————————— Heading
300 S. 17th Ave.
Oklahoma City, OK 73111

July 13, 1989————————————————————————————— Date

Mr. Frank Smithers
Photon Applications, Inc. ———————————————— Inside
555 North Lake Street   Address
Kansas City, MO 64118

Dear Mr. Smithers:————————————————————— Salutation

Colon—

——————————————————————————————————
——————————————————————————————————   Single-space
——————————————————————————————————   paragraphs

——————————————————————————————————
——————————————————————————————————   Double-space
——————————————————————————————————   between
——————————————————————————————————   paragraphs

——————————————————————————————————
——————————————————————————————————

Sincerely yours,————————————————————— Complimentary
                                                                        close
Comma—   *Edwin Campbell* ————————————— Signature
Dr. Edwin Campbell ————————————————————
Director of Research   Identification

Figure 15-1   **Block or flush left letter form**

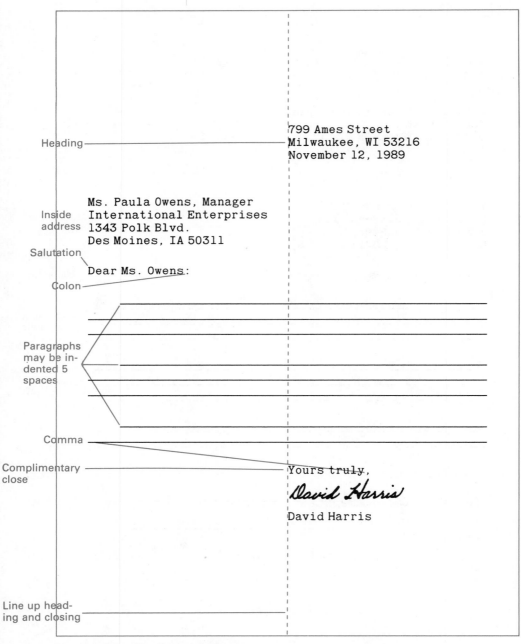

Heading ——————————————————— 799 Ames Street
                                        Milwaukee, WI 53216
                                        November 12, 1989

Inside   Ms. Paula Owens, Manager
address   International Enterprises
            1343 Polk Blvd.
            Des Moines, IA 50311

Salutation
            Dear Ms. Owens:
Colon

Paragraphs
may be in-
dented 5
spaces

Comma

Complimentary ——————————————— Yours truly,
close
                                        David Harris
                                        David Harris

Line up head-
ing and closing

Figure 15-2   Semiblock letter form

```
4673 Westlake                                                    ———— Heading
Fort Worth, TX 76132

April 17, 1989


DMC Machine Corporation
741 Washington                                                   ———— Inside
Salt Lake City, UT 84101                                              address

Subject: Recall of the SMX Lathe ————                            ———— Subject line
                                                                      substitutes for
_____                                 salutation
_____
_____

_____
_____
_____
_____


Walter Collins
Walter Collins ——————                                           ———— No complimen-
                                                                      tary close
```

Figure 15-3   Simplified letter form

the bottom. You may have to type the letter more than once in order to achieve the proper balance.

The third form, shown in Figure 15-3, is called simplified or streamlined because it omits two traditional parts of the standard letter: the salutation and the complimentary close. A subject line substitutes for the salutation, and the signature and typed name appear at the bottom without a complimentary close. The simplified form is useful if you don't know how to address your reader, or if the letter will have many readers. On the other hand, to some readers this form may seem abrupt and somewhat rude, so use it only if the form is standard practice for your company.

All letters should be typed on standard 8.5 × 11 white or off-white bond paper. Do not use onionskin or erasable paper. The second page of a letter should be on plain paper even if you use letterhead stationery for page one. At the top of each page after page one, type the reader's name, the date, and the page number, like this:

Smithers
July 13, 1989

page 2

Parts

A standard business letter has the following basic parts.

1. Heading The heading includes the writer's street address, city, state, ZIP code, and the date. On company letterhead stationery, the address is already printed, so you need type in only the date. Notice that the writer's name does *not* appear in the heading.

2. Inside address Include in the inside address the reader's name (preceded by Mr., Ms., or other title such as Dr.), title if appropriate (Publications Manager, Vice President of Sales), department, company name, street address, city, state, and ZIP code. Sometimes you must include a building number or internal mail stop. Duplicate the inside address on the envelope. (See section 2.3 for information about finding addresses of companies.)

3. Salutation Address the reader as "Dear (person's name)" and end the salutation with a colon. You can address people you know well by first name only. Otherwise, use either a title (Dear President Jones) or Mr. or Ms. If you know that a woman prefers Mrs. or Miss, use that. If you are uncomfortable with Mr. or Ms. in a salutation or do not know the reader's sex, use the reader's full name: "Dear Chris Harkness." Letter salutations can easily become sexist if you don't write to a specific, named individual, so if you can't determine a specific name, use an appropriate title such as "Dear Sales Manager." Avoid salutations like "Dear Sir" or "Gentlemen"; these terms can no longer be used in a generic sense. The simplified form in Figure 15-3 omits the salutation and substitutes a subject line. Subject lines can also appear in letters that use a standard salutation.

4. **Body** For easy reading, keep sentences and paragraphs fairly short in the body of the letter, and include only one idea per paragraph. Single-space within paragraphs and double-space between them. Section 15.3 tells what to include in the body of a letter. Try to keep a letter to one page, two at most.

5. **Complimentary close** Traditional closings are "Sincerely," "Sincerely yours," or "Yours truly." The complimentary close begins with a capital letter and ends with a comma; other words are not capitalized. If you know the reader well, you might choose a less formal closing like "Cordially" or "Regards." The simplified form omits the complimentary close.

6. **Signature and identification** Letters are always signed, but below the signature you should type your name. Many of us have only semilegible signatures, and you want to make sure that the reader knows how to spell your name. If you write as a representative of your organization, type your title below your name. If you are a woman and prefer to be addressed as Mrs. or Miss, indicate that with your typed name: "(Mrs.) Carol Green." Sometimes, too, the company name appears in this identification; follow your company style in this matter.

Sincerely yours,
ELECTROVISION, INC.

Elizabeth Swanson

Elizabeth Swanson
Engineering Products

7. **Optional information placed in the lower left corner**

- If someone else types the letter, the writer's initials appear first in capital letters, followed by the typist's initials in lowercase letters.

  LMR/st   or   LMR: st

- If you enclose additional material with the letter, you can indicate that by

  Enc.   or   Encl: resumé

For multiple enclosures:

Enc. 2   or   Encl: Proposal
                                    Background data sheet

- If you send copies of the letter to other people, it's a courtesy to let the primary reader know.

  cc: A. Wright
        T. Liu

8. Other optional information

- If you want to remind the reader of previous correspondence or a numbered contract or document, you can insert a reference line just before the salutation:

  Re: Order #E43487

  or you can include this information in the opening paragraph.

- You can also insert a subject line just before the salutation. (Remember that in the simplified form, the subject line replaces the salutation.)

  Subject: Cost Estimates of the Halifax Project

  Again, this information can appear in the opening paragraph.

15.3 TYPES OF LETTERS

All letters—no matter what the purpose or content—fall into three general categories: (1) letters that *ask*, (2) letters that *answer or tell*, and (3) letters intended to *persuade*. Table 15-1 lists many of the types of letters you may be called on to write. In letters that *ask*, you appeal to the reader to respond to a request; in letters that *answer or tell*, you convey information, appreciation, or complaint; and in letters intended to *persuade*, you attempt to convince the reader that your position is right.

In addition to these specific purposes, letters also build and maintain relationships in the business world. It's quite possible to learn to know a person well over the years simply through exchanging letters to carry on the routine business of your companies. If your letters are both professional and cordial, your correspondents will consider themselves your friends and will think more highly of your organization.

Sometimes letters overlap categories, but in general their purposes fall into one category or another. The following sections discuss in detail the content of one type of letter from each of the three categories in Table 15-1. In addition, Chapter 16 has examples and information about job search letters, and Chapter 21 discusses transmittal letters.

Table 15-1 COMMON TYPES OF BUSINESS LETTERS

Letters that ask	Letters that answer or tell	Letters intended to persuade
order	response to request,	sales
request	complaint, or	job application
complaint	invitation	invitation
collection	follow-up information	recommendation
invitation	announcement (good	request
	or bad)	complaint
	report	
	cover or transmittal	
	congratulations	
	appreciation	
	remittance	
	acceptance	
	resignation	

Letters That Ask: A Request Letter

In a request letter you are asking your reader to do you a favor, so you should carefully plan both the content and organization. Your letter should be direct and short—one page if possible. Fill out a Planning Sheet (see section 1.6) to remind yourself about purpose, reader, topic, and tone. Keep your key words in mind as you draft the letter, and use as many as you can. Whatever you say, make it as easy as possible for your reader to respond.

1. Opening paragraph   In one or two sentences, the opening paragraph must state your subject and purpose and establish a satisfactory tone. Rather than introduce yourself or give lengthy explanations here, get to the point: be as specific as possible about what you want to learn. Here's an example of an effective opening paragraph:

> Dear Ms. Asher:
>
> I am writing to ask for your help in identifying the skills needed by students who want to become professional technical writers. Central University plans to set up a certificate program in technical writing, and we want that program to meet the needs of local industries. Please help us plan that program by answering the following questions.

Here is an example of a student's opening paragraph:

> Dear Mr. Bradley:
>
> Since I am considering the purchase of a high-quality four-track cassette recorder for my home studio, please send me information on your Studiomaster Studio 4.

Note the friendly, positive tone in both examples. Part of their success comes from the frank appeal for help and the shift at once from the "I" of the opening to the "you" of the rest of the sentence. One expert on business letters says to remember that the first paragraph belongs to the reader; your purpose is to introduce the reader to the topic. You can also write good openers if you use action verbs like "request," "ask," and "send."

2. Middle paragraphs Middle paragraphs should explain the request and give the specifics of what you need. If you ask for more than three items or pieces of information, put them in a numbered list for easy reading. If appropriate, frame your request as a question or a series of questions to make your reader's job in responding easier. Here is an example of the middle paragraphs from a student letter that uses a list (Cook 1985):

> As a student of technical writing at Green Hills State University, I am working on the investigation and evaluation of surge protectors. A large part of my report will be based on manufacturer-provided specifications. Those specifications I will be most concerned with are
>
> 1. maximum spike dissipation
> 2. maximum spike voltage
> 3. maximum spike current
> 4. clamping spike voltage
> 5. surge current clamping ratio
> 6. clamping response time
> 7. noise attenuation
> 8. frequency range
> 9. filter network
>
> Even if you are able to supply me with only a few of the above specifications, it will be a great help.

Note that the paragraph asks for *very specific* information. It also explains the reason for the inquiry and how the information will be used.

Another student asks specific questions in the middle paragraph (Takagi 1984):

> I am writing to you because of the high ratings the Monark 872 received in the *Consumer Reports 1984 Buying Guide* issue. Would you please answer these three questions about the Monark 872?
>
> 1. How is resistance applied to the flywheel?
> 2. What is a high-momentum flywheel, and what are some of its advantages?
> 3. Could you supply me with the name and address of a distributor or a local retailer of the Monark 872?

Don't word your request in such general terms that your reader doesn't know how to reply. If you say, "Please send me information on radial tires," you could be looking for the history, specifications, sales records, durability, or road test results of radial tires. You probably won't get the information you need, and you may be lucky to get any information at all.

3. Closing paragraph Use the last paragraph to motivate the reader to respond to your request. If you need the information by a certain date, give that date in this paragraph. Be sure to allow enough time to respond, though; otherwise, you risk alienating your reader by sounding too demanding. It's better to give a specific date like "October 14" than to say "within three weeks," so your reader doesn't have to calculate the date. And be courteous enough to give the reason for your cutoff date.

If possible, the closing paragraph should show that there will be some benefit to the company or person addressed if they respond, perhaps by your interest in purchasing the product or by your offer to send a copy of the finished report. Also indicate your appreciation of an answer. You don't have to gush, but a simple "thank you" is appropriate. Remember that you want to end the letter without unprofessional begging but with a positive tone. Unless you are writing to a private individual, a stamped, self-addressed envelope is usually not necessary.

Here are examples of good closing paragraphs:

> I would much appreciate receiving brochures on General Electric Model #WWA8364V and Hotpoint Model #WLW3700B by October 15 so that I can include the information in my report. I will be grateful for any assistance you are able to give me (Gunn 1985).

> My project is to be completed by November 14, 1989, and I would appreciate any help you could give me in answering these questions. I would be happy to send you a copy of my analysis when it is completed.

Letters That Answer or Tell

Letters that answer or tell may either present information or respond to some kind of request from your correspondent. The two situations may be rather different, even though both are letters that tell.

An informational letter A letter conveying factual information that is not intended to persuade can be straightforward and positive in tone. Because you are writing directly to a real human being, make the letter as personal as possible, using names and personal pronouns like "I" and "you" and short active sentences. However, you should probably keep the tone semiformal rather than informal since you want your information to stand objectively on its own. You can again follow a three-part organization, as does the example in Figure 15-4.

A response letter As you work in your technical profession, you will frequently need to respond to requests of different kinds. Response letters

WQC | Water Quality Control
Indianapolis, Indiana 46206

March 30, 1989

H. J. Rogers
Interstate Water Monitoring Agency
8800 River Parkway
Central City, Ohio 45801

Dear Hal:

Purpose

Your agency will be interested in the success we had during the last month with a carbon absorption system for emergency treatment of contaminated runoff water. This surface runoff came from improperly stored drums of industrial chemical waste and threatened both Watson Creek and the nearby water wells of private dwellings.

Our sampling of water in the drainage basin indicated the presence of 142 different chemical compounds, including organic compounds like benzene, toluene, ketones, xylene, styrene, and polychlorinated byphenyls (PCBs).

Details

We constructed three devices to meet the immediate problem:

1. Trenches to intercept the surface runoff
2. A catchment basin to collect the runoff from the trenches
3. A carbon absorption unit to treat the contaminated water in the catchment basin.

The carbon absorption unit was constructed from two watertight trash dumpsters (20' × 10' × 5') placed on a bed of crushed rock. The contaminated water was pumped into the bottom of the first dumpster and released through a perforated pipe into 8000 lbs. of crushed limestone with an aeration ring at the bottom to remove volatile organic compounds. The overflow from this dumpster went into the second dumpster and flowed through two separate cells of activated carbon. The treated water was then released into Watson Creek.

(Continued)

Figure 15-4   An informational letter

Rogers   page 2
March 30, 1989

 James Chemical Laboratories conducted chemical
analyses of the released water at four different locations Closing—what
and found a 99 percent reduction of organic compounds use is to be
after treatment. We were pleased with the quick results we made of infor-
attained with this carbon absorption system on an mation
emergency basis.

 Hope you find this information useful.

       Sincerely,

       Tess

       Tess H. Wade
       Water Quality Control

Figure 15-4 Continued

differ from informational letters because the reader looks to you to fulfill some specific need. Your letter will be judged by that reader in three ways:

 1. on content. Does the response answer the questions and give the information needed?
 2. on tone. Is the tone both friendly and businesslike? Is the response gracious rather than grudging?
 3. on form. Does the letter follow standard business form? Is it correct and clear?

 In Figure 15-5 a customer relations supervisor responds to a student's request for technical information for a college report. Notice how the opening paragraph both acknowledges the letter of request and sets a friendly tone. The writer can only partly fulfill the request for information, but the first paragraph in the body tells what he can do, emphasizing the attempt to be helpful. In the second body paragraph, instead of simply saying "We don't have that information," the writer suggests an alternate source. Finally, the writer closes with an action request of his own—and the positive "Good luck" as a final word.

 A different kind of response letter is shown in Figure 15-6. Because the content of this letter is so positive ("you got the job!"), the writer probably did not need to worry about the tone. Yet this letter is carefully crafted to be welcoming and to center on the reader and the reader's needs. Note the careful mix of "we" and "you" along with the concentration on

U.S. SUZUKI MOTOR CORPORATION

October 4, 1985

Michael A. Murray
199 So. 15th Street
San Jose, CA   95112

Dear Mr. Murray,

Thank you for your interest in Suzuki motorcycles. We will
be more than happy to assist you with technical information
for your comparison report.

Please find enclosed brochures, spec sheets and photocopies
from the Owner's Manual that should provide you with the majority of the information you have requested.

Unfortunately, U.S. Suzuki does not publish some of the information you have asked for. Horse power, quarter mile performance, top speed, and fuel consumption are the specific items
that we will be unable to provide you with. I would recommend,
however, that you search through the motorcycle magazines in
the library for their test reports on these models as they
should provide the balance of the information you need.

If it would not be too great an inconvenience, we would be
pleased if you would send us a copy of your results when it
has been completed.

Good luck on your report!

Sincerely,

U.S. SUZUKI MOTOR CORPORATION

Mike Clark
Customer Relations Regional Supervisor

MC/kks

enclosures

Figure 15-5   A response letter

Silicon Compilers

2045 Hamilton Ave. • San Jose, CA 95125
408/371-2900

May 5, 1986

Marian Cochran
1253 S. 7th, #42
San Jose, CA 95112

Dear Marian:

It is my pleasure to offer you the technical publications
summer internship, with a salary of $8.00 per hour. This
position will begin on May 12 with a completion date of
July 15, 1986. May 12 through May 23, you will be working
on a part time basis. From May 27 through July 15, 1986
you will be full time, working approximately 40 hours per
week.

We look forward to your joining us at Silicon Compilers in
what we hope will be a mutually advantageous internship.
Please sign and return the enclosed copy of this letter
as your acceptance of our offer. If you have any questions
please call me at (408) 371-2900.

Sincerely,

Paula Bell
Manager Technical Publications

I accept the position at Silicon Compilers Inc.

_____   _5/_____
        Candidate's Signature   Date

Figure 15-6   A response letter

the specifics of the contract. The student's response was "Isn't this a *nice letter!*"

Letter Intended to Persuade

You may think immediately of selling when you think of persuasion, but many times you must sell an idea or course of action rather than a product. Persuasion requires special sensitivity to the reader's position relative to the writer. You may want to review sections 1.2 on reader evaluation and 1.4 on tone before you tackle the draft of a persuasive letter.

Figure 15-7 shows a good example of a persuasive letter written by a student. Notice that the reader is given much specific information: the specifics of the student's needs, the background of the situation, and reasons why the request should be granted. Note too that the tone is respectful; this letter is written "up," and the student takes care with his language and approach. This letter was successful: through the combination of tone, logic, and the appeal to a sense of fair play, the reader was persuaded to admit this student to the program.

15.4 THE PROCESS OF WRITING LETTERS

Planning

In planning a business letter, first decide on your purpose: what you want this letter to accomplish. After consulting the list in Table 15-1 and reviewing the material in Chapter 1, fill out a Planning Sheet like that shown in section 1.6.

Always try to write a letter to a specific, named individual, even if you have to call the company to find out who that individual is. Try to learn as much as you can about the primary reader, categorizing him or her by type, job title, and background. (See section 1.2.) If you should send copies to secondary readers, plan the letter to meet their needs as well— for example, by avoiding technical jargon if secondary readers are generalists or managers.

Since letters are personal, direct communications from you to the reader, your tone is very important. You can easily offend a reader by being rude, condescending, curt, or servile. Determine your position relative to your reader, asking yourself if you are writing up to a superior, down to a subordinate, or horizontally to a peer. Also determine whether the content of the letter is good news or bad news; the content will dictate both your tone and the method of organization you should adopt.

For example, here is part of a request letter that has a tone problem. The letter is written to volunteer CPR (cardiopulmonary resuscitation) instructors asking for their help in teaching CPR to 5,000 people.

567 South Eighth Street
Greenfield, NY 13905
March 31, 1988

Prof. Thomas Herman
Chairman
Aeronautics Department
Greenfield State University
Greenfield, NY 13903

Dear Professor Herman:

I want to become an aeronautics major at Greenfield State. I have applied for the program three times before, but I believe I have not been accepted because of my GPA. I realize that my GPA is not fantastic, but please note that it has been increasing since I started in Fall 1986. Last semester my GPA was 2.83.

Purpose of letter

I have been hoping to study aeronautics since I was in high school. Before I graduated from Northrop Community College in June, 1988, I received a letter from Phyllis Hale, Assistant Evaluations Coordinator at Greenfield, stating: "You will be given full credit for all of the courses you have taken at Northrop." Because of this letter, I sincerely thought I would soon be studying aeronautics, so I moved up here to Greenfield from Middletown.

Instead, during the last two years, I have been taking as many of the classes required for an aeronautics degree as possible without actually being in the program. Now I have run out of courses to take. This semester I took some math classes just to fill my schedule, but I don't want to do that anymore. I want to start taking aeronautics classes.

Supporting details

I have also been turning my interest in aeronautics into reality. For one year I worked as a lineman and A&P mechnic here in Greenfield. In a technical writing class I am now learning to do report writing, which will be helpful in my planned career as a maintenance supervisor. I am also writing a formal report for that class on business jets, which is both fun and interesting.

I hope you will understand and help me with my dilemma. Please help me achieve my goal of becoming a maintenance supervisor by admitting me to the aeronautics program.

Action request

Sincerely,

Ken Frith

Ken Frith

Figure 15-7   A persuasive letter

> . . . We need 250 CPR instructors to teach at the seven sites listed be-
> low. . . . We need instructors to sign up for four-hour shifts. We will be starting
> at 7:00 A.M. both days and will finish at 5:00 P.M. on Saturday and 3:30–
> 4:00 P.M. on Sunday. You may sign up for as many shifts as you wish. We
> will try to give you your choice of site, but it may not be possible.
>
> All instructors MUST attend an orientation, as the classes won't be
> taught in the traditional modular method, and all instructors MUST be able
> to show proof of retraining. Only three orientations will be held, and your
> attendance at one is MANDATORY. . . .

Since this letter is written to experienced instructors, the readers are at
least peers of the writer, but notice the brusque and rather rude repetition
of the words "need," "must," and "mandatory," giving the impression that
the readers are subordinates (and not very bright ones at that).

If the writer wants to attract volunteers, the tone needs to be friendly
and persuasive, recognizing how much these instructors will contribute.
It could be done like this:

> . . . The only way we can train 5,000 people in this lifesaving skill is
> with your help: we need 250 CPR instructors to teach at the seven sites listed
> below. . . .
>
> Can you help by signing up for one or more four-hour shifts on Saturday
> and Sunday, October 18 and 19? We will be starting at 7:00 AM both
> days. . . . We will do our best to give you your choice of sites.
>
> These classes won't be taught in the traditional modular method, so
> retraining sessions are planned for the following three dates: October 2, 11,
> and 14. After retraining, you'll have an opportunity for a skill checkout with
> an instructor trainer, and you'll be issued a new instructor card.
>
> Call (215) 333-2614 for details and to reserve a spot in a retraining
> session. You've already contributed much as a volunteer instructor, and we
> hope you can help us again.

Gathering Information

You may have at hand all the information you need before you start your
letter. If not, spend some time verifying details. Be very specific, whether
you are asking for something, giving explanations and information, or
trying to persuade the reader. Letter-writing costs are high because of the
time involved in writing them, and you can avoid second letters of clarification if you are accurate and complete the first time.

Organizing

Good or neutral news and positive persuasion If your letter conveys good
news, urges the reader to undertake positive action, or simply presents
objective information, you should use this three-part organization:

1. Tell the reader the main point of the message.
2. Substantiate that main point with details.

3. Close with the action you want the reader to perform, or repeat the main point you want to make.

This direct approach works well with busy readers. They want to know at once the purpose and main point of your letter, and they will not object to a straightforward presentation.

Figure 15-7 showed how this basic organization works in a persuasive letter. The first paragraph opened with a direct statement of what the writer wanted—"to be an aeronautics major." In the next four sentences, the writer explained the problem: his grade point average. Paragraphs two, three, and four gave details to support the request: background, the current school situation, and the writer's efforts to improve his preparation. The last paragraph contained a direct appeal for help and repeated the long-range goal of becoming a maintenance supervisor, thus reminding the reader of the importance of the request.

Bad news and negative persuasion If your letter must present bad news or persuade a reader *not* to do something (or to do something he or she doesn't want to do), you need to change the method of organization. Your letter may still have the three basic parts, but they must be sandwiched between expressions of goodwill. You shouldn't begin with a direct statement of the purpose and main point. Sometimes the news is too harsh ("You didn't get the job"), and at other times the reader is hostile and needs to be met on some common ground. With bad news you need to prepare the reader before you state the main point. Here is the organization to follow:

1. Set up a common ground with the reader or make a positive comment.
2. Give background.
3. Tell the reader the main point of the message.
4. Support the main point.
5. State the action you want the reader to perform.
6. Close with a positive statement or one that conveys goodwill.

What types of letters will contain bad news and negative persuasion? Most often, these will be letters of complaint or replies to complaint letters (especially when the complaint must be denied). Sometimes you must write a letter about your inability to meet a schedule or about problems in production or personnel. As you move up in your chosen career, you may have to write negative letters to job applicants. In all these cases, you need to combine a roundabout method of organization with careful consideration of tone and style. Negative letters most often call for a formal or semiformal style and for a tone that is positive or neutral, but always courteous. Remember the image of yourself and your company that you want to project. Below is a rejection letter to a job applicant—a difficult letter both to write and receive. Notice that the bad news is cushioned by expressions of goodwill and that the tone is courteous and encouraging.

> Dear Mr. Willis:
>
> Thank you for your interest in Carbide Engineering. We enjoyed meeting you at the plant on July 12 and were impressed with your background in CAE workstations.
>
> While you were one of the three finalists for the engineering position, we are sorry to tell you that the position has been filled from within the company. This in no way reflects on your credentials for the job, which are outstanding. We will keep your resumé in our active file should another opening occur.
>
> We wish you success in your career goals.

Writing, Reviewing, and Editing

Once you have determined an appropriate method of organization, write a draft of your letter. Make sure that you write the letter in terms of what's important to the reader. This is sometimes called the "you" attitude. For example:

> USE: You can expect delivery of the complete order by August 20, 1989.
> NOT: We will ship your order on August 17, 1989.
> USE: Your application is currently being reviewed by engineering managers, and you will hear from us within the next two weeks.
> NOT: We have sent the application to our engineering managers, and we will notify you of any action within two weeks.

When your draft is written, review it by checking the content against your Planning Sheet. Edit for technical accuracy, for punctuation, and for correct spelling of names as well as other words. The acceptance of your message often depends as much on the correctness of your letter as it does on the content, so you must take extra care with the details. Follow standard forms for salutations and closings, but avoid tired clichés like "In re your letter" or "Pursuant to the matter of." Instead, write in direct, uncomplicated sentences. See Chapter 8 for help on avoiding clichés. Have a fellow student read the draft. Then type the letter, following one of the three standard forms. Make sure to allow adequate margins, framing the letter on the page. Proofread the final copy for accuracy before you sign it. Since a letter is a communication from you and your company directed personally to a reader, your reputation as a professional rides on each letter you write.

CHECKLIST FOR WRITING LETTERS

1. Which standard form shall I choose?
   _____ flush left
   _____ modified block
   _____ simplified
2. Have I included all the standard parts and the necessary subparts?
   _____ heading
   _____ inside address
   _____ salutation
   _____ body
   _____ complimentary close
   _____ signature and identification
   _____ typist identification (if needed)
   _____ enclosure (if needed)
   _____ cc (if appropriate)
   _____ reference or subject line (if appropriate)
3. Is this a
   _____ letter that asks?
   _____ letter that answers or tells?
   _____ letter that persuades?
4. Is the content
   _____ good news?
   _____ bad news?
   _____ neutral news?
5. What is my position relative to the reader? Am I writing
   _____ up to superiors?
   _____ down to subordinates?
   _____ horizontally to peers?
6. Have I carefully reviewed and edited to eliminate errors?

EXERCISE 15-1   Evaluating Request Letters

Directions: Study the letters A, B, and C to determine what they do well and what problems, if any, they contain. Look at the form, content, and tone. Be prepared to discuss them in class.

Letter A

September 24, 1989

Sanyo Electric Inc.
200 Riser Road
Little Ferry, NY

Attn: Microwave Oven Customer Service

Dear Sirs:

I am a university student doing a comparative study on microwave ovens for a semester project. I would appreciate it if you could send me some information regarding your microwave ovens. I want to know about the advantages of your brand, the warranty, safety, cooking features, and cost.

This information will enable me to compare your ovens with other brands and to compile a report recommending the best oven to meet the criteria mentioned above. This study is due the first week in November, and I would appreciate it if you could send me some information by the third week in October.

I would like to thank you in advance for your cooperation. The information will be beneficial to me now as well as in the future when I need to purchase a microwave oven.

Sincerely,

Carol

Carol A. Rekidder
60 Lake Drive
Kearney, NB

Letter B

September 17, 1989

Condor Computer Corp.
PO Box 8318
Ann Arbor, MI 48017

Subject: Condor 3 Database Software

I am researching and reporting on various database software on the market today. This report will be the foundation of several purchase decisions.

Could you please send me the following information on your Condor 3 software package:

1. Available technical reference material
2. System compatibility data

Could you also send any other information that is available on this subject.

Thank you,

Gregory T. Hart

Gregory T. Hart

Letter C

```
2330 Walnut Street
Springfield, MA
February 25, 1989

Apple Computer
20525 Mariani Avenue
Cupertino, CA 94014

Dear Sir or Madam:

I need some information about Apple computers. This information
will be used in a report comparing different brands of personal
computers.

This report is part of the technical education curriculum at the
university. I will be comparing different brands of computers by
price, expandability, software expansion, color or graphics
capability, and service contracts offered.

I hope to bring out the advantages of the Apple over the other
computers. With the help of the needed information, I shall be
able to make consumers aware of the features of the Apple.

I would really appreciate your cooperation in this matter. It
will help me if the information is sent right away.

                                    Sincerely yours,

                                    Lee Anson Judge
```

EXERCISE 15-2   Evaluating a Persuasive Letter

Directions: Analyze the following letter (see page 316) for tone, organization, and total effect. Determine how you would respond to such a letter and write a short paragraph explaining your response. Submit the paragraph to your instructor and be prepared to discuss your response in class.

EXERCISE 15-3   Writing a Request Letter

Directions: Write a letter of request to get information for your term project. Follow the suggestions in section 2.3 to find addresses of specific companies and obtain the names of individuals at those companies. Send the letters after review and revision. If this is a collaborative project, assign one or two people to write request letters. Before sending the letters, review them as a group for accurate content and correct form.

CITY-WIDE

Waste Collectors
P.O. Box 3114
Shreveport, LA 71001

February 3, 1989

Mr. Joseph Polaski
200 Plum Street
Shreveport, LA 71106

Dear Mr. Polaski:

Something troublesome is about to happen to your garbage collection service. The City Council is considering getting rid of City-Wide Waste Collectors and putting your home collection service in the hands of a company with no proven track record here.

That means you could be paying more for your service each month and at the same time losing the City-Wide drivers our customers speak so highly about, getting a new kind of billing statement in your mailbox, and suffering through a long period of confusion while another company tries to figure out how to do our job.

As a valued customer, you can help stop this action by immediately:

Sending the enclosed postage-paid card to City Hall. Do it today!

Writing or calling your councilperson or the mayor.

Attending the City Council meeting on February 15 to tell the Council you like our service and want to keep it. The Council meets on the second floor of City Hall.

Writing letters to the editor of the city newspaper.

If the city makes a decision in your favor and renews the City-Wide contract, you will get at no increase in cost:

1. Continuation of your unlimited pickup service

2. Free curbside motor oil recycling

3. A year's supply of garbage bags

If you want to stay with City-Wide, please make your wishes known. Contact the mayor and your councilperson—let them know you want the contract with City-Wide renewed. Do it today!

Sincerely,

Allen Harkness

City-Wide Waste Collectors

EXERCISE 15-4 Writing a Follow-up Letter

Directions: After three weeks (or when your instructor suggests), send a second letter of request to those companies that have not responded.

EXERCISE 15-5 Writing an Informational Letter

Directions: Choose a subject from your major field that you understand well. Write a brief informational letter to your writing professor explaining that subject in an objective manner and using language that your professor can follow even if he or she is not a specialist in that field.

EXERCISE 15-6 Writing a Persuasive Letter

Directions: Write a letter to the chairperson of your major department discussing some problem or concern of yours. Make the letter persuasive, giving the background of your concern and the reasons why you advocate a specific action.

REFERENCES

Cook, Ronald. 1985. Letter to a manufacturer. San Jose State University.
Frith, Ken. 1988. Adapted from a persuasive letter. San Jose State University.
Gunn, Elizabeth. 1985. Letter to a manufacturer. San Jose State University.
Takagi, Leslie. 1984. Letter to a manufacturer. San Jose State University.

CHAPTER 16

Resumés and Job Search Letters

The most important writing of your career may well be the resumés and letters you write in search of a job. Certainly getting your first career position is important, so you'll want to learn all you can about writing an effective resumé and job search letter. What's more, you need to keep polishing those writing skills throughout your professional career. Current labor studies indicate that people change jobs every three to four years and can expect about four major career moves in a lifetime. Each job

318

change or career move requires that you "market" yourself with what you write. Marketing is exactly what you are doing as you search for a job: writing persuasively to sell your skills, abilities, experience, and potential to a prospective employer. And because competition can be fierce—with 100 or more applicants for a single job slot—you must stand out from your competitors and convince the reader that you are worthy of an interview. It's not easy, but with the right resumé and job application letter, you can do it. In this chapter you'll learn about job search documents.

Even though the information on writing resumés and letters comes at the beginning of this chapter, you may want to read section 16.4 first to learn how to research potential employers and plan your job search documents.

16.1 DEFINITIONS OF JOB SEARCH DOCUMENTS

The primary job search document consists of two parts: a resumé and an accompanying letter of application or request. A *resumé* is a short and highly organized summary of your education, activities, and work experience. Its purpose is to tell a prospective employer of your skills and abilities. The word resumé comes from a French word meaning "to sum up," and resumés are sometimes called fact or data sheets.

A *job application letter* is a short business letter directed to a specific person at a company that (1) tells of your interest in a specific job, (2) highlights applicable information from your resumé, and (3) requests an interview. The purpose of the letter is to obtain an interview. Job application letters are sometimes called cover letters or letters of transmittal, but while a standard cover or transmittal letter is subordinate to the main document, a job application letter is at least as important as the resumé.

Other letters that are part of the job search include

- follow-up letters
- thank-you letters
- acceptance letters
- refusal letters or those withdrawing from consideration

Students sometimes ignore these additional letters, but writing them marks you as a professional and can help you secure the job you want, both now and in the future.

Because resumés and job application letters differ significantly in form and organization, they will be discussed separately in this chapter. You should plan and write your resumé before you write a job application letter. When you compose your resumé, tailor its form to your own needs; do *not* simply follow one of the forms given here.

16.2 RESUMÉ FORM AND ORGANIZATION

Most resumés contain the same six basic components:

1. personal data
2. job or professional objective
3. educational information
4. work experience
5. skills and abilities
6. references

The order and arrangement of these six components will differ depending on the kind of resumé you write. In this chapter you will learn about two kinds—chronological and analytical—and the advantages and disadvantages of each. You may want to try writing both kinds to see which works better for you.

Both chronological and analytical resumés should be short and succinct, preferably confined to one page or at the most two. The typical resumé gets a first reading of ONLY 15–20 *seconds*; therefore, if you ramble, the reader may never get to your key information. That very short reading time also means that the form must be easy to read. You need to arrange information carefully on the page: use white space as a frame and highlighter and choose boldfaced type, underlining, capital letters, and lists to direct the reader's eye. Always put the most important information at the top of the page. You need not write complete sentences in a resumé (in fact, you probably shouldn't), but you should emphasize with verbs what you can do or have done. Your resumé should be *absolutely free of errors or smudges*; that paper represents you to the reader, and spelling or typing errors may well disqualify you at once. You don't need to pay for professional typesetting, but make sure that the copies made from your original are clear and on good quality white or off-white paper. You need not use fancy or colored type; make it your best professional effort without fancy touches.

Chronological Resumés

A chronological resumé, also called a traditional or "obituary" resumé, is relatively easy to write because it simply tells what you have done in the past. Figure 16-1 shows the chronological resumé of a graduating senior (Martinez 1986).

Components of a chronological resumé   Each of the six major components is allotted a separate section on the page, with the most important or relevant information placed near the top of the page.

    1. Personal data. Always put your name, address (including ZIP code), and phone number (including area code) at the top. If you have a

JULIE MARTINEZ
4326 N. Elm Street
New London, IA 51003
(319) 296–3329

CAREER
OBJECTIVE

An entry level position in the field of marketing, preferably in the areas of product development and research.

EDUCATION

UNIVERSITY OF IOWA, Iowa City, IA
Bachelor of Science degree in marketing with a minor in English, expected December 1988
GPA: 3.5/4.0

Pertinent course work:

Marketing Management	Marketing Channels
Industrial Marketing	Marketing Research
Consumer Behavior	Business Logistics
Business Statistics	Programming in BASIC
Technical Writing	and FORTRAN

EXPERIENCE

MARKETS, INC., Burlington, Iowa

1/86–
present

Market Research Interviewer
o Conduct surveys under diverse conditions and among varied consumer groups including college students, retail store and motion picture patrons, and housewives.
o Compile data from various studies to form basis for a national research project on consumer spending potential for microwave ovens.
o Write biweekly reports on research projects.

6/85–
12/85

THE DAILY IOWAN, Student Newspaper, University of Iowa, Iowa City, IA

Advertising Manager

o Increased the number of regular advertisers from 9 to 21 over a period of two semesters.
o Eliminated paper's budget deficit.

EXTRA–
CURRICULAR
ACTIVITIES

Member, American Marketing Association, University of Iowa; served as Vice–President (two terms)

Tutor, Statistical Analysis course (two terms)

Member, Future Farmers of America

REFERENCES

Available upon request.

Figure 16-1 Chronological resumé for graduating senior

temporary and permanent address, include both; it's important that a prospective employer be able to contact you easily. At one time personal data also called for information like height, weight, age, marital status, children, health, hobbies, and interests. Today, however, most resumés omit that information. For one thing, it's illegal for employers to discriminate on the basis of sex or age, and for another, that information is often judged irrelevant to your job potential. You may choose to include some bits of personal information, but do so *only* if they will enhance your application and if they bear directly on the job you want.

Personal information beyond your name, address, and phone number should not go at the top of the resumé but at the bottom. Examples of personal information that might be appropriate include: (1) the fact that you are single if the job calls for extensive travel, (2) your fluency in a second language if the job involves international contact, (3) your status as a parent if the job calls for working with children or involves nurturing qualities. Evaluate personal information very carefully to ensure that it will enhance your application. If it won't, leave it out.

2. Job or professional objective. Career advisers do not agree about whether or not you should clearly state the job you want on the resumé. If you make the objective too specific ("applications programmer"), you may eliminate yourself from consideration for related positions. If you make the objective too broad ("challenging position using my abilities in business"), the statement means nothing. Some experts thus recommend putting the objective only in your application letter. Remember, though, that the job objective provides a focus for the resumé. In many ways it's like a thesis sentence—with the details of the resumé supporting the objective.

It may be best to tailor your resumé to the situation. If you are responding to a specific job opening for which you qualify, stating the objective specifically is to your advantage. But if you're prospecting—to see what possibilities exist—you might want to be more general or list multiple objectives. In all cases, you should have researched a company *before* you write and submit a resumé so you know what kind of jobs are available there. Section 16.4 gives some advice about this. You may want to create more than one resumé, each with a different objective and a slightly different emphasis. The added work and expense is worth it if you find the job you want. Learning to use a word processor (see Chapter 9) will make adapting your resumé to specific jobs much easier. If you are entering the job market, you may want to list both present and long-term objectives. For example:

CURRENT OBJECTIVE: To create clear, concise documentation and training aids for computer system and software end-users

PROFESSIONAL GOAL: To supervise and/or manage a group producing technical documentation and training materials

3. Educational information. If you are a graduating senior or a recent graduate, education should be near the top of the page because your education is probably more relevant to the job objective than is your work experience. In this section you should list in *reverse chronological order* the schools you have attended, with city and state, inclusive dates of attendance (1985–89), and degrees received or the date a degree is expected. Add your grade point average (GPA) or class standing if it is better than average. If your GPA is better in your major, indicate that. This section can also include such educational highlights as honors, awards, research studies, projects, particularly relevant course work, offices, and activities. Such activities are important to employers because they show your active involvement in college life. (While you're in college, it pays to do more than just attend class. It is never too late to join organizations that interest you.) Do not include high school activities on your resumé unless you are still in college or the high school work is unusual or relevant to the job.

4. Work experience. List your work experience, again in *reverse chronological order,* leading off with your current or most recent job. Identify each job with a job title and tell what your primary duties were. Instead of writing complete sentences, condense the information into short phrases, using active verbs and specific nouns. For example:

- *Trained* new employees on etching and masking procedures
- *Wrote* documentation for EZY interface
- *Reported* monthly sales plan figures and variances from prior years to upper management
- *Assigned* duties and projects to a staff of 7–10 salespersons

Name the company, identify the city and state, and give the inclusive dates of employment. You may also want to name your supervisor. If you have had many part-time jobs unrelated to your career objective, you can group them ("summers 1985–87 fast-food cashier at McDonalds and Burger King"). Explain any large gaps in your employment history ("1986–87 travel in Europe").

5. Skills and abilities. Some chronological resumés add a section highlighting special skills. These could include fluency in a second language, knowledge of specific programming languages, skills such as welding or wirewrapping, or experience with systems like specific word processors or laboratory equipment. Determine what to include by testing the skill's relationship to your job objective, and omit this section unless you have pertinent skills to list.

6. References. Most employers want you to have references, but they do not expect to find them listed on the resumé. A simple line such as "References available on request" is sufficient. You may think that you need to include the references to fill up the page, but by themselves references don't mean anything to the employer. It's more important for you to find solid information from your education and work experiences to fill the page.

Even though you don't list references, you *should* type a list of them, including name, job title, place of business, address, and phone number, and you should bring several copies of that typed list with you to an interview. References should be people who know the quality of your work and can speak enthusiastically about it. Have three to five names, including at least one work manager, and for new graduates, one professor. Be sure to ask these individuals ahead of time if they would be willing to give you a good reference. You don't want your sources to be surprised by a request for information.

Arrange these six parts of the resumé on the page so they are easy to read. Figure 16-2 shows a typical chronological resumé for an experienced technician (Fuller 1987). Notice how the Experience section has replaced the Education section (the reverse of Figure 16-1) because the technician has a track record of professional experience.

Advantages and disadvantages of a chronological resumé The advantages of a chronological resumé are that (1) employers are familiar with this traditional format, (2) the chronological organization traces your past and tells the reader what you have done, and (3) the resumé is easy to compose. Disadvantages are that (1) readers must figure out from your past what you are capable of doing because the resumé itself does not indicate your potential, (2) your experience may not support your job objective, and (3) the traditional aspect may make your resumé look like all the others.

Analytical Resumés

An analytical or skills resumé emphasizes your qualifications and skills, especially as they relate to your stated job or professional objective. This kind of resumé requires you to think through your past and isolate the three to five skills or abilities that will make you a valuable employee in a particular career slot. Those skills and abilities can come from your education, paid or volunteer work experience, school or extracurricular activities, hobbies, or research projects. An analytical resumé includes the same six components as a chronological resumé, but their emphasis and organization differ. In an analytical resumé you want to emphasize your abilities and accomplishments in the order of their importance, regardless of their chronological sequence. Study the analytical resumé in Figure

MICHAEL FULLER
3034 Lincoln Avenue
Minneapolis, MN 55403
Phone (612) 342–0718

OBJECTIVE: Position as an Electronic Technician that will utilize my
               five years experience. Seeking a swing or graveyard shift.

EXPERIENCE:
               TECHNICAL CORP. Semiconductor Division, Minneapolis, MN

1/84 to   ASSOCIATE ENGINEER
present   Responsible for breadboarding, testing, and
               troubleshooting integrated circuits for Bipolar Analog
               Design Engineering.
               — Developed and wrote test programs using the following:
                 o LTX Test System/LTX Basic—for Plantronics amplifier
                 o Hewlett Packard 9825A Test System/HPL Basic
               — Implemented bench tests and analyzed data from
                 amplifiers, logic processors, and D/A and A/D
                 converters.
               — Designed and built two—sided PC Boards.
               — Developed, organized and implemented a DBASE II
                 tracking system for the development of new IC products.
               — Created a parts inventory system (IBM—XT).
               — Used curve tracers, oscilloscopes, distortion
                 analyzers, logic analyzers.
               — Built test jigs. Checked layout plots.
8/83–1/84   PQR ELECTRONICS CO.—Milwaukee, WI

               ELECTRONIC DESIGN TECHNICIAN—Bipolar Analog Design
               Engineering
               — Helped design linear data conversion ICs.
               — Implemented bench tests of converter products.
               — Developed linearity test program for A/D converter
                 using the H—P 9825A.
               — Debugged at IC level with a micromanipulator probe
                 station.

               PROGRAMMING LANGUAGES:
               Fortran, Assembly Language IBM 360/370, HPL Basic, IBM PC
               Basic, 8085 Assembly Language, and CADAM

EDUCATION
1985—   UNIVERSITY OF MINNESOTA—Minneapolis, MN
present   Major: General Engineering   GPA: 3.0/4.0

1983—85   INSTITUTE OF TECHNOLOGY—Milwaukee, WI
               A.S. Degree—Electronics Engineering Technology, July 1985
               GPA: 3.9   Awards: Dean's List 4 quarters, Certificate
                                     of Achievement

1982—83   MACALESTER COLLEGE—St. Paul, MN
               Major: Math/Computer Science 30 Credits GPA: 3.5

REFERENCES: Available upon request

Figure 16-2 Chronological resumé for experienced worker

16-3 to see how the writer has grouped her skills and abilities to support the job she seeks (Nelson 1986).

Components of an analytical resumé Here's how to treat the six sections in an analytical resumé.

1. Personal data. As in a chronological resumé, the key data at the top should include your name, address with ZIP code, and phone number with area code. If you have a temporary and a permanent address, list both to make it easy for the employer to contact you. Other personal data may be located at the bottom, but your space is usually limited. One reason for including personal data is that it might introduce a topic of mutual interest for discussion at an interview, but observe the cautions about personal data discussed in the section on chronological resumés.

2. Job or professional objective. Clearly stating your objective is crucial to an effective analytical resumé because it acts like a thesis sentence with the resumé itself supporting the objective. If you are simultaneously applying for several jobs, you may have to tailor the job objective to each company. Objectives can be stated in many ways:

- A position in financial planning or management leading to a career in corporate finance
- Teaching woodworking in junior or senior high school, with an interest in coaching baseball or softball
- Technical editor
- Seeking a position in developing software systems. Flexible in area of assigned work but most interested in system programming, computer network systems, microprocessor applications, and graphics

3. Skills and abilities. In an analytical resumé this section is the most important, for it supports the job objective and sells you to the employer. At the left margin, you list three or four particular skills or qualifications and follow each skill with specific accomplishments that back it up. If you can show how you saved money with a new process, reorganized for more efficiency, and designed or implemented a successful operation, you will demonstrate your value. Skills, abilities, or qualifications can be listed in several ways:

- by names of job functions: manager, salesperson, designer, planner, coordinator, writer, programmer
- by adjectives describing the skill: supervisory, analytical, mechanical, organizational, research

PAMELA NELSON
342 Drywood Lane   Home: (316) 342–1203
Wichita, KS 67207   Work: (316) 287–9090

OBJECTIVE: A position as a technical editor/writer

SKILLS

Organizational
o Took over nonfunctional word processing center and established all
  center policies and procedures.

Technical
o Performed hardware and software installations.
o Developed proficiency in diagnosis and troubleshooting of software
  problems.
o Wrote BASIC programs for the Wang and IBM PC.

PROFESSIONAL ACCOMPLISHMENTS

Administrative/Management
o Advertised for, interviewed, and hired word processing operator.
o Prepared WP center's 1986 operating budget.
o Developed new overtime policy for office support staff.

Writing/Editing
o Served as production editor for monthly regional newsletter.
o Rated "Outstanding" three consecutive years for editing skills.
o Wrote two users' manuals, one for authors and one for novice WP
  operators.

Creative Problem Solving
o Converted a 20–page typeset form for use on word processing
  equipment, which reduces production time by 50–75% and revision
  time by 95%.
o Developed new method of booking corporate conventions that
  produces more detailed documentation in less than half the time of
  the old method.

EMPLOYMENT HISTORY

Sr. Word Processing Technician Donald Corporation, Wichita, KS
January 1984 to present

Purchasing/Advertising Donald Corporation, Wichita, KS
Administrative support of nine staff and two department heads.
July 1983–January 1984

EDUCATION

MA Degree in English with Writing Concentration (June 1988)
Carnegie–Mellon University, Pittsburgh, PA   GPA: 4.0 (to date)

BA in Music (Voice Performance) June 1982
North Texas State University, Denton TX

REFERENCES available on request.

Figure 16-3 **Analytical resumé of a person changing careers**

- by applications of abilities: art, budgets, promotion, production, troubleshooting, human relations

The more specific you are with your backup information, the more believable you will be. For example, if you list a skill such as "budgeting," you need to say something like "Planned monthly budgets for Goldmark Department Store. Prepared daily and monthly cash flow records and reports. Improved cash flow reporting time by 50 percent." Then be prepared during your interview to back up any claims you make with specific evidence. What you are doing here is confidently driving home your accomplishments to the reader.

4. Work experience. You may have already covered your work experience in the skills section, but many effective analytical resumés add a brief summary of employment (in reverse chronological order). For example:

EMPLOYMENT
Winnebago County Accounting Office Rockford, Illinois	9-87—present
Swanson Bookkeeping Service Elmhurst, Illinois	1-86—9-87
Mutual Insurance Company Chicago, Illinois	12-84—12-85
Agricultural Research Institute DeKalb, Illinois	Summer 1984

Explain any significant gaps: "August to December 1984: Travel in Mexico."

5. Education. Since some of the skills and abilities you list may well come from your education, this section is usually treated briefly (and in reverse chronological order) with a simple listing of schools attended, location, dates of attendance, and degrees.

EDUCATION
Vassar College, Poughkeepsie, NY   B.S. Biology 1988
City College of New York, New York, NY
    45 hours of biology and computer science
Willamette University, Salem, OR   B.A. Psychology 1980

6. References. As with the chronological resumé, obtain promises of references from at least three or four sources, and make the bottom line of your resumé say "References available on request." Take several copies of your typed list of references to the interview.

Figure 16-4 shows an example of an analytical resumé for a student about to graduate (Magnuson 1986). Notice how specific he is in listing his computer-related experience.

William Magnuson

Permanent Address:
244 Jackson St.
Houston, TX 77004
(713) 213-6687

Campus Address:
385 S. 10th St.
#309
Houston, TX 77004
(713) 277-1092

OBJECTIVE	Computer Programmer/Analyst
EXPERIENCE	Languages: Pascal, FORTRAN, BASIC, 8080 assembler Operating Systems: MP/M II, North Star DOS Word Processors: Wordstar, Wang, Spellbinder, Format II Editors: XEDIT, EDT Hardware: CDC Cyber 174, North Star "Horizon" and "Advantage," DEC PDP-11/45, VAX, Wang VS-100, Onyx Work: Developed payroll, quarterly report, profit and loss report, and journal programs. Debugged and modified in-house programs. Courses: Data Structures, Combinatorics
Instructional	Conducted training services on North Star, Wang, and Plexus systems. Tutored math (up to calculus), electronics, and English.
Technical	Performed maintenance work on North Star computers. Installed upgrade kits in Apple II+. Courses: Microcomputer Repair, Solid State Circuit Design.
Operational	Performed data entry in an accounting office, an insurance department, and an agricultural lab. Backed up data bases, wrote and managed operations of computer systems.
Communication	Wrote reports and managed operations of computer systems. Documented List Management report processor for Highlands Insurance. Wrote formal technical report of 30 pp. on compact disk players. Courses: Technical Writing, Speech

EMPLOYMENT
    Montgomery High School District, Westfield, TX 9/88–present
    Central Bookkeeping Service, Bryan TX 1/86–8/88
    Highlands Insurance, Highlands, TX 10/85–6/86
    Institute for Science, Houston, TX 6/83–8/85

EDUCATION   BS: Mathematics and Computer Science
             Anticipated in December 1987
             Texas Southern University, Houston TX   G.P.A. 3.73

References available on request.

Figure 16-4 Analytical resumé for graduating senior

Advantages and disadvantages of an analytical resumé The advantages of an analytical resumé are that it

1. tells an employer what you think your greatest strengths and abilities are
2. relates specific skills to the job objective
3. allows you to combine educational, work, and unpaid experience in one section
4. may draw the reader's attention because it is not as common as the chronological resumé

One disadvantage of an analytical resumé is that its nontraditional organization makes it more difficult to write than a chronological resumé. Thus, if you do not clearly state your job objective and specify your skills, this resumé will not be effective. On the other hand, if you clearly support the job objective, an analytical resumé can be a powerful persuasive tool.

CHECKLIST FOR WRITING A RESUMÉ

1. Have I researched job opportunities and this company?
2. Which form of resumé is appropriate for this job?
   _____ chronological
   _____ analytical
3. Do I have complete information for each of these parts?
   _____ personal data
   _____ job objective
   _____ education
   _____ work experience
   _____ skills and abilities
   _____ references
4. Have I listed education and work experience in reverse chronological order?
5. Have I used active verbs and specific nouns?
6. Is my resumé easy to read? Does it lead the reader's eye to important information? Will it tell key information about me in 15–20 seconds?

WRITING ASSIGNMENT #9 Writing A Resumé

Directions: Choose a real job for which you are qualified or will be qualified when you graduate, and write either a chronological or analytical resumé that will support your application for that job. You can find job descriptions:

- in the classified section of your newspaper
- at your college career center
- by calling local companies

Make this a real resumé of your education, experience, and skills. Follow the suggestions in this chapter, but remember not to reproduce the forms shown exactly; your resumé must be distinctively yours. Keep the resumé to one page. Attach your Planning Sheet (see section 1.6) and the job description to your resumé when you submit it for review. If possible, meet with three to four other students for peer review of your resumé while it is in draft. Use their suggestions to revise it before submitting it to your instructor.

16.3 APPLICATION LETTER FORM AND ORGANIZATION

The letter accompanying a resumé is a standard business letter that follows the general organization of a letter of request (see Chapter 15). It should be only one page long, typed single-spaced on good quality bond paper, be well placed on the page, and have absolutely *no* errors, erasures, or whiteouts. Before you send your letter, have it edited by at least two people whose judgment you trust.

If you are responding to an ad or job posting, address your letter to the person and department indicated in the announcement. Otherwise, try to send your letter to the specific manager for whom you would be working rather than to the personnel department. You may have to call a company to learn the specific name and job title. If you must send your letter to the personnel department, obtain a specific name in that department. Do not simply address your letter to the corporation. Think about the reaction you have to a piece of mail addressed to "Occupant" as opposed to a letter addressed to you by name.

Like a letter of request, an application letter has three parts.

The Opening Paragraph

In the first paragraph you tell specifically what job you are seeking, including the title of the position and where you heard about it. You want this paragraph to attract the reader's attention and single your letter out from others, so you need to tailor it to the company and the job; you can't write a standard opening paragraph that will work for 20 different companies. Experts in job search recommend openers like these:

- If the position was advertised:

Your ad for the position of ＿＿＿＿＿＿＿＿＿＿ in ＿＿＿＿＿＿＿＿
is of special interest to me, and I would like to apply.

- If someone known to the reader told you about it:

Tom Jasper, Vice-President of Sales and Marketing at Starr Manufacturing, suggested that I contact you about the engineering aide position you have open in your quality assurance department.

• If you have already talked to this person:

As you requested in our telephone conversation on Friday, October 5, I am enclosing my resumé for your consideration. I feel my background in biotechnology could fill XXX Company's need for a laboratory assistant.

• If you are building on past skills and experience:

Having doubled the sales in my region during the past three years, I am looking for a challenge like the one offered by your organization. Your advertised opening for a sales manager with a record of increasing sales appears to be suited to my capabilities and interests.

• If you are prospecting for an unadvertised job:

Because of my three years experience as _____ , I believe I am qualified for the position of _____ at _____ .

or

Can _____ use someone with five years experience in designing _____ ?

or

I would like to apply for a position in _____ department with the goal of becoming _____ .

Middle Paragraphs

Paragraphs in the middle of the letter should highlight the abilities and skills that will be of most use to your employer. This information is already summarized in your resumé, but your letter should emphasize qualifications that fit this particular job at this particular company. For a new graduate, such qualifications might include courses of study, work experience, grades, or outstanding accomplishments. Try to see yourself as a potential employer would, and talk specifically about what you have to offer. Mention the company name in this section instead of just talking about "your company," and tell why you want to work there. One executive says (Rogala and Liswood 1980): "Make me feel like *I'm* your first choice, not one among fifty. The cover letter should be directed toward my company and its needs. Don't make it a fill-in-the-blank exercise." In other words, make sure that your cover letter *builds* on the resumé, tying the more general resumé to the very specific job.

The letter in Figure 16-5 was written by a person seeking to change careers (Schaeffer 1984). Notice how she relates the experiences in her background to the specific job she is applying for.

327 Romeo Avenue
Manchester, NH 03104
October 26, 1989

Ms. Ann Hartley
Personnel Specialist
CDS Co., Inc.
812 North Third Street
Buffalo, NY 14213

Dear Ms. Hartley:

I would like to become a writer in CDS's System Information
organization. My experience and competence as both a writer and a
user of computer manuals and programs should help me meet the
specific needs of your technical writing department.

As a documentation specialist, I prepared software manuals for a
new product by revising, updating, and editing four manuals
concurrently. I refined statements applicable to the old product
and, by using ICF, made them applicable to the new product. I
rewrote all statements referring to gender. I also inspected all
acronyms and created a policy for their uniform use in all manuals
in the library.

As a programmer, I wrote and documented programs for a
semiconductor firm, organizing instructions in a user—friendly
format using structured programming techniques. These programs
were effective in running the company's Production Plan Reports
and are still used.

As a procedures analyst, I wrote procedures and instructions for
data entry operators to edit and update sales information. I
conducted walk—throughs and supervised the trainees to ensure
comprehension of the procedures and correct application of the
instructions.

My practical experience supplements the theory I learned at
school in technical writing, technical editing, and printing
classes. As a result, I know when to research something, when
something looks questionable, and when something was treated
differently 200 pages earlier.

You will find more detailed information about my education and
work experience in the enclosed resumé. I can supply you with a
portfolio and letters of recommendation.

I am willing to relocate, and I can make arrangements for one or
more interviews at any of your locations. Please call me at (603)
342—6678 or write to me at the above address, and I shall be happy
to discuss my qualifications and their application to CDS's System
Information organization.

Sincerely,

Barbara Schaeffer

Barbara Schaeffer

Enclosure

Figure 16-5   Job application letter

The Closing Paragraph

In the closing paragraph you come to the point of the letter: your desire for a job interview. Make it easy for the reader to contact you by giving your phone number and times when you can be reached. (Your phone number also appears in your resumé, of course, but it pays to repeat it here.) Be specific but brief in this paragraph, and take care to project a tone of self-confidence without being pushy. Closing paragraphs might be written like these:

- I would very much like an opportunity to discuss my qualifications with you, and I will be available for an interview after June 7, 1989. Would you please contact me any weekday after 3:00 P.M. at (216) 282-4587? I am looking forward to hearing from you.

- I look forward to discussing with you my qualifications and the research department's career opportunities. I hope that you will find a match between your recruiting needs and my professional interests. To schedule an interview, please contact me at (515) 324-2919. You can reach me at this number on Tuesday and Thursday afternoons. Thank you for your consideration.

- If the experience listed above would be valuable to your department, please contact me for an interview. My telephone number is (713) 461-4130.

- I would like to join your test lab because I think I can make a real contribution. May I meet with you to discuss my qualifications in more detail? I will call you in two weeks to schedule an interview, or you can reach me at (404) 332-1180.

One of the biggest challenges in writing the application letter is telling about yourself without constantly beginning sentences with "I am" or "I have." While the examples above do not avoid this problem entirely, notice the techniques you can use:

- Begin with a clause or phrase.

If my experience fits your needs . . .
To contact me you can call . . .

- Shift from yourself as subject to your experience or education.

My classes in package design . . .
The experience in agricultural inspection . . .

- Concentrate on the company's needs.

Your job announcement asks for someone with experience in . . .
Paul Smith told me that the mapping department needs . . .

Figure 16-6 shows a job application letter from a student about to graduate from college. Notice how specifically he relates his experience to the needs of the employer.

202 Dentwood Drive
Jacksonville, FL 32209
March 10, 1989

Ms. Dawn Simonds
Personnel Director
Pyramid Advertising
832 Main Street
Jacksonville, FL 32207

Dear Ms. Simonds:

I would like to apply for a position in your art department with
the goal of becoming a graphic designer for major accounts. I am
currently taking courses to earn a Bachelor of Science degree in
Graphic Design in May 1989 at Jacksonville University, which has
one of the best graphic design programs in Florida.

Throughout my college education my grade point has averaged 3.6/
4.0. I have gained valuable experience by writing for my college
newspaper and by selecting materials and organizing page layouts.
For this work I received a certificate of Outstanding Achievement
in Journalism. Some of my courses at Jacksonville University
include advertising, illustration techniques, lettering and
typography, and printing techniques, including use of a process
camera.

My work experience covers many facets of design. At Rosen
Electric I used CAD programs for graphics, which would be helpful
should Pyramid Agency now or in the future have a computer
graphics system. As a retail sales clerk for Albright's I
analyzed the effectiveness of my department's advertisements.
Working as a free-lance designer, I have designed and printed
both greeting cards and business cards.

I have always been interested in the advertising business, and I
am confident that I would enjoy working for Pyramid. I am enclosing
a resumé, and at an interview will be glad to show you a portfolio
of my work.

Please call me at (904) 342-9876 to set up an interview. I look
forward to meeting with you.

Sincerely,

Mark Wolder

Mark Wolder

Enc.: Resumé

Figure 16-6   Job application letter

16.4 THE PROCESS OF WRITING RESUMÉS AND JOB SEARCH LETTERS

Planning

Before you write either a resumé or a job application letter, you need to spend some time analyzing yourself, your abilities, and your short-term and long-term goals—and how you plan to achieve those goals. Begin by listing 10 achievements that gave you both personal satisfaction and a sense of accomplishment. These achievements can come from work, school, or personal experiences. Then analyze them to see where your interests and abilities lie. Next, write down each of the jobs you have had, and list three to five specific things you accomplished on each job. Finally, on a separate sheet of paper write down personal considerations, asking yourself questions like these:

- Where do I want to live? Am I willing to relocate? to travel on the job?
- Do I want to work for a big corporation? a small company?
- Can I work at night? on weekends? more than 40 hours a week?
- What's most important to me in a job?

Answering these questions will help you determine what kind of job you want and will help you construct your resumé and job application letter.

Now is also a good time to begin a folder of information about yourself. Include your college transcripts, but add a list of your professors and the courses you took at all the schools you attended. Keep notes on each job you've had—your duties, your title, your supervisor's name and title, and the address of the company. Keep a list of the addresses where you have lived, and the time you lived there. Note organizations you've joined and the dates. In other words, keep track of all the things that happen to you. If you keep this folder current, you'll always have at hand the data needed for job applications, resumés, and security clearances. In the long run such a folder can save you considerable time and effort.

Gathering Information

The next step in the job search is learning about a prospective employer. Several months before you need a job, begin gathering information on companies. Attend job fairs at your college or in nearby cities and collect brochures from employers who attend. Find out what your college or university career planning and placement center can do for you; register for their training sessions and campus interviews. Research specific companies through company brochures and annual reports, or through articles in business and technical periodicals. Study the *Occupational Outlook Handbook* published by the U.S. Department of Labor to see where the jobs are expected to be. Begin to read newspaper ads, especially on Sundays, when many companies advertise heavily, to learn what kinds of jobs are being advertised in your field. When you read, remember that only 10 percent of available jobs are actually advertised, so other jobs are available. In addition to investigating the sources listed above, you should tell your friends, relatives, and acquaintances that you're looking for a job.

Organizing and Writing

Details on organizing and writing resumés and job application letters appear in sections 16.2 and 16.3. Be sure to infuse both documents with your own personality. Do not simply choose an organization or phrase from the samples and plug in your personal information. As you read the samples, you may be uncomfortable with the tone of some of them, and if you react this way, the tone is wrong for you. The way you write and organize resumés and application letters should reflect you at your professional best. That's why it's also crucial to avoid errors in punctuation, spelling, or content. Errors of any kind in these documents reflect badly on your competence. Job application letters should be individually typed. Buy a new typewriter or printer ribbon and use the best and clearest typeface you can find. Take care even with mundane matters like the envelope—make it accurate and neat. Remember that out of every 100 applicants, 99 don't get the job. The more careful you are with even little matters, the better your chances are of being the *one* that is hired.

Reviewing

Besides reviewing for accuracy, you should review your resumé and job application letter for accessibility, asking "How easy is it to find the important information?" Remember that what you submit will have many competitors, and you must make your point in 15–20 *seconds* of reading time. Plan for quick and easy reading by using white space, list markers, bold type, and underlining as highlighters. Ask friends to review these documents for you—reading very quickly at first to see if they find the important information, then reading more carefully to check for details.

16.5 ADDITIONAL JOB SEARCH LETTERS

Your part of the job search process is not finished when you send off your letter and resumé; often, you must write a number of other letters before you are settled into a new job. All these letters should be short, friendly, and polite, and they should be directed to specific, named individuals.

If you submit a job application letter and resumé and have had no response within three or four weeks, write a follow-up letter in which you mention your previous letter and the date you sent it, repeat your request for an interview, and enclose another copy of your resumé. One manager told me that he *expects* follow-up letters. He will not consider anyone who is not interested enough in the job to follow up on the original application. In order to do this, you will, of course, keep copies of all your letters and maintain a file of responses.

If you are granted an interview, respond immediately and enthusiastically. Then, within two days after the interview, write a brief thank-you letter to the person who interviewed you. If you had more than one major interview, write to each key person. Since only 1 percent of job seekers take the time to write thank-you letters, your letter will be noticed favorably. Use this letter to reinforce your value to the company and to follow up on any questions or comments that arose during the interview. Mention people's names and the company name. Repeat your interest in the job.

The easiest letter to write is the one accepting a job; don't neglect it. Again, be personal, enthusiastic, and convey your thanks, but don't gush. Confirm details such as your starting date and moving arrangements.

A harder letter to write is one in which you must reject a job offer or ask to be withdrawn from consideration. In your happiness at accepting one offer, don't risk antagonizing another company that has seriously considered you. Who knows—you may want to work for it in a few years. Thank the company and the individual for their time and consideration, and offer your regrets. Do not say anything negative about the company you are rejecting. You need not tell them what offer you have accepted; you can simply say, "I have decided to accept another offer." Close with an expression of goodwill. One type of rejection letter is shown in Figure 16-7, where the writer repeats his thanks for the interview, even though he had already sent a thank-you letter.

CHECKLIST FOR WRITING JOB SEARCH LETTERS

1. Have I determined the specific goals and skills to highlight in my application letter?
2. Have I studied the company so I know its background and needs?
3. Does my letter follow a standard format? Is it correct in every detail of spelling and punctuation?
4. Does the opening paragraph tell what job I am seeking? Is it tailored to the company?
5. Do the middle paragraphs highlight the skills and abilities that fit this particular job?
6. Does the closing paragraph motivate the reader to grant me an interview?
7. Have I made it easy for the reader to contact me?

```
                              3139 North 47th Street
                              Milwaukee, WI 53218
                              June 13, 1989

Mr. Donald Cook, Administrator
GRA Corporation
2486 Anderson Drive
Raleigh, NC 27608

Dear Mr. Cook:

     I was pleased to receive your offer of employment as an
electrical engineer with GRA. I greatly appreciate the
opportunity I had to visit your facilities, and I enjoyed talking
with members of your engineering departments. I have reviewed
your literature on work types, benefits, and salary, and found
them quite satisfactory; however, at this time I am unable to
accept your offer.

     Thank you for the interest you have shown in my career. You
may be sure that I will keep GRA in mind in the future.

                              Sincerely yours,

                              Robert P. Johnston

                              Robert P. Johnston
```

Figure 16-7   **Job rejection letter**

8. Do I need to write any of these other job search letters?
    _____ follow-up
    _____ response to an interview
    _____ thank-you
    _____ acceptance
    _____ rejection
    _____ withdrawal

EXERCISE 16-1   Analyzing Job Application Letters

Directions: Study job application letters A, B, and C for content, tone, and correctness. Be prepared to discuss any problems or strong points.

WRITING ASSIGNMENT #10   Writing a Job Application Letter

Directions: Following the form of a standard business letter, write a job application letter for a real job, either the job for which you prepared your

Letter A

Paul Hillyer
PO Box 1467
St. Louis MO

DEXI Corporation
Professional Staffing
5100 Henry Drive
St. Louis, MO

Dear Sir:

I am currently employed as a development engineer at the software division of Entry Company. My primary interest is in the development of software systems which involves a knowledge system programming, microprocessor, graphics, and/or computer network systems. I think that my experience is relevant to your needs, and that my record in adapting and learning would be valuable to you.

I have heard of the positive working environment at your company and would appreciate being considered for a challenging position allowing further career growth.

If you have any questions, please contact me at (314) 442-9012 during normal business hours, or leave a message at (314) 342-3254.

Thank you very much and I look forward to hearing from you.

Yours sincerely,

Paul Hillyer

Paul Hillyer

Letter B

```
Jackie Ash
113 Wanton Way
Pittsburgh, PA

May 15, 1989

To Whom It May Concern:
I would really like to work there. The people I met on the
tour were very nice, plus I'd like to know more about your
company. I would like a job researching or technical
writing. I can write and I can do research if you tell me
what it is that you're researching. I can get along with
people.

I can work 8 to 5, and I sort of like to have weekends and
holidays off. Actually I will take any job you have open,
because the bottom line is, I really need the money.

I will send you my resume upon request.

                        Thanking you in advance,

                        Jackie Ash
                        Jackie Ash
```

resumé or another job advertised or listed at your career center. Do not copy the sample letters in this chapter. Your letter must sound like you and reflect your skills and experience. Submit the letter for review in draft form to a small group of your fellow students. Listen carefully to their comments and revise your letter as needed, concentrating on specific content, effective tone, and correct form.

Letter C

```
4356 64 Street
Milwaukee, WI 53102
January 5, 1989

A. P. Washowicz
Manager, Production
Clean—Line Research, Inc.
875 North Avenue
Milwaukee, WI 53108

Dear Mr. Washowicz:

I am responding to your ad in the Milwaukee journal on
January 3, 1989, for an assistant engineer on a research
vessel. I will soon graduate with a degree in mechanical
engineering, and I have had experience working with both
Caterpillar diesel engines and hydraulics.

Although I don't have the three years experience you ask
for, I am a hard worker and learn quickly. I took college
classes in hydraulics and fluid mechanics. My resumé gives
details.

Please contact me at the above address for an interview.

Sincerely,

J. Dolan
```

REFERENCES

Fuller, Michael. 1987. Adapted from a resumé supplied by Sue Birdwell, Xidex
    Corp.
Magnuson, William. 1986. Adapted from a resumé written by Steve Teraji, San
    Jose State University.
Martinez, Julie. 1986. Adapted from a resumé written by Lina Melkonian, Career
    Planning and Placement, San Jose State University.
Nelson, Pamela. 1986. Adapted from a resumé written by Pamela Nelson, San Jose
    State University.
Rogala, Judith and Laura Liswood. 1980. The briefcase. *San Jose Mercury*. 29 July.
Schaeffer, Barbara. 1984. Adapted from a letter written by Barbara Schaeffer, San
    Jose State University.

CHAPTER 17

Instructions and Manuals

How many times have you clenched your jaw in frustration while trying to assemble a new bookcase, clean the gunk from your coffeemaker, or figure out your income tax? Nearly all of us have been victims of bad instructions at some time, so we realize how crucial good instructions are and how carefully they must be written. Every day you follow instructions of some kind—whether at home, at work, or at school. You try a new recipe for Chicken Marengo; you install a new oil filter in your car; you register for an evening class; you fill out those income tax forms. In each case you perform a series of actions in a specific sequence in order to complete the task successfully, and most of the time you learn what you must do from written instructions. As a form of teaching, instructions are one of the most important kinds of technical writing.

When you teach, you tell a reader either how to *do* something (instructions), or how something *happens* or *is done* (procedures). Both ways of "telling how" are sequentially organized: one step follows another in the order of performance. But notice the difference in purpose: instructions *teach* performance of specific activities, and procedures *give information* about ongoing or repetitive activities. For example, you might write instructions for properly disposing of organic industrial wastes like oil, toluene, and benzene; in this case the reader will be carrying out the disposal. You could also write a procedural description of how such wastes are generally handled; in this case the reader simply wants information and is not going to do the job. Chapter 18 deals with procedures. This chapter explains instructions; in a final section it also explains how instructions and other techniques and forms are combined in manuals.

17.1 DEFINITIONS OF TERMS

Instructions are a form of writing telling a reader how to do something—either how to find a place or person (also called "directions") or how to perform some physical operation (also called "orders"). Instructions can range from three or four simple lines to huge volumes dealing with operation, maintenance, and repair of complex machines. Figure 17-1 is an example of a simple but useful set of instructions included as part of an advertising brochure for radial tires. In contrast, think of the detailed sets of instructions needed to operate and maintain an industrial complex like an oil refinery. In a complicated operation like that, each task must be clearly identified and then subdivided into component parts, and instructions must be designed for each part. The needed instructions will fill many volumes, and parts will go to different types of readers. Because of their distinctive form, simple instructions are often found in a separate document like a brochure or booklet. Complex instructions form the major part of manuals for installing, testing, operating, maintaining, and repairing equipment.

Readers come to instructions seeking direction and advice; they want to know how to do a job. Usually they are impatient to begin working and won't take time to read the instructions from beginning to end before they start. What's more, the saying "When all else fails, read the instructions" is too often true; as a writer you must plan for spot reading, hasty reading, and reading after the fact (when the reader has already done part of the job and may be in trouble). With that in mind, this chapter suggests ways to help you organize and write good instructions. Because they are often read *while* the reader is performing the action, instructions must be carefully organized, easy to follow, and very clear. Those requirements dictate the form used in successful written instructions.

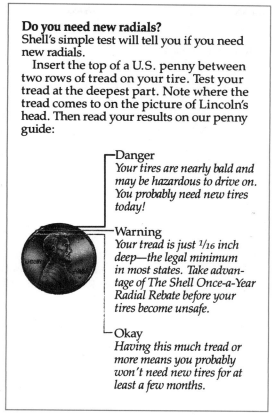

Do you need new radials?
Shell's simple test will tell you if you need new radials.

Insert the top of a U.S. penny between two rows of tread on your tire. Test your tread at the deepest part. Note where the tread comes to on the picture of Lincoln's head. Then read your results on our penny guide:

Danger
Your tires are nearly bald and may be hazardous to drive on. You probably need new tires today!

Warning
Your tread is just 1/16 inch deep—the legal minimum in most states. Take advantage of The Shell Once-a-Year Radial Rebate before your tires become unsafe.

Okay
Having this much tread or more means you probably won't need new tires for at least a few months.

Figure 17-1   Simple instructions

17.2 THE FORM AND ORGANIZATION OF INSTRUCTIONS

Good instructions generally consist of two parts: (1) an overview, and (2) a set of individual steps that follows a sequential method of organization. The amount and variety of detail in the overview will vary depending on the reader's expertise and the complexity of the operation. The detail given in each step will also vary depending on whether the reader is doing the task for the first time or has performed similar tasks in the past.

The Title and Overview

The title defines the operation and its purpose, so it should include the words "How to" or "Instructions for" and the name of the operation. The title should be specific enough that readers know at once they have the right set of instructions. Titles might read

> How to Assemble the Easy-Gro Spreader
> Cleaning Instructions for Automatic Coffee Makers

An overview serves as an introduction to the instructions. Readers *should* study both the overview and the detailed steps before they begin, but many of them ignore the overview and begin immediately with Step 1. Sometimes they get in trouble doing this, but they seldom blame themselves; instead, they blame the "lousy instructions." Your job as a writer, then, is to make overview information instantly accessible to impatient readers.

All overviews should include these four items:

1. an opening statement that, like the title, identifies the purpose and content of the instructions. An opening statement (like this one from IBM 1981) might read

> The Hardware Maintenance and Service Manual is the publication you use to isolate and repair any failure of a Field Replaceable Unit (FRU).

2. a list of the materials needed. Materials are the items that will be used up while the job is being done, such as paint, oil, nails, and plastic tubing.

> For cleaning wood furniture you will need:
>
> • gum turpentine
> • boiled linseed oil

3. a list of the tools or equipment required. Tools and equipment are items that are reusable and transferable from one job to another.

> Tools needed to hang wallpaper:
>
> **1.** smoothing brush   **5.** blades
> **2.** seam roller   **6.** plumb line
> **3.** pasting brush   **7.** trim guide
> **4.** razor knife

4. a summary of the steps involved or a statement of the number of steps.

> X-ray film developing includes these five major steps, all performed in dark-room conditions:
>
> **1.** developing
> **2.** rinsing
> **3.** fixing
> **4.** rerinsing
> **5.** drying

You can help the reader keep on track by numbering the steps in the overview and then repeating those numbers as you explain the individual steps. You can also help the reader by using the same grammatical form for naming each step—the technique called parallelism described in detail in section 6.2.

Some instructions will need other information in the overview, such as:

- warnings and precautions that apply to the whole operation
- definitions of terms, especially for the first-time operator
- discussion of the skills or expertise needed and, for beginners, perhaps assurance that the job can be accomplished with minimal skill
- directions on how to organize materials and set up equipment
- special conditions required, such as temperature or humidity range or a dust-free environment
- the time required to complete the whole task or certain segments of it
- troubleshooting information
- for very formal instructions, the theory of operations

The Individual Steps

The individual steps are the core of a set of instructions. They should be in sequential (chronological) order—that is, in the order in which they must be performed. If there are many steps, you should group related items under fewer than 10 umbrella terms so that the reader's short-term memory is not overwhelmed. (See section 3.2 for more on short-term memory.) Mark each major step by letters or numbers, by bullets, by paragraphing, or by the use of boldface type. Letters and numbers are best because they help the reader follow the sequence. Start each step with the verb that tells the reader what to do. For example (Thames n.d.):

To remove old finish:

1. *Place* the piece of furniture in a horizontal position so remover won't drip or run down.
2. *Apply* remover to one small section at a time—a leg, rung, or table leaf. *Protect* other parts with newspaper.
3. *Apply* a heavy coat of remover, brushing in only one direction if you use a wax-containing remover.
4. CAUTION: Immediately *wipe off* any remover that drips on parts, or stubborn spots will form.

Notice that the last instruction begins with the word "Caution." Precautions warn of possible damage, and they should be inserted wherever the reader needs to be alerted. The two terms used to indicate precautions are

WARNING: Given to show danger of injury or death to the operator or others nearby

CAUTION: Given to prevent damage to equipment or materials

Highlight warnings and cautions by boxing them or using bold or red type to attract attention. If you box the precautions, center the word "WARN-ING" or "CAUTION" above the entry. For example (Varian 1986):

WARNING

The internal cryoarrays remain cold for some time after power is turned off. They could cause severe frostbite if handled when cold. Never disassemble the pump and handle unless you have determined they are at room temperature.

CAUTION

Be sure not to bend the capillary tube at the braze joints when cleaning the hydrogen vapor bulb assembly because the tube will break easily.

In addition, you can use a heading called a Note to explain or add information. This heading is usually put at the left margin. For example,

NOTE: Both rods are inserted from the same side of the bracket.

Help the reader by illustrating your words with plenty of graphics: photographs, exploded drawings, line drawings, and the others explained in Chapter 12. Place the graphics next to or right after the explanation, and choose close-ups that show someone's hands performing the operation if possible. Study the instructions in Figure 17-2, written for lab technicians. Notice how much the simple line drawings help explain the steps. Also notice how substeps are grouped under the 10 major steps.

Lay out the steps on the page for easy readability. Don't try to save space by jamming them together; use white space as a frame around each step, making the steps easy to find when the reader is in the middle of the operation. Use a type size that is large enough to read easily.

For clear instructions, write in the present tense, using the second person (you), active voice, and imperative (command) mood. That way you will be talking directly to the individual reader in a straightforward way. Imperative mood means that, as writer, you are giving orders, not information.

COMMAND: Remove the four screws on the battery case.

INFORMATION: The four screws on the battery case are removed.

or

Installer removes the four screws on the battery case.

*M*icro**Trak**™

Chlamydia trachomatis
Culture Confirmation Test

Test Procedure—
Coverslip Technique

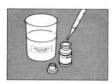

1 Reconstitute reagent
- Remove seal and rubber stopper
- Add 3.0 ml deionized* water
- Replace stopper
- Swirl gently to dissolve
- Equilibrate at room temperature for 15 minutes

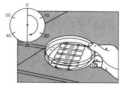

6 Incubate
- Incubate at 37° C for 30 minutes in **moist** chamber

2 Fix monolayer
- Remove medium from vial
- Immediately add ethanol
- Leave for 1 to 10 minutes
- Aspirate ethanol

7 Aspirate excess reagent
- Remove from incubator
- Aspirate excess reagent from monolayer

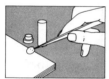

3 Remove coverslip from vial
- Remove coverslip from vial using forceps
- Touch edge of coverslip to blotting paper

8 Rinse monolayer
- Remove coverslip from slide using forceps
- Rinse in deionized* water for 10 seconds
- Touch edge of coverslip to blotting paper

4 Place coverslip on slide
- Place coverslip cell-side up on labeled slide

9 Mount specimen
- Place drop of mounting medium on clean, labeled slide
- Place coverslip, cell-side down, on top of drop

5 Stain monolayer
- Moisten cells with deionized* water or PBS (if dry)
- Blot excess
- Add 30 μl of reagent to monolayer
- Ensure entire area is covered

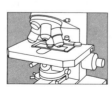

10 Examine
- Examine monolayers using fluorescence microscope

*or distilled

Figure 17-2 Instructions with line drawings for technicians

Note too that you can help the reader by carefully specifying "the four screws," indicating that there are only four screws. If you said "remove four screws," the reader might wonder if there were others not to be removed.

17.3 THE PROCESS OF WRITING INSTRUCTIONS

Planning

Before you can write instructions, you must understand clearly what readers will be able to do after reading what you write. Clarify this for yourself by writing either (1) a purpose statement or (2) a list of performance goals. You can also use these statements in the overview section of your instructions.

PURPOSE: to explain how to properly unpack and install **XXX** brand microwave oven

PERFORMANCE
GOALS: After reading these instructions, the reader will be able to

1. unpack the oven from the shipping carton
2. safely install the oven in a desired location
3. unlock the oven scale
4. install the glass tray
5. hook up the oven to grounded electrical service

Notice that by thinking through performance goals, you also begin to break the task into its component parts.

Next, learn as much as you can about your potential readers, because reader background and type will determine the vocabulary you can use and the skill level you can expect. Ask: What do they already know? What skills do they have? What terms do they understand, and what terms must be defined? Do they want to be told every small detail, or do they simply want an outline of the general steps? Once you've characterized your readers, back up one step from the knowledge you think they have, and assume slightly *less* knowledge and skill. Those readers who are proficient will skip your added explanation, and those who need it will be grateful to have the details.

For instructions, the reader's purpose is clear: readers want to be able to perform the action. Many instructions are written for generalists and first-time users and you must explain any unfamiliar terms. In Figure 17-3, for example, you can see part of a glossary included in a brochure on how to hang wallpaper written for generalists. With these definitions, readers will more easily understand the individual steps.

Such instructions must also specify all the tools, equipment, and

Note: The term "wallcoverings" in today's usage refers to any materials used to cover walls. "Wallpaper" is a traditional term used only for paper wallcoverings.

Adhesive — A paste for applying wallcovering to a wall.

Backing — A fabric, paper or synthetic material to which wallcoverings are laminated for strength and support.

Booking — The technique of folding the top and bottom of a wallpaper strip to the center, paste side to paste side. This allows the material to "relax" for several minutes so that it can assume its final dimension from the water in the paste. For booking time, refer to manufacturer's instructions.

Broad Knife — Wide putty knife used as a trim guide and also for scraping old wallpaper from walls. It may be used to spread spackle in order to fill holes or cracks in walls.

Butted Seam — To "butt" a seam, edges of wallcovering strips are fitted edge to edge without overlapping.

Deglossing — The breaking down of a glossy painted surface so that wallcovering will adhere to the surface properly.

Drop Cloth — A plastic sheet used to cover furniture or flooring for protection during pasting and papering.

Flock — A wallcovering with a velvet-like surface design.

Grasscloth — A handmade wallcovering, made by gluing woven grasses onto a paper backing.

Gre-Sof — A product used primarily for cleaning surfaces before refinishing or applying wallcoverings.

Figure 17-3 Glossary of terms needed for instructions

materials needed to perform the operation, so beginners will have everything they need before they start. If you leave out a piece of equipment, the reader can justifiably call your instructions lousy.

Experienced readers—those who have performed similar operations—already know the basic equipment needed and are familiar with certain steps of the operation. They do not want definitions and excessive detail. The set of instructions in Figure 17-4 is written for technicians, and the writer assumes their familiarity with the parts of the pilot generator and the tools needed to perform the test. These instructions would not be useful to a person who did not understand pilot generators. Thus, if you're unsure of the expertise of your readers, write for generalists and give them plenty of detail.

Remember to limit your topic in writing instructions. You need not describe the object, give theory, or even explain why the operation is carried out in a certain way. This type of added information clutters the instructions. The one exception is safety; it pays to explain *why* readers must do (or not do) something.

FIELD TEST FOR SAFE IGNITION
(Turn Down Test)

WARNING

WITH PILOT GAS REDUCED TO LOWEST POINT WHICH WILL GENERATE THE MIILLI-VOLTAGE REQUIRED TO OPEN VALVE, THE MAIN BURNER MUST LIGHT OFF SMOOTHLY. IF IGNITION IS DELAYED DURING TEST, IMMEDIATELY SHUT OFF GAS. WAIT 5 MIN. BEFORE CONTINUING TEST.

1. Disconnect one pilot generator wire from valve terminal. Jumper thermostat terminals of valve.

2. Open pilot gas valve. Light pilot and adjust for MAXIMUM pilot flame. Check appearance to be sure flame is in proper position to ignite burner.

3. Open main burner gas cock. Touch and hold the loose pilot generator wire to valve terminal. Burner should light off smoothly within a few seconds.

4. Reduce pilot flame by adjustment valve until flame around cartridge is about 1/2 of maximum. Wait two minutes for generation to stabilize. Touch and hold the loose pilot generator wire to valve terminal. If valve opens, burner should light off with no delay. Again reduce pilot gas flow slightly, wait two minutes, and test. Repeat until pilot flame is too low to produce sufficient millivoltage to open valve. Reposition pilot generator if necessary.

5. Remove jumpers from valve and reconnect all wires in their original positions. Make all wiring connections clean and tight. Readjust pilot gas flow for blue, non-blowing flame surrounding the generator cartridge. Be sure the main burner flame does not hit generator cartridge or snorkel.

Figure 17-4 Instructions for technicians

CAUTION
DO NOT OPERATE THE MICROWAVE WHEN IT IS EMPTY.
Microwave energy can damage the magnetron tube if the oven is operated empty for an extended period of time.

Even if you are writing very simple instructions for first-time users, your tone should be courteous and objective. Don't talk down to your readers or try to impress them with big words. In other words, don't let how you say it obscure what you say. To help you plan good instructions, use a Planning Sheet like that in section 1.6.

Gathering Information

You need to know and understand the operation yourself before you can write good instructions. This means you must find and read specifications, descriptions, and procedures. As you read source material, determine the purpose of the source; for example, differentiate between general instructions (how to change the oil in an automobile) and specific instructions (how to change the oil in a 1987 Ford Escort). If you are writing general instructions based on those specific to a certain product or piece of equipment, you will have to adapt and revise details. If you are using instructions written for a product or operation similar to your model, don't simply copy that form and organization and plug in your own information—tempting as that might be. Think through the purpose, define the reader, and make sure that the approach is good before you imitate it, and even then, be careful not to plagiarize. Don't use the same wording or major ideas and pass them off as your own.

Besides written material, a good way to get the information you need is by interviewing experts—people who have designed or tested the equipment or are otherwise familiar with the operation. The most important and best way, though, is to observe, examine, or actually perform the operation yourself.

Organizing

Instructions are relatively easy to organize because the individual steps follow a sequential order; that is, what should be done first, what next, and so on. The trick is to *observe* the actions and anticipate what the reader must know before starting each one. Be careful to include all steps, and put them in the right order. You should not send the reader forward and then backtrack.

> NOT: Apply two coats of thin sealer 24 hours apart. Smooth with 3/0 steel wool after the last coat has dried 24 hours.
> INSTEAD: Apply one coat of thin sealer. Wait 24 hours, and then apply a second coat. Wait 24 hours; then smooth the surface with 3/0 steel wool.

Writing, Reviewing, and Editing

Make your instructions easy to follow by keeping items of equal importance parallel—as in Figure 17-5.

Use numbers, letters, or subheadings to identify each major section, and begin with the verb indicating the action. As you write each step, think about the best way to illustrate what you are saying, and plan for graphics as you go. Notice how the exploded diagram in Figure 17-5 makes the parts and their relationships clear.

In summary, follow these techniques to communicate most directly and clearly to the reader:

1. Write in the present tense and the imperative mood, usually beginning each step with the verb that tells what to do. Write directly to the reader.

CS 8 FA CRYOPUMP OPERATOR'S MANUAL

6.5 PRESSURE RELIEF VALVE CLEANING/REPLACEMENT

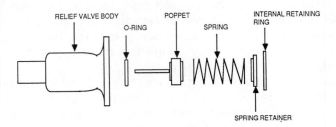

1. Remove the internal retaining ring located inside the relief valve body. Be careful not to lose parts as the assembly is spring-loaded.

2. Remove the spring retainer, spring, poppet valve and O-ring from the valve body.

3. Carefully remove the O-ring from its groove in the poppet valve.

4. Inspect the O-ring and discard if damaged.

5. Thoroughly clean the inside of the relief valve body making sure all surfaces are free of foreign matter and particulates.

6. Remove and clean the screen on the inside of the pump body. It is held in place by friction and should be replaced if damaged.

NOTE: The poppet pin guide and its retaining ring on the inside of the pump relief valve should not normally require removal. Visually inspect and replace if damaged or foreign matter is lodged around this area.

7. Reinstall the inside screen.

8. Apply a light film of vacuum grease to the poppet O-ring.

9. Install the O-ring on the poppet valve making sure it is in its groove and is not twisted or otherwise misapplied.

10. Attach the spring to the back side of the poppet valve so that the spring is held on the poppet recess groove.

11. Place the spring retainer (lip side toward spring) on the spring and insert the assembly in the relief valve.

12. Install the internal retaining ring into its groove in the relief valve body.

END

Figure 17-5   Parallel construction in instructions

2. Use the active voice so that the subject of the sentence is also the doer of the action.

SAY: Tighten the three wing nuts securely.
NOT: Wing nuts are tightened securely.

3. Alert the reader to contingencies or conditions by putting that information first in the sentence and highlighting any warnings.

WARNING: If the burner does not ignite within 5 seconds, immediately shut off the gas.

To distinguish the top of the panel from the bottom, locate the two holes drilled for casters in the bottom edge.

This construction puts the contingency at the beginning as a warning and the command itself near the end where it receives emphasis.

4. Give your instructions a clear title that identifies them as instructions and tells what the reader will learn. (See section 6.1 for more on writing titles.)

5. Do not omit articles like "a," "an," and "the" in order to save space. Articles are important to help the reader understand what you are saying. Table 17-1 summarizes these points.

Table 17-1   TECHNIQUES FOR WRITING INSTRUCTIONS

Writer's purpose	Reader's purpose	Method of organization	Tense	Person	Voice	Mood
to tell how to do some action	to be able to perform some action	sequential	present: "Open the valve."	second: "You attach the heat shield to the vent."	active: "Insert the filter."	imperative (command): "Cover the surface."

The best way to review your instructions is to have a typical member of your target audience perform the steps as you have written them. Watch the operation in progress, noting when the tester backtracks, hesitates, or raises questions. At this time do not give oral explanations to supplement your instructions; instead, note that what you have written is unclear. Later ask your tester where the problems occurred. Revise the instructions to clarify murky points. Then find *another* tester to try out the new version. Follow the same procedure to clarify and rewrite.

CHECKLIST FOR WRITING INSTRUCTIONS

1. Have I given my instructions a clear title that identifies them as instructions and tells what they are for?
2. Have I divided the instructions into an overview and individual steps?
3. Does my overview include
   _____ a list of materials needed?
   _____ a list of tools or equipment required?
   _____ a summary of steps involved or statement of the number of steps?
4. Do I need to include
   _____ warnings and precautions applying to the whole operation?
   _____ definitions of terms?
   _____ a discussion of skills needed?
   _____ directions for organizing materials and setting up equipment?
   _____ a list of special conditions required?
   _____ the time needed for completion?
   _____ troubleshooting information?
   _____ a theory of operations?
5. Is my breakdown of the individual steps
   _____ arranged chronologically?
   _____ organized by numbers, letters, or subheads?
   _____ written in parallel structure?
6. Do individual steps
   _____ give warnings or cautions where needed?
   _____ use graphics to clarify?
   _____ stand out by the use of white space?

EXERCISE 17-1   Analyzing Simple Instructions for Generalists

Directions: Sometimes you can best see the problems in instruction writing by analyzing instructions written for simple operations that most people know how to do and take for granted. The following examples come from an assignment (Shanley 1986) to tell "How to Make a Peanut Butter and Jelly Sandwich." Parts of five student responses are shown. Analyze each example to determine its good and bad points, and be prepared to discuss your conclusions in class.

Example 1

3. Spreading the filling (filling = peanut butter and jelly)
   Take a knife and dip into the peanut butter jar. Extract a wad of peanut butter about 1/2 as big as your fist. Place the wad in the middle of one of the slices. Take the knife and spread the peanut butter over the center

toward the edges. Apply only enough pressure to spread the peanut butter without tearing the bread. Spread it so that the thickness of the peanut butter is about the same at the edges as at the center of the bread . . .

Example 2

Stick the knife in the peanut butter and bring out some peanut butter on the knife. Spread the peanut butter on one side of a slice of bread. Put enough on the bread until the desired amount. Clean the knife. Put the knife in the jelly and bring out some jelly. Put this jelly on one side of the OTHER slice of bread. Spread jelly until desired amount. Take both slices of bread and put them together so that the peanut butter side and the jelly side of the bread touch. The tops of the bread should be together, but if they aren't, the sandwich should still be good.

Example 3

VI. Hold the knife at the dull end and dip the sharp end into the peanut butter jar and into the peanut butter.
VII. Remove about 4 handfuls of peanut butter and spread on bread.
VIII. Repeat steps VI through VII with the jelly.
IX. Now that the peanut butter and jelly are on one or both slices of bread, put two pieces of bread together with peanut butter and jelly in between.

Example 4

A. Open Bread and Remove Two Slices
   1. This step is to prepare the bread used to contain the peanut butter and jelly. Opening the bread refers to opening the wrap that keeps the bread fresh.
   2. To open the bread wrap, one must have the ability to remove the fastener that holds the bread wrap closed. Materials include bread wrapper and a plate. The plate is used to hold the bread and to contain any crumbs that may result from handling of the bread.
   3. Removal of Bread
      (a). Holding the bread loaf in the left hand, grasp the fastener at one end of the wrapper and remove.
      (b). Be careful that the bread does not spill out of the wrapper.
      (c). Reach into the open wrapper and grasp two (2) slices of bread.
Note: If you skip the end pieces of the loaf, your bread will remain fresh in storage.
      (d). Pull the bread through the open end of the wrapper and place on the plate in a side-by-side orientation . . .

Example 5

I. Overview
   A. Definition. A peanut butter and jelly sandwich is a combination of peanut butter, jelly, and bread. By definition sandwich means to enclose, to be wedged between, to encase. In this case, the peanut butter and jelly are encased by two slices of bread.

   B. Intended Reader. These instructions are for the person who wants to
      make a peanut butter and jelly sandwich.
   C. Knowledge and Skills Needed
      1. How to use a knife to spread peanut butter and jelly.
      2. How to unscrew a cap-jar lid.
   D. Brief Overview of the Steps
      1. Peanut butter is removed from its container and spread on bread
         with a knife.
      2. Jelly is removed from its container and spread on bread with a
         knife.
      3. Sandwich is assembled.
II. Steps in Making the Sandwich
   [Steps are detailed in this section.]

EXERCISE 17-2 Writing Good Instructions for Generalists

Directions: Write a set of instructions for one of the following activities.
Be sure to follow the form and organization recommended in this chapter.
Specify a purpose and reader.

how to wash a car	how to iron a shirt
how to cut up a chicken	how to load a dishwasher
how to trim a Christmas tree	how to change linens on a bed
how to fillet a fish	how to carve a jack o' lantern
how to build a campfire	how to shovel snow
how to split wood	how to prune a fruit tree
how to balance a checkbook	how to wean a calf
how to get from where you live to your technical writing classroom	how to sew a button on a shirt

EXERCISE 17-3 Analyzing Instructions for Customers

Directions: Analyze the following instructions written for customers of a
large wholesale warehouse, and write a paragraph or two telling specifically what is good and what needs to be rewritten. Then rewrite as needed.

PROCEDURES FOR HANDLING SHORTAGES OR DAMAGES

SHORTAGE AT TIME OF DELIVERY
A. The driver will issue a credit request for merchandise that is verified by
   the driver as being short.
B. The driver may issue a credit request for average case cost for merchandise
   shortage which is not specifically identified. Credit will be issued for the
   cost of an average case of merchandise. After this credit is issued there
   will be no further adjustment made.
NOTE: The driver must report any shortage in excess of five (5) cases to our
office immediately.

DAMAGE AT TIME OF DELIVERY
A. The driver will issue a credit request for identified damage only at time of delivery.
B. The driver will issue a credit request for damaged units within a case, rather than picking up the entire case.

DAMAGE AFTER DELIVERY
A. Merchandise found damaged after the time of delivery is returnable only if the case label is intact and the merchandise is in its original case. There will be no credit given for units which have been price marked or damaged at the store. The code number, invoice number, and the date on the case label must be on the request for pickup of damaged merchandise. The damaged product will then be picked up upon the next delivery.
B. Items over 30 days old will not be picked up.

EXERCISE 17-4 Writing Instructions for a Paper Airplane

Directions: Write a set of instructions telling a fellow student how to make a paper airplane. In this case, do not use graphics but rely on the written word to convey your ideas to the reader. Limit your materials to basic school supplies such as regular bond paper, a ruler, a paper clip. Make two copies of the instructions. Turn in one copy to your instructor and exchange the second copy with a classmate. On a launch day designated by your instructor, you will attempt to fly your classmate's airplane. Before that, you should follow the instructions to create the plane, and keep track of any problems you have. Write a memo to the writer evaluating the instructions; point out good points and any problems. Turn in a copy of the memo to your instructor.

WRITING ASSIGNMENT #11 Writing Instructions in Your Major Field

Directions: Choose an operation in your major field of study that you know how to perform well. Write a set of instructions to tell a generalist how to perform this operation. Use source material, but do not simply copy someone else's information. Exchange instructions with a classmate. Then write a memo to your instructor explaining the good and bad points of the instructions you read. Turn in a copy of the instructions along with the memo.

17.4 COMBINED FORMS IN MANUALS

As a person involved in a technical or scientific field you will spend much time reading manuals—and perhaps writing them—because manuals provide the factual and theoretical backup for machines and procedures. Manuals answer questions like

- How does it work?
- How do I use it?
- What are its parts?
- How do I install it?
- How do I maintain it?
- How can it be fixed?

You may be familiar with the owner's manuals for your automobile, your stereo, or your household appliances, and you may realize that these manuals are written for generalists. The owner's manual for your car, for example, tells you how to operate the heater and air conditioner, and gives you information about tire inflation and oil changes. But much about your car is not in that manual; in fact, a dozen or more other manuals probably exist to answer all the questions that designers, installers, and mechanics might pose.

Definition of a Manual

A *manual* is a document that gives both information and instructions. Many of the writing techniques discussed separately in this book come together in manuals: they usually combine definitions, descriptions, analyses, and explanations (all of which give information) with instructions and procedures (which tell how to do something or how something works). They are written for many purposes.

Parts of Manuals

In the past, manual writers usually assumed a common background and vocabulary among their readers, including knowledge of industry or trade jargon and basic concepts. Today, however, a common background can no longer be assumed because of the increased numbers of "do-it-your-selfers" and the movement of high-technology equipment like computers into the general marketplace. Now manuals must be written for first-time users and generalists as well as for experienced operators, technicians, and specialists. Thus, many manuals must provide (1) instruction for novice or first-time users—or "tutorials," (2) instruction by demonstration for more experienced users—sometimes called the "cookbook" approach, and (3) reference information for both that is presented in a compressed and easily accessible manner.

Tutorials   Tutorials take the reader by the hand, defining terms, clarifying concepts, describing and illustrating every move, and explaining the result of each action. Tutorials also provide positive feedback to assure novices that they can succeed and that the operations, if taken a step at a time, are not difficult. In other words, tutorials begin "at the bottom" with elementary concepts and definitions and such simple operations as how to turn on a machine. Usually a tutorial will take a user through a sample or typical operation from beginning to end to illustrate the procedure, but will not go into all the possibilities of that procedure.

Demonstrations   Experienced operators or installers will be frustrated by a tutorial's detailed approach; they prefer a demonstration of a whole process "from the top down." This group of readers already knows in general what to do and wants instructions in what is sometimes called the "cookbook" format. If you think about how most recipes are written, you can understand the difference between a tutorial and a demonstration. A recipe written for an experienced cook assumes a great deal of kitchen knowledge. For example, consider these instructions: "Add two egg whites, beaten stiff but not dry." Such an instruction assumes that the reader knows how to separate the yolk from the white, how to beat eggs (with a spoon? fork? wire whisk? electric beater?), and what "stiff" means as opposed to "dry." By eliminating the details that such explanations would call for, demonstrations can cover much more material and make that information more available to the experienced reader. Many successful manuals are now "task-oriented"—organized, in other words, not by the parts of the machine, but by the specific tasks the reader must accomplish.

Reference   A good manual should also make it easy for the reader to find information about specific topics—information that the reader may want out of sequence from the steps in a procedure. You can provide such information in many ways; for example, with

- a detailed index
- tabs or dividers for quick access to sections
- tables and other kinds of graphics like graphs and exploded drawings
- lists
- a detailed table of contents

In writing reference information you need to put yourself in the reader's place and ask where you might look for specific information and under what kinds of headings. Then you need to make sure that information is in the right location and cross-referenced if necessary.

Graphics in Manuals

Good manuals use graphics extensively because graphics can condense information and show relationships clearly. Because manuals combine several kinds of writing, different sections will also have different purposes. Thus, you need to understand the special purposes of particular graphic aids and choose accordingly (see Chapter 12).

The challenge in manual writing is to meet all reader needs in one compact book: to provide tutorials for the novice, demonstrations for the experienced user, and easily accessible reference material for both. If you have a manual to write, your first task is to define the purpose and determine your readers. Examine existing manuals both from your company and other companies, but don't automatically follow the format they use. Respond to the unique needs of your particular manual and reader, using the techniques you have learned in this book to write clear definitions, descriptions, explanations, instructions, and procedures.

EXERCISE 17-5   Analyzing a Manual

Directions: Examine a manual from your major field or from the general consumer field. Analyze how well it

- defines terms
- describes objects and procedures
- gives instructions
- helps the novice user
- helps the experienced user
- provides reference information

Write a short evaluation report and be prepared to discuss strengths and weaknesses of the manual in class.

REFERENCES

IBM. 1981. *Hardware maintenance and service manual.* No 602S075. Personal Computer Hardware Reference Library. New York: IBM Corp.

Shanley, Kim. 1986. Parts of student responses to a classroom assignment. San Jose State University.

Thames, Gina. n.d. Adapted from *Furniture restoration.* New York: Cornell University Cooperative Extension Service.

Varian Associates. 1986. *Cryopump operator's manual for CS 8 FA.* No. 87-400437. Santa Clara, CA: Varian Vacuum Products Div.

CHAPTER 18

Procedures

When a famous professional quarterback has surgery to repair a ruptured disk, the news media explain how such surgery is carried out—complete with diagrams of the spinal column. When a nuclear power plant approaches meltdown, we learn how such plants provide power and how they control the process. These are process descriptions, or procedures, providing information about how something works or the way something happens.

Procedures like these are simplified explanations for generalists, but they can also be much more complex when written for operators, technicians, and specialists. A scientist, for example, might go into great detail explaining to peers how proteins are transferred from the blood to tissue, how the ozone layer expands or contracts, or how swamps maintain an ecological balance. In technical fields, manuals about installation, maintenance, and repair often contain sections explaining how machines and parts of machines work.

Procedures are closely related to instructions because both kinds of writing "tell how." But procedures simply present information, whereas instructions specifically tell the reader how to do something. Instruction writing is covered in Chapter 17. In this chapter you will learn the form and the process of writing procedures.

18.1 DEFINITIONS OF TERMS

Procedures are a form of writing that explains how something happens or how it is done. The reader of the procedure is not likely to perform the operation but seeks to understand how it works. In a procedure you describe the steps of a process, much as you would describe the parts of a physical object, so procedures are often called "process descriptions." For example, Part 1 of this book is called "The Process of Technical Writing," but it could also be called "A Procedure to Follow in Technical Writing." Likewise, section 18.3 of this chapter is titled "The Process of Writing Procedures" because it gives information about a method that experienced writers use in explaining how something happens or is done.

The difference between physical descriptions and procedures lies in the method of organization. Descriptions are usually organized spatially, while procedures are organized chronologically. A procedure can be complete in itself in a letter, memo, brochure, or short report, or it can be part of a longer document like a proposal or feasibility report. When procedures are combined with definition and physical description, a reader can learn from one document what something is, how it looks, and how it functions. Combinations like these often appear in various kind of manuals.

Procedures fall into three general groups: (1) those accomplished by machines or systems of machines, (2) those that occur in nature, and (3) those done by people. Table 18-1 lists typical procedures in each category.

Table 18-1 TYPES OF PROCEDURES

Machine Procedures	Natural Procedures	Human Procedures
how airbrakes stop a truck	how organic matter decays	how beer is made how a plot of land is surveyed
how a chain saw cuts wood	how fog forms how cancer cells	how an oil painting is
how a sewage system operates	replicate how shorelines	cleaned
how a CPU powers on and self-tests	change shape	how a calf is dehorned

Notice how similar a procedure done by humans is to a set of instructions. Both involve people, but whereas instructions tell *how to do* something, procedures simply give information, telling *how something is done*. Readers of procedures also have a different purpose; they want information, not instruction.

18.2 THE FORM AND ORGANIZATION OF PROCEDURES

Procedures have three parts: (1) an introduction that provides an overview, (2) a step-by-step description of the action, and (3) a short conclusion that summarizes, discusses advantages and disadvantages, or relates the procedure to a larger ongoing project. Before you write a procedure, carefully assess your potential reader, because the reader's background and purpose will determine how much specialized vocabulary you can use, whether you can assume any familiarity with a similar procedure, and how much detail you should include. Readers can range from generalists to specialists. Use a Planning Sheet like that in section 1.6 to help clarify your purpose and reader.

The Title and Introduction

The title should both explain the purpose and identify the procedure. You can explain the purpose with an "ing" form of the verb and identify the procedure with key words:

- Plotting a Cross-Country Flight
- Fluoridating Municipal Water Supplies

or you can use the word "How" and key words:

- How a Laser Works
- How Scientists Predict Earthquakes

or the word "process" or "procedure" and key words:

- Start-up Procedure for XXX Minicomputer
- The Process of Electroplating

Notice that the title is different from one you would give a set of instructions. Instruction titles should begin with "How to" or "Instructions for."

The introduction should include four components: the purpose of the procedure, definitions of key terms, a list of tools and equipment, and an overview of the main steps. In addition, there may be optional components. Examples of each required component are given below.

1. **Purpose** Begin with an opening statement that identifies the procedure and its purpose. For example

- Coldheading is a widely used method of making screws; in coldheading a head of predetermined size and shape is formed or "upset" on one end of a cut blank of rod or wire (Clarke 1978, 167).
- Crossdating is a method of dating wood (including trees, posts, and structural beams) by comparing the ring patterns in the older wood with the patterns in more recent wood (Phipps and McGowan 1981, 11).

2. Definitions   Be sure to define the process and the key terms essential to the process. Remember that your readers want information, so the procedure must begin with clear word definitions. For example (Corbalis 1986):

> The electrocardiograph (EKG) is a very important test done in hospitals, doctors' offices, and other medical facilities; it helps in diagnosing and monitoring the effects of drugs, therapy, and illness on the heart. It is also used as an auxiliary test in the diagnosis and treatment of many other conditions, such as pregnancy and lung disease. The EKG is a graphic representation of the electrical activity of the heart. It is represented in the form of a time versus amplitude graph.

For help in writing good definitions, see Chapter 10.

3. Equipment   Tell the reader what tools, equipment, or special conditions are required. You can do this in a list, as you would in instructions, or you can write a paragraph explaining what is needed and what each item is used for. You might do it this way (Bardellini 1980):

> The only tools needed to perform a seismic refraction survey are the following:
>
> - a seismograph—an instrument capable of measuring the time it takes a signal moving through a given medium to get from point A to point B.
> - geophones   —small sensitive receivers of the signal (sound or percussion wave) as it passes by.
> - seismic cable —the means by which the geophones pass the information on to the seismograph.
> - shot point   —the point at which the signal emanates. The signal is produced by various means, for example, a hammer, explosives, or a mechanical device.

4. Overview   To provide an overview, indicate the main steps of the procedure. The best way is to name each major step, presenting the steps in a list if you have more than three. At least tell how many major steps there will be, so the reader can follow them as they unfold. For example (Leonard 1985):

The basic process of laser action involves four steps:

1. excitation
2. spontaneous emission
3. stimulated emission
4. amplification

Optional components   In addition, you should add any of the following information that would apply:

- where and when the procedure takes place
- any special training or advanced preparation required
- how long the procedure takes
- the theory on which the procedure is based

Figure 18-1 shows how the theory behind a procedure might be explained. Note also how the theory grows out of the introduction and leads into the step-by-step description. Words in capital letters are defined in the glossary.

Readers seeking information will probably read all of a well-written

---

Surgical Lasers Explained (Process of Laser Action)

A surgeon wants a tool that will cut and ablate (remove) tissue structures, coagulate vessels, and simultaneously sterilize the process. The laser, because of three special qualities, meets the surgeon's requirements. The three qualities that differentiate laser radiation from conventional radiation sources are listed below and illustrated in the figure.

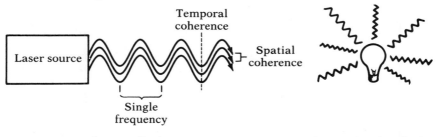

1. Single FREQUENCY. All laser radiation has the same frequency. This makes visible laser light MONOCHROMATIC. Some lasers produce invisible radiation; nevertheless, the beam is pure.

2. Temporal Coherence. All the wave fronts in a laser beam are in phase.

3. Spatial Coherence. The distance between waves in a laser beam remains virtually constant. This produces directed, or COLLIMATED, light as opposed to dispersed light.   *Continued*

Figure 18-1   A procedure with theory

Pure frequency allows the surgeon to act on specific tissues. Temporal coherence allows the surgeon to deliver constant energy to the target tissue. Spatial coherence allows the surgeon to accurately focus the beam.

The process of laser action gives the laser these qualities. Although the laser beam is generated automatically within the laser device when it is turned on, a basic understanding of the process should give the surgeon added skill and confidence.

The basic process of laser action involves four steps:

1. Excitation
2. Spontaneous Emission
3. Stimulated Emission
4. Amplification

These steps all occur within the laser mechanism.

Step 1: Excitation

The first step is initiated when power is supplied. It is based on fundamental properties of matter.

In a normal state atoms exist as minute particles of matter each with a nucleus and a specific number of electrons. These electrons assume definite energy levels. In the absence of external energy, the electrons assume the lowest possible energy level. However, when energy is supplied, electrons can take higher energy levels. The energy absorbed must exactly equal the difference in energy between levels; there are no partial steps.

Electrons in elevated energy levels are called excited electrons. Thus, pumping energy into atoms excites electrons and initiates the excitation step.

Step 2: Spontaneous Emission

Usually within one-hundredth of a MICROSECOND an excited electron will regain a lower energy level. Since the atom has less energy after the transition than before, energy must have been released. The released energy is exactly equal to the energy gained during excitation. The energy is given off as a PHOTON, a minute particle of light. This process is called spontaneous emission.

All the photons spontaneously emitted by similar atoms have the same frequency. They travel in all directions, however. Therefore, spontaneous emission alone will not create a collimated beam.

Step 3: Stimulated Emission

The step most crucial to laser beam generation is stimulated emission. In this step a photon strikes an electron that is already excited. When this happens, the energy of the photon is not absorbed by the atom as in excitation. Instead, the atom simultaneously releases the energy of the photon and the energy it would have released by spontaneous emission. The result is two photons with equal frequency. Not only do they have equal frequencies, but they also travel in the same direction.

Figure 18-1 **Continued**

Step 4: Amplification

As photons produced by stimulated emission travel, they can strike other excited electrons. If the process continues, the number of photons traveling in the same direction with identical frequency amplifies until a beam is produced.

Unless certain conditions exist, however, amplification cannot occur. Because excited electrons are very short-lived, a photon normally has little chance of striking one. The laser mechanism provides the necessary conditions.

Three basic components of the laser mechanism involved in the process of laser action are illustrated in the figure.

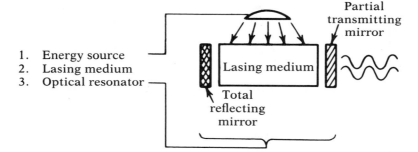

1. Energy source
2. Lasing medium
3. Optical resonator

The energy source pumps energy into the LASING MEDIUM and excites electrons. Enough energy must be supplied to keep more electrons excited than unexcited. This condition is called a POPULATION INVERSION.

With abundant excited electrons, amplification can proceed. Photons traveling along the axis of the OPTICAL RESONATOR are reflected by the mirrors at each end. Then they travel back through the inverted population amplifying the beam in their direction.

After sufficient amplification, part of the collimated beam passes through the partially reflecting mirror at one end of the optical resonator. The reflected beam travels back through the medium stimulating enough photons to make up for the beam leaving the resonator. Thus, a sustained laser beam is produced.

Summary of Laser Action

Bringing all the steps together, the process of laser action occurs as follows. The energy source pumps energy into the lasing medium. Electrons within the medium become excited and they form a population inversion. At this stage conditions for stimulated emission exist, and a beam amplifies along the axis of the optical resonator. Part of this beam passes through the partially reflecting mirror and is available to the surgeon. The process is described by the word *laser*; it is the acronym of Light Amplification by Stimulated Emission of Radiation.

Figure 18-1   Continued

introduction; they are less likely to skip to Step One than are readers of instructions. Even so, your introduction should be entirely functional; in other words, do not give details just because they are interesting to you.

The Step-by-Step Description

After the introduction, break the process into its component steps, keeping the number of major steps under nine. Within each step you can subdivide; you will have to do that if the process has more than one thing happening at a time. Arrange the steps in a chronological method of organization, and help the reader keep on track by devoting a separate paragraph or section to each major step. To show the sequence, you can number the steps, identify each step with a subheading, or if the steps are simple, use transitional words and phrases such as "next," "then," or "finally." Unlike the steps of instructions, those of procedures are usually written as fully developed paragraphs.

The Conclusion

A good conclusion will help the reader understand either by summarizing the steps or by relating the procedure to a larger operation. With a long and complicated procedure, the reader needs a summary and reinforcement of the major steps. If the steps are relatively few and easy to understand, a better conclusion might be to

- discuss the advantages and disadvantages of the procedure
- show how this procedure is related to other work
- compare it to similar procedures
- list the results of the procedure
- indicate its importance or significance
- recommend this procedure or one procedure over another

In short procedures, the conclusion may be eliminated.

In the following document, part of a longer comparative report designed for generalists, a student writer (Neher 1984) explains the procedure of multitrack sound recording. Notice that the introduction defines the procedure, discusses needed equipment, and lists the three principal steps. Each step is then discussed in detail.

The Multitrack Sound Recording Process

Multitrack recording is the process of recording two or more audio signals onto separate tracks on a magnetic recording tape.

The main purpose of multitrack recording is to simultaneously or consecutively record multiple sound sources so that they are synchronized on the recording tape, and so it is possible to individually adjust the level, equalization, and special effects on each sound source as it is being recorded or mixed down.

Multitrack recordings are generally made in a soundproof environment, such as a recording studio, so that no extraneous noises are recorded on the tape.

In making a multitrack recording, a recording engineer usually operates the recording equipment, while at least one performer is needed to produce the sounds. The sounds can be produced with musical instruments, voices, other recordings, or sound effect devices.

The equipment needed to make a multitrack recording is a multitrack tape recorder with a mixing section, and at least one microphone for each sound source. During recording, each sound source is picked up by a microphone, which converts the sound into an electronic audio signal. Each signal is fed into the mixing section of the recorder where the input level and tone are adjusted. The processed signals are then recorded onto their own separate channels or tracks on the recording tape. The resulting multitrack recording can then be mixed down onto a standard stereo tape for distribution.

The principal steps in making a multitrack recording are

1. Setting up and adjusting the recording equipment.
2. Making a master recording.
3. Mixing down the master recording.

1. SETTING UP AND ADJUSTING THE RECORDING EQUIPMENT

The recording engineer and the performers work together to set up the appropriate microphones for each sound source. The engineer decides the type and placement of these microphones. While the performers produce sample sounds, the engineer adjusts the input level of each signal as it is fed into the mixing section of the recorder. Distortion results if the input level is set too high, and background noise results if the level is set too low. The engineer also adjusts the equalization or tone of each signal.

2. MAKING A MASTER RECORDING

Once the recording equipment is set up and properly adjusted, the master recording is made. The engineer turns on the recorder and signals the performers to begin. During the performance, the engineer monitors the record levels and tone adjustments to make sure they stay within acceptable bounds. When the performance is over, the engineer stops the tape and rewinds it. If, after playing back the tape, the engineer and performers are satisfied with its quality, the tape is ready to be mixed down. Otherwise, the recording process may need to be repeated, or individual performers may do their parts over (overdubbing).

3. MIXING DOWN THE MASTER RECORDING

Since most tape players can only play one or two tracks, master tapes having more than two tracks need to be mixed down onto more common two-track tape for distribution. The engineer uses a multitrack recorder and a second, mixdown recorder to accomplish this task. The master recording is played on the multitrack recorder, and the audio signals that are produced are routed through the mixing section. The signals are combined into a composite two-track signal, which is then recorded by the mixdown recorder. During this process, the engineer can again adjust the volume, equalization, and relative balance of all the tracks being mixed down from the master tape.

CONCLUSION
     Multitrack sound recording can produce high-quality demonstration tapes of specific musical techniques and sounds. The advantage of multi-track recording is in the degree of control the engineer has over the sound quality.

18.3 THE PROCESS OF WRITING PROCEDURES

Planning

In planning a procedure you need to keep in mind your reader's purpose in turning to this piece of writing: the reader wants information. That means your purpose as writer is to *explain*, not to give instructions. You will write clearer procedures if you remember that the reader will not be using what you say to perform the operation. It's confusing for readers if you mix the forms of instructions and procedures, so you need to have the differences clearly in mind.

     You must also know who your readers will be and how much background information they already possess. One way to determine potential readers is by asking *where* this procedure will appear. If it is part of an article that will go to a professional journal or a general purpose magazine, you can profile readers by looking at back issues and evaluating the vocabulary, level of detail, and breadth or depth of coverage in the published articles. If your procedures will appear as part of a report, you should find out where the report is going and to how many people. Fill out a planning sheet like that in section 1.6.

     Many procedures are written to acquaint generalists with technological advances or to explain the results of scientific research. If you are writing for generalists, follow the suggestions in section 1.2 to keep the procedure interesting and easy to follow:

- avoid specialized language and define terms
- use comparisons and anecdotes
- avoid equations and formulas
- keep sentences relatively short
- use graphics

     But procedures may also be written for specialists and technicians—those who have extensive background in the general field but who may be unfamiliar with the procedure you are explaining. For these readers, you can rely on a shared vocabulary and on understanding of basic concepts. The following two paragraphs are from a procedure written for specialists in air pollution measurement and control (EPA 1986, 3.0.9). Notice that the writer assumes understanding of basic terms and processes.

> *Nondispersive infrared spectroscopy* utilizes infrared light in a limited range of the electromagnetic spectrum. The light is not scanned or "dispersed" as with scanning laboratory spectrometers. In general, the light is filtered to select light wavelengths that will be absorbed by the molecules that are to be measured. The light passes through a gas cell that contains the flue gas extracted from the stack. A portion of the light from the lamp passes through a cell containing a reference gas that does not absorb the filtered light. A detector senses the amount of light absorption in the sample cell relative to the signal from the reference cell. Through proper calibration, the detector responses are electronically converted to pollutant concentration readings. A variant of this technique, called gas filter correlation spectroscopy, uses a reference cell that absorbs 100% of the light in the molecular absorption region of the pollutant.
>
> Infrared analyzers have been developed to measure gases such as SO_2, NO, NO_2, HC1, CO_2, and CO. The commercially available monitors differ primarily in the design of the detector and the level of rejection of interfering gases.

The tone you use in procedures will affect the reader's response. Readers who seek information and explanations should never be made to feel stupid because they don't already know. Your job as a technical expert and a communicator is to share your information objectively and confidently. The style you choose will influence your tone, so be aware that you *choose* a style—don't let it simply happen. Most technical procedures will be semiformal or formal in style, unless you are writing for children or popularizing science and technology; in those cases you can use an informal style.

Gathering Information

Accurate information is essential for procedures. Your readers come to you for explanations, expecting what they read to be correct and verifiable. That means you must go to original sources for your own information. You must read specifications, talk to experts, and observe the process in action whenever possible—taking notes as the various steps occur. You may have to spend considerable time gathering information: taking measurements, timing operations, taking things apart, and studying the results of experiments. True specialists know much more about the subject than they can practically include in a written procedure. The additional information allows them to select and shape the material to meet the reader's needs more effectively.

Note that when you write procedures, you are explaining not a one-time phenomenon but a typical occurrence. Therefore, in gathering information you may have to examine many specific occurrences in order to generalize to the typical procedure. For example, if you are writing "How

Proposals Are Written to Secure Government Contracts," you must look not just at one sequence of events that resulted in a contract but at many such sequences.

Organizing and Reviewing

The three-part organization of a procedure combines hierarchical and chronological organization. The overall structure of a procedure is hierarchical; you divide the procedure into manageable parts with an introductory overview, a body, and a conclusion. Within the body, however, the method of organization is chronological; that is, each step is discussed where it appears in the time sequence. Occasionally, the procedure is also causal—A causes B, which causes C, and so on. Even here, though, the organization is sequential and is best understood as a series of steps. Thus, a procedure is relatively easy to organize if the actions occur in straight-line order.

If several actions occur simultaneously, the explanation in words becomes more complex, and you might need to supplement your explanation with a flow chart. In Figure 18-2 the flow chart for installing a new computer system helps the reader understand the simultaneous actions that would occur in the 30 days between the pre-installation meeting and the point at which the system is functional.

With a visual representation like this, you can take up one component at a time. You might discuss the simultaneous events in a left-to-right order:

- data preparation and data input
- office preparation and system installation
- training and customer education classes

Alternatively, you could break the procedure into six parts:

- data preparation
- office preparation
- training
- data input
- system installation
- customer education classes

Once you have determined the organization you wish to use, review the details to ensure that you have not forgotten any steps. To check that your order of presentation is clear, ask someone unfamiliar with the procedure to look at your outline. Only after this kind of careful review are you ready to begin writing.

Writing

When you write, remember that you are giving information, not instructions. That means you should:

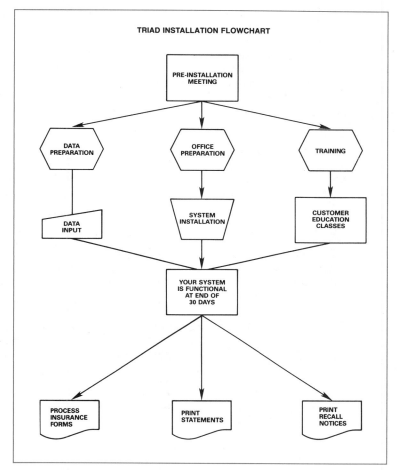

Figure 18-2 Flow chart to help explain a procedure (Triad installation)

1. Write in the present tense and in the indicative (information-giving) mood. The present tense is most appropriate for the continuing actions described in procedures.

> WRITE: In planning a cross-country flight, the pilot locates the course line and draws it on the chart.
> NOT: Locate the course line and draw it on the chart.
> WRITE: Convection heat transfer occurs when a fan blows heated air around food in an oven.
> NOT: Convection heat transfer occurred when a fan blew heated air around food in an oven.

2. Keep the procedures objective and impersonal, using nouns and the third-person pronouns "he," "she," "it," and "they."

- When the rotor is spun, the gyroscope acquires a high degree of rigidity, and its axle keeps pointing in the same direction, no matter how much the base is turned.
- To use the locking clip, the operator first threads the vehicle's seat belt through the correct frame and buckles it. Then he or she pulls the shoulder belt portion until the lap portion is as tight as possible.

3. Use the active voice whenever you can. The active voice is effective when the procedure is performed by a person:

- The materials testing technician (MTT) takes direct readings on the hammer gauge against vertical surfaces such as walls.
- Before the washing process begins, the operator selects the water level, cycle, water temperature, and speed of the motor.

The passive voice may be used when the procedure is performed by a mechanism or by natural forces:

- After the light has been gathered by the objective lenses and processed by the prisms, the image is focused and magnified by the ocular lenses.
- As the water enters the tube, air is forced into the air pressure dome.

Even in these cases, though, you can usually use the active voice:

- The objective lenses gather the light reflected from an object, and the prisms process that light. Then the ocular lenses focus and magnify the image.
- As the water enters the tube, it forces air into the air pressure dome.

4. Write a title that clearly identifies what you have written as a procedure. Use function words and key content words like "How a Passive Solar System Works" or "The Process of Passive Solar Heating." (See section 6.1 for more on writing titles.)

Table 18-2 summarizes the techniques for procedure writing.

Table 18-2   TECHNIQUES FOR WRITING PROCEDURES

Writer's purpose	Reader's purpose	Method of organization	Tense	Person	Voice	Mood
to tell how something happens or is done (information)	to learn how it works or how it happens	sequential chronological hierarchical	present: "The operator OPENS the valve."	third: "A technician AT-TACHES the heat shield" or "The heat shield IS ATTACHED."	active or passive: "A mechanic INSERTS the filter" or "The filter is inserted."	indicative (giving information): "The surgeon covers the surface."

In addition, think of ways to enhance the words you are writing. Many procedures make good use of graphics, either to supplement the words or to shape the text. When you write for generalists, you will probably want to include graphics. In Figure 18-3 the flow chart is cartoonlike because the writer wants to make the presentation enjoyable as well as informational. Even so, the information is serious and important, and the tone is objective, not condescending.

If the procedure is complex, graphics can also enhance explanations for technically oriented readers. In Figure 18-4 on page 380 the writer summarizes the steps involved in digital radar landmass simulation (DRLMS). Notice how the simple graphics help clarify the explanations.

Reviewing, Editing, and Rewriting

To test the effectiveness of your procedure, have both experts and novices read it. Experts can tell you if what you say is accurate and if your chronological sequence is appropriate. Novices can tell you if your explanation makes sense to someone unfamiliar with the procedure. Apply what your reviewers say and rewrite any inaccurate or unclear passages. Make sure that you define any terms your readers don't know, and edit for correct punctuation, spelling, and sentence construction.

CHECKLIST FOR WRITING PROCEDURES

1. Does my title clearly explain the purpose and identify the procedure?
2. Does the introduction include
   _____ the purpose of the procedure?
   _____ definitions of the procedure and any key terms?
   _____ a list of needed equipment and supplies?
   _____ an overview of the major steps or an indication of the number of steps?
3. Should I also include
   _____ the time and place of the procedure?
   _____ special training or advanced preparation required?
   _____ the time needed for the procedure?
   _____ the theory on which it is based?
4. Are the individual steps
   _____ arranged chronologically?
   _____ marked by number, letter, subhead, or paragraphing?
   _____ subdivided if necessary?
   _____ usually written in paragraphs?
5. What type of conclusion is best for this document?
   _____ summary
   _____ advantages and disadvantages

A Water Treatment Plant

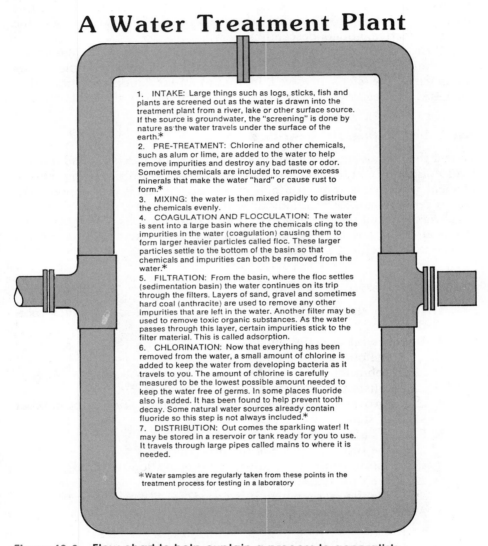

1. INTAKE: Large things such as logs, sticks, fish and plants are screened out as the water is drawn into the treatment plant from a river, lake or other surface source. If the source is groundwater, the "screening" is done by nature as the water travels under the surface of the earth.*

2. PRE-TREATMENT: Chlorine and other chemicals, such as alum or lime, are added to the water to help remove impurities and destroy any bad taste or odor. Sometimes chemicals are included to remove excess minerals that make the water "hard" or cause rust to form.*

3. MIXING: the water is then mixed rapidly to distribute the chemicals evenly.

4. COAGULATION AND FLOCCULATION: The water is sent into a large basin where the chemicals cling to the impurities in the water (coagulation) causing them to form larger heavier particles called floc. These larger particles settle to the bottom of the basin so that chemicals and impurities can both be removed from the water.*

5. FILTRATION: From the basin, where the floc settles (sedimentation basin) the water continues on its trip through the filters. Layers of sand, gravel and sometimes hard coal (anthracite) are used to remove any other impurities that are left in the water. Another filter may be used to remove toxic organic substances. As the water passes through this layer, certain impurities stick to the filter material. This is called adsorption.

6. CHLORINATION: Now that everything has been removed from the water, a small amount of chlorine is added to keep the water from developing bacteria as it travels to you. The amount of chlorine is carefully measured to be the lowest possible amount needed to keep the water free of germs. In some places fluoride also is added. It has been found to help prevent tooth decay. Some natural water sources already contain fluoride so this step is not always included.*

7. DISTRIBUTION: Out comes the sparkling water! It may be stored in a reservoir or tank ready for you to use. It travels through large pipes called mains to where it is needed.

*Water samples are regularly taken from these points in the treatment process for testing in a laboratory

Figure 18-3 Flow chart to help explain a process to generalists

IN NORTH AMERICA, OUR GOVERNMENTS HAVE SET STANDARDS
OR GUIDELINES FOR DRINKING WATER. WHEN WATER LEAVES A
TREATMENT PLANT IT IS AS CLEAN OR CLEANER THAN REQUIRED.

PRE-TREATMENT

2

Cl₂ GAS   LIME   ALUM

1

INTAKE

3 MIXING   4 COAGULATION AND FLOCCULATION

5
FILTRATION

7 DISTRIBUTION

Cl₂ GAS

6 CHLORINATION

Figure 18-3   Continued

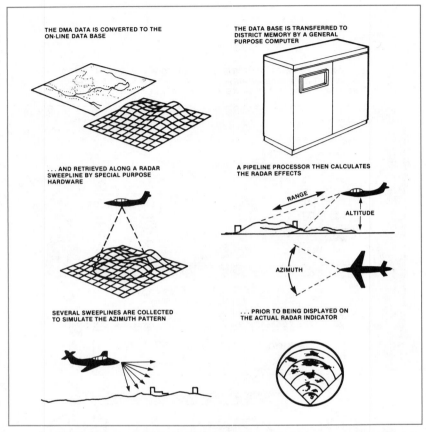

THE DMA DATA IS CONVERTED TO THE
ON-LINE DATA BASE

THE DATA BASE IS TRANSFERRED TO
DISTRICT MEMORY BY A GENERAL
PURPOSE COMPUTER

... AND RETRIEVED ALONG A RADAR
SWEEPLINE BY SPECIAL PURPOSE
HARDWARE

A PIPELINE PROCESSOR THEN CALCULATES
THE RADAR EFFECTS

RANGE

ALTITUDE

AZIMUTH

SEVERAL SWEEPLINES ARE COLLECTED
TO SIMULATE THE AZIMUTH PATTERN

... PRIOR TO BEING DISPLAYED ON
THE ACTUAL RADAR INDICATOR

Figure 18-4   **Simple drawings to help explain a procedure: digital radar landmass simulation**

_____ relationship to other work
_____ comparison with a similar procedure
_____ results of the procedure
_____ its importance or significance
_____ a recommendation

EXERCISE 18-1   Analyzing a Procedure

Directions: The following procedure was written by an industrial design student (Liang 1986). The intended reader is an office manager who wants to use a desktop copier but does not understand the process involved in producing the copies. Determine whether this procedure has all the needed parts and whether it would be clear to the reader. Be prepared to discuss your evaluation.

COPYING PROCESS

Electrophotographic copying system is based on electrostatic charging and photoconductivity; it is a complex copying process which office copiers today are based upon and can be classified into six distinct steps:

CHARGE: A corona discharge caused by air breakdown uniformly charges the surface of the photoconductor.

EXPOSE: Light reflected from the image discharges the normally insulating photoconductor and produces a latent image, a charge pattern on the photoconductor that mirrors the information to be transformed into the real image.

DEVELOP: Electrostatically charged and pigmented polymer particles, called toner, are brought into the vicinity of the oppositely charged latent image; they adhere to the latent image, transforming it into a real image.

TRANSFER: The developed toner on the photoconductor is transferred to paper by giving the back of the paper a charge opposite to that on the toner particles.

CLEAN: The photoconductor is cleaned of any excess toner using, for example, coronas, lamps, brushes or scraper blades (Burland 1986).

FUSE: The image is permanently fixed to the paper as it passes between two heated rollers or under fusing heat.

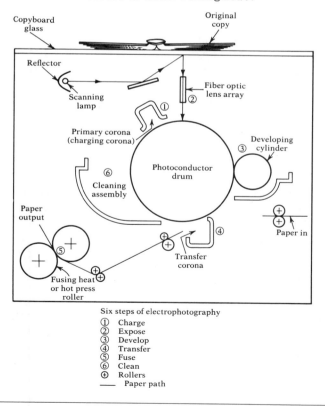

Six steps of electrophotography
① Charge
② Expose
③ Develop
④ Transfer
⑤ Fuse
⑥ Clean
⊕ Rollers
—— Paper path

Figure A   Desktop copiers: a procedure to analyze.

WRITING ASSIGNMENT #12   Writing a Procedure for Generalists

Directions: Following the suggestions in this chapter, write a procedure for one of the following:

Procedures done by people:

how bricks are made	how cloth is woven
how coal is mined	how CPR is given
how a car is jacked up	how wine is made

Procedures done by machines:

how a squirt gun works	how air brakes work
how a fire extinguisher works	how a snorkel works
how a toaster works	how a vacuum cleaner works

Procedures in nature:

how sound is transmitted	how a skunk emits its charge
how rust forms	how wood becomes petrified
how frost forms	how leaves turn color in autumn

EXERCISE 18-2   Writing a Procedure as Part of a Report

Directions: As part of your formal research or feasibility report, write a procedural description. Check your outline to see if you have placed it effectively. If not, revise your organization. Seek review of the procedure. If you are working as a team, assign one person to write any needed procedures. Have that person work with the graphics specialist to supplement the procedure with a flow chart or other illustration.

EXERCISE 18-3   Explaining a Procedure to Young Readers

Directions: Choose a procedure that is carried out in your major field and write a brief explanation of it for junior high school students. Use simple definitions and tie it to their experience with some kind of analogy. For help with definitions, see Chapter 10.

REFERENCES

Bardellini, David. 1980. *The process of seismic refraction.* Informal report, San Jose State University.

Clarke, Donald, ed. 1978. Screws. *Encyclopedia of how it's made.* New York: A & W Publishing.

Corbalis, Mary. 1985. *How an electrocardiograph works.* Informal report, San Jose State University.

EPA. 1986. *Quality assurance handbook for air pollution measurement systems*, vol. 3. Research Triangle Park, NC: Environmental Monitoring and Support Laboratory.

Leonard, David. 1985. *An evaluation of Nd:YAG surgical lasers for Allied Hospital.* Formal report, San Jose State University.

Liang, Alice. 1986. *Desktop copiers.* Formal report, San Jose State University.

Neher, Jonathan. 1984. *The multitrack sound recording process.* Informal report, San Jose State University.

Phipps, R. L., and J. McGowan. 1981. *Tree rings: Timekeepers of the past.* No. 361-618/114. Alexandria, VA: U.S. Geological Survey.

CHAPTER 19

Abstracts and Summaries

Abstracts and summaries condense material from a longer, major document, thus allowing busy readers either (1) to grasp the main ideas without reading the whole thing, or (2) to review, for clarification, the main points they have just read. Both abstracts and the several kinds of summaries form a package with a major report or proposal and depend on the main document for their content. Because they must be comprehensive as well as succinct, they are not easy to write, but readers of technical documents depend on them for help in understanding complex subjects. If you follow the procedure given in this chapter and practice writing the various forms, you can learn to create clear abstracts and summaries. This chapter has three main sections, as listed above.

384

19.1 DEFINITIONS OF TERMS

"Abstract" and "summary" are general terms, each having several variations. Abstracts generally appear in proposals and reports, while summaries can appear in nearly all forms of technical documents.

Abstracts

An *abstract* is a summary of a longer technical document and is written primarily for a technical reader like a specialist or technician. It is independent of the main document, appearing either in a collection of abstracts or on a separate page at the front of a proposal or report. Abstracts are of two types: descriptive and informational.

Descriptive abstracts   A *descriptive abstract* tells what the report is about in general terms but does not include the main points of the information. It is very short—usually three or four sentences—and is written for technical readers. Since it does not include the key points of the report, a descriptive abstract has somewhat limited value. Therefore, if the abstract is the only summary provided, a growing trend is to add to the description of the report the key information in it, thus creating an informational abstract. However, if the document includes both an abstract and an executive summary, the abstract will often be descriptive. See section 19.2 for an example of a descriptive abstract and an explanation of how to write one.

Informational abstracts   An *informational abstract* is a summary of the report that includes the major facts, the conclusions, and the recommendations. It follows the general organization and style of the report itself but omits examples, illustrations, and references. Abstracts do not quote from the main document but summarize and paraphrase; therefore, they do not use quotation marks. Informational abstracts are also short, usually no more than 200 to 300 words. They should, however, be fully intelligible pieces of writing, so when you write one, do not leave out the articles "a," "an," and "the" in order to save words. Section 19.2 includes an example of an informational abstract.

Key Words

Since many abstracts are now filed in computer data bases, you may need to supply at the end of your abstract from three to six *key words* that would help a reader locate that abstract in an index or data base. In choosing key words you need to think like a potential reader, deciding what words your readers would use to search for the information in your abstract, and (by extension) in your report or proposal. Think of the computer data base as an index in the back of a book in which your abstract is included. What cross-reference words would you supply? Develop the habit of listing the key words below your abstract under the heading Key Words.

Summaries

Technical writers use three types of summaries to condense information for readers: (1) executive summaries, (2) report and document summaries, and (3) overviews. Each type meets a slightly different need, and some documents include all three.

Executive summaries   *Executive summaries*, like informational abstracts, condense the main information in a report or proposal—including facts, conclusions, and recommendations. They differ from informational abstracts in that their intended readers are executives or managers: people who are not oriented to technical matters but to planning, allocation of resources and personnel, costs and profits, and the broader concerns of the organization. Therefore, the executive summary helps the decision-making process by providing the manager with an accessible and concise condensation of the report. It may also include the following information about the project or the document:

- the purpose
- the scope
- the methods
- definition of key terms and concepts
- relationship to larger management concerns

An executive summary often covers more material than an abstract, always looking at the subject from the manager's point of view. Necessary technical terms are defined and explained; those that the manager doesn't need for understanding are omitted.

Executive summaries should be written after the body of the report is finished. They are usually placed immediately after the table of contents. Because of the necessary additions, they may be longer than abstracts, but they are usually confined to one or two pages. Like abstracts, executive summaries should be concise, but without leaving out connective words. See section 19.2 for an example.

Report and document summaries   When you write on technical subjects, your job is to help your reader understand complex subject matter in every way you can. One good way to help is by repeating and summarizing the main points at the end of a document or a section of a document. You can find an example of such a summary in Figure 18-1.

Because these summaries appear in the concluding section, they are sometimes called Conclusions, but a true summary includes no material not in the document itself. A true Conclusions section, on the other hand, draws logical inferences from the material and presents these to the reader as findings—sometimes continuing to recommend specific action. Typically, these Conclusions sections first summarize, then conclude, and then recommend. These types of conclusions are discussed in Chapter 6. Another way to summarize is with a checklist like those at the end of each chapter in this book.

Overviews Overviews and previews are like abstracts in that they come *before* the main document. Generally, overviews simply present the main points, which helps the reader see both the organization and the key words in those main points. Sections and subsections can also have their own overviews—each one becoming more detailed. Overviews are explained in Chapter 6. You can also find examples at the beginning of each chapter in this book.

Table 19-1 summarizes these differences.

19.2 THE FORM OF ABSTRACTS AND EXECUTIVE SUMMARIES

General summaries and overviews vary in form and organization, but abstracts and executive summaries have a traditional form and format:

- They are written in paragraph form, using complete sentences.
- The title of the original document and its author's name appear at the top.
- Sometimes the page is labeled "Abstract" or "Executive Summary." If the abstract is attached to a report, no further identification is needed.
- If the abstract appears in a listing of abstracts, the location of the article or report is also given: the journal or publication, volume, publisher, and date.
- To assist in indexing, many abstracts are followed by key content words.

The following informational abstract was written by a student (1984) and summarizes a long comparative report.

Gayle Dolby. *Evaluation of Four Word Processing Software Packages for Minicomputer Systems Suitable for Use in a Technical Publications Group*

The ABC Corporation is considering purchase of a mainframe minicomputer system and has narrowed its selection to the Hewlett-Packard (HP) 3000 and the Digital Equipment Corporation (DEC) VAX 11/780 system. This report evaluates four word processing software packages, two of which run on the HP 3000 system, and two of which run on the DEC system. Following a feature-by-feature comparison and an evaluation of the

Table 19-1 MAJOR TYPES OF SUMMARIES

Purpose	Primary reader	Location	Readers' use	Length
Descriptive abstract to describe the contents of the main document	specialist or technician	in indexes of abstracts or attached to front of report	to see if they want to read report	short—3–4 sentences
Informational abstract to summarize the key information in the main document	specialist or technician	in indexes of abstracts or attached to front of report	1. to see if they want to read report 2. to see major points	fairly short—200–250 words, or less than 10 percent of the main report
Executive summary to aid in decision-making by condensing key information and providing background for recommendations	manager or generalist	attached to front of report	to understand major points in order to make decisions	1–2 pages
Report and document summaries to review the main points	all types	at end of report or document	to understand through repetition of major points	fairly short, but varies
Overview to preview the main points	all types	at beginning of report or document	to see key information and organization that will follow	fairly short, but varies

strengths and weaknesses of each package, Text and Document Processor (TDP)/3000 is recommended for use on the HP 3000 system and WORD-11 is recommended for use on the DEC system.

> Key words: word processing; software; minicomputer systems; publications; technical writing.

The two examples in Figure 19-1, written by professionals in the soil sciences, are a descriptive abstract and an executive summary written for the same document. By comparing the two, you can see how the writers meet different types of reader expectations and needs.

Another example of a descriptive abstract appears in Figure 22-1; informational abstracts appear in Figure 21-2 and in the student comparative report that follows Chapter 21.

19.3 THE PROCESS OF ABSTRACTING AND SUMMARIZING

For Abstracts and Executive Summaries

Often you will be asked to write either an abstract or an executive summary of your own writing, and occasionally you must summarize someone else's document in this way. Fortunately, you can follow the same seven steps in either case.

1. Determine the purpose of the main document and identify the primary readers and their level of technical expertise. Choose the type of summary you will write based on this information: if the readers are technical specialists, you will probably write an abstract, but if the readers are managers, you will write an executive summary.

2. Get an overview by reading the whole document carefully but rapidly. This means that if you are abstracting your own writing, you must first finish writing the main document.

3. Read the document a second time, marking or listing subheadings, key words and topic sentences. In a well-written document, subheads and topic sentences will help you understand the organization.

4. Determine the key information from the subheadings, key words, and topic sentences. Write a rough outline of those major points.

5. Using the outline, write the abstract or executive summary in complete sentences. Do not omit articles or adjectives in your attempt to keep it short. Try to keep an abstract between 200 and 300 words (or less than 10 percent of the original), but do not leave out important information. Try to keep an executive summary to one or two pages, again in proportion to the document. Keep your reader's needs clearly in mind. Unless you are writing both an abstract and an executive summary for the same document, write an informational abstract. As a true summary

for technical
readers

Descriptive Abstract

Felix Y. Yokel, Richard L. Tucker, Lymon C. Reese. *Soil
Classification for Construction Practice in Shallow Trenching.*
NBS Building Science Series 121. U.S. Dept. of Commerce. March
1980.

description of
report contents

Construction practices in trenching and data on potential causes
of trenching accidents are reviewed. A study is made of the soil
properties and site conditions that must be identified in order to
determine the stability of shored and sloped excavations against
cave—ins. Two possible alternate soil classification methods are
recommended. The methods are simple enough to be used by
construction foremen and at the same time use parameters which
can be measured or identified without ambiguity. The classification
methods are supplemented by appropriate field tests and correlated
with allowable side slopes and lateral soil pressures on shoring.

Key words: Braced excavations; construction; excavation;
geotechnical engineering; retaining structures;
shoring; slope stability; soil classification; soil
pressure; soil testing; trenching.

for managers

Executive Summary

reasons for a
soil classifica-
tion system

criteria

As a result of a National Bureau of Standards' study of the
problems associated with excavation safety, it has been concluded
that there is a need for a simple soil classification system that
can be used by field supervisors to make rapid decisions on sloping
or shoring requirements. The soil classification system should
meet the following criteria: it should be comprehensive (cover
essentially all the conditions that could be encountered); it
should consider (at least implicitly) all the critical conditions
that can cause instability; it should not be ambiguous (two
persons classifying a site should arrive at the same conclusion);
it should be usable by construction supervisors and OSHA
compliance officers in the field without the assistance of an
engineer.

summary of
the two sys-
tems

Two alternative systems are proposed for consideration:
The "Simplified Classification System" and the "Matrix
Classification System." The simplified system requires fewer
decisions by construction supervisors, but it also somewhat
narrows the number of choices available to contractors. The
matrix system is more sophisticated and would require better
training of construction personnel and OSHA inspectors. The
systems are summarized in Tables 1, 2, and 3. The tables contain
definitions of soil types as well as stability requirements
associated with the various soil types identified. Lateral loads
that shoring systems should be designed to resist are defined by
the "lateral weight effect" and further explained in Figure 1.
Recommended allowable configuration of sloped excavations
associated with the steepest allowable side slopes defined in
Tables 1 and 3 are shown in Figure 2.

Figure 19-1 **A descriptive abstract and executive summary for the same
report**

The soil classification systems are also correlated with provisions for spaced sheeting (skip shoring). Simple field identification methods are recommended, which can be used to determine the soil type (or class) in the field. These consist of a visual–manual identification method, supplemented when necessary by in–situ strength tests using a pocket Penetrometer[1] or a hand operated shearvane and by a drying test which is performed on an undisturbed sample and used to determine whether the soil is fissured and whether it is cohesive or granular. In case of doubt or dispute these simple field tests can be further corroborated by traditional in–situ and laboratory tests.

explanations of terms

[1]The pocket Penetrometer is a small (vest–pocket sized) commercially available device that measures in–situ shear strength of cohesive soils.

footnote

Figure 19-1   Continued

that includes key information, the informational abstract is more useful to the reader than the descriptive abstract's simple explanation of the contents of the report.

6. Read what you have written and check it against the document to make sure it covers all key information. Edit and correct any errors.

7. Extract from three to six key words and list them below your abstract.

For Summaries and Overviews

You will usually write a summary of your own work for the concluding section of documents like reports and procedures. It's important to finish the document first so you can then go back and pick up the main points to include in the summary. Use your outline and the subheadings of your document to guide you in choosing the main points. Write the summary in paragraph or list form. Double-check your work by comparing your summary with your outline and topic sentences.

You'll place an overview in an introduction so the reader sees it *before* reading the main document, but it may be easier to write it *after* you have written your report, procedure, instructions, or proposal. That way you'll know exactly how much to include and which main points to stress. For more on this, see Chapter 6.

CHECKLIST FOR WRITING ABSTRACTS AND SUMMARIES

1. Who is my primary reader?
_____ technician or specialist

_____ manager
_____ operator or generalist
2. What types of abstract and summary will be most appropriate for my purpose and that reader?
_____ descriptive abstract
_____ informational abstract
_____ executive summary
_____ document summary
_____ overview
3. Do I need to include key words? If so, which ones?
4. In writing my abstract or summary, did I
_____ read the original document to get a sense of the whole?
_____ reread, noting the title and subheadings to get the main ideas?
_____ mark or highlight key words?
_____ mark or highlight the topic sentence in each paragraph?
_____ follow the general organization of the document and write the summary in a page or less, omitting examples and illustrations?
_____ below the summary, list from three to six key words?
_____ check the summary against the original document for accuracy?

EXERCISE 19-1   Practice in Writing an Abstract

Directions: Read the following short report (Assn. of Monterey Bay Area Governments 1984, 1,5) and write (1) a descriptive abstract and (2) an informational abstract of the same material. The intended reader of this report is a generalist.

Groundwater Management Study of Pajaro Basin Completed

The Pajaro Basin Groundwater Management Study, sponsored by the Association of Monterey Bay Area Governments (AMBAG), examines groundwater conditions, uses, demands, recharge, and management. The report was prepared by H. Esmaili & Associates with guidance from the Technical Advisory Committee, which represented municipal, local, and regional interests.

Background

Water flows into the Pajaro Basin from Monterey and Santa Cruz Counties. The east side of the basin rests against the San Andreas Fault and the west against the Pacific Ocean. To the north is the Soquel-Aptos Groundwater Basin, and to the south, the Salinas Valley Groundwater Basin.

In most coastal valleys, groundwater is recharged within stream channels by water of excellent quality. In the Pajaro Basin, rainfall percolating in the surrounding hills is the principal source of recharge, and the Pajaro River is the main source of poorer quality groundwater.

The City of Watsonville is the only major user of surface water in the region. Groundwater, drawn for the most part from depths of less than 700 feet, supplies over 90% of the total water demand in the basin.

The following groundwater-related problems have been identified:

- an overdraft of between 6,000 and 18,000 acre-feet per year;
- seawater intrusion, extending up to three miles inland at some locations;
- water quality problems related to pollution, and;
- potential loss of natural recharge due to development.

The emphasis of this study was on methods for protecting and enhancing natural recharge, and on new sources of water to halt the seawater intrusion.

Study Findings

Groundwater Table Levels. The amount of water pumped from the groundwater table annually is expected to increase slightly during the next 15 years. Groundwater level declines in the inland portions of the basin do not justify any major water projects at this time.

Seawater Intrusion. The seawater intrusion problem can be addressed by reducing or halting pumping from wells in the affected areas. If pumping were stopped in the seawater intruded areas, 6,500 acre-feet of replacement water would be needed to control the intrusion. The only water supply projects which would meet estimated supplemental water needs in the basin are Pescadero Creek Dam, the San Felipe Project, Arroyo Seco Dam, and deep aquifer wells.

If no action is taken, seawater intrusion will lead to the need to drill deeper wells, relocation of wells inland, and finally to the removal of the land from agricultural production.

Loss of Natural Recharge. The sand hills surrounding the Pajaro Basin, in their undeveloped state, have unusually low runoff rates. This corresponds to a high recharge rate of the aquifer. Conversion of these watersheds to residential and agricultural use greatly increases runoff rates.

The increased runoff has resulted in erosion, sedimentation, and damage to drains and sensitive lowland habitats. Runoff due to conversion to residential use amounts to the equivalent of one quarter to one fifth of the annual rainfall.

The study indicates that an integrated water management approach for the sand hills could improve runoff control and recharge protection.

Recommendations

The study recommends a project to supply 6,500 acre-feet of water per year to halt the seawater intrusion. It recommends the following drinking water supply alternatives in the following order:

1. Deep aquifer wells in the Pajaro Valley
2. The Arroyo Seco Project
3. The San Felipe Project
4. The Pescadero Creek Dam.

The study recommends selected management of specified groundwater problems in the Pajaro Valley by a proposed groundwater management agency.

This agency would guide a project to supply 6,500 acre-feet of water per year to the seawater intruded areas.

Four additional small studies should be made in the sand hill areas to assist in the development of specific programs to deal with runoff and lost recharge. Recharge source protection measures should be taken in these areas.

Additional groundwater monitoring and a groundwater pollution control program are advised.

Costs

Damage to roads and drainage facilities currently costs the individual counties from $200 to $300 per homesite in the sand hills areas. Inaction in dealing with seawater intrusion could cost from $255,000 to $725,000 each year. The costs for the management approach to problems in the Pajaro Basin are estimated to be $256,000 annually for administration. There would also be an initial cost of from $105,000 to $210,000 for research and the construction of a monitoring well.

EXERCISE 19-2   Writing an Abstract of Your Own Work

Directions: When you have finished your term project, write an informational abstract to attach to the report. Even though it is your own writing, describe the report objectively, as though someone else had written it. Submit the abstract for either peer or instructor review. If this is a team project, assign one person to write an abstract of the finished document. Review the abstract in your team to ensure its accuracy.

EXERCISE 19-3   Writing an Abstract and an Executive Summary of Your Own Work

Directions: If you will have two types of readers for your term project report (for example, a professor in your technical field and your writing instructor or fellow students), write both an abstract and an executive summary—one for each type of reader. If you are doing a collaborative feasibility study, assign one or two team members to write the abstract and executive summary. Have other team members review it for accuracy and reader appropriateness.

REFERENCES

Dolby, Gayle. 1984. *Evaluation of four word processing software packages for mini-computer systems suitable for use in a technical publications group.* Formal report, San Jose State University.

Yokel, Felix Y., Richard L. Tucker, and Lyman C. Reese. 1980. *Soil classification for construction practice in shallow trenching.* NBS Building Science Series 121. Washington: U.S. Dept. of Commerce, March.

CHAPTER 20

Proposals

What picture pops into your mind when you hear the word "proposal"? Is it the old-fashioned one of a young man on his knees before a simpering maiden offering himself in marriage? Perhaps not. These days marriage is more likely to be a joint decision arrived at by equals, and that meaning of the word seems out of date. In the world of manufacturing, business, and research, though, proposals are both current and very important.

In fact, thousands of people in technical fields earn their salaries by seeking out, planning, and writing proposals. These proposals can either be to provide services or to supply equipment and materials; they can be in any field, including research, development, sales, training, and evaluation. Proposals range from simple suggestions for a change within a department to complex bids for the design, construction, and launch of an

396

orbiting space station. With all this variety, it's likely that at some time in your career you will be asked to write a proposal. This chapter will help you.

20.1 DEFINITIONS OF PROPOSAL TERMS

A *proposal* is a written offer to solve a problem or provide a service by following a specified procedure, using identified people, and adhering to an announced timetable and budget. The elements of this definition are shown graphically in Figure 20-1.

If you study each part of this definition, you'll have a good idea of what's involved in writing a proposal.

1. A proposal is an *offer or a bid*—in other words, a sales or persuasive document. Proposals are much like resumés in that both sell your skills and abilities; however, proposals focus on your ability to do a single task better than anyone else. Because a proposal is a sales document representing you and your organization, you must take special care to make it attractive, organized, and error-free.

2. The proposal offers to *solve a problem or provide a service*. This means you must clearly state the problem, after making sure that your understanding of it is the same as your reader's. You may need to contact the reader in advance to clarify the problem, or to understand exactly what the reader's needs are.

3. The proposal spells out a specific *procedure* or plan of attack for solving this problem or providing this service. The point is to convince readers that it's the best procedure or plan. You may have intense competition for this job, with many other companies or individuals bidding against you, and you need to be innovative, clear, and complete in telling *how* you'll solve this problem or *what* the service will be.

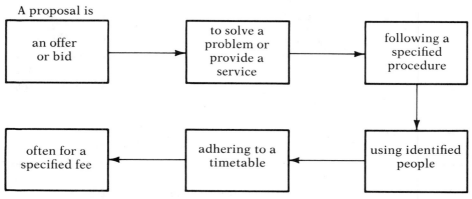

Figure 20-1   Elements of a proposal

4. In a proposal, you identify the *people* who are going to do the job and explain their expertise. To show that you (and your group) are better than the competition, you discuss past accomplishments, experience on similar projects, and successful completion of previous jobs within time and budget limits. You can also provide resumés of key personnel to assure readers that the leaders know what they're doing.

5. An effective proposal includes a carefully planned *timetable.* You break the task into its segments and project the time needed to complete each segment. To do this, you will need some background in a similar project or advice from an expert. You must consider contingencies and make sure that the projected timetable is both possible and competitive.

6. If the project is for a *fee*, the proposal must forecast both the hourly cost of workers and the cost of equipment and supplies. You must figure in a profit but maintain a competitive edge. It's not easy to do all this, but it can be done. The first few times you prepare proposals, seek help from experienced technical people in your field. When you win a contract, keep careful track of costs as they occur and use those figures to help write the next proposal. If you don't win, find out why and file that information so you'll be better prepared next time. Only a small percentage of proposals are accepted, but you can also learn from your rejections.

20.2 TYPES OF PROPOSALS

Proposals can be classified in two ways:

    1. by origin—whether unsolicited or solicited
    2. by the intended reader—whether internal or external

In both classifications, proposals can be presented either informally or formally; form and organization are discussed in detail in section 20.3.

Unsolicited and Solicited Proposals

If you see a problem that needs solving, a change that needs to be made, or a service that needs to be performed, you may write an *unsolicited* proposal. For example, suppose that your job frequently takes you away from your desk so that you can't answer your phone. As a result, callers must leave messages with the department secretary, meaning that you can't respond quickly. You suggest to your manager that an answering machine would help you work more efficiently. "Put it in a proposal," you're told. Or suppose that your company specializes in detection and prevention of methane gas migration. You learn that a local developer plans to construct a large park and outdoor amphitheater over a filled city dump. You recognize the potential for methane gas problems, and you write a proposal for monitoring the construction procedures to ensure safety in the park. When your proposal is unsolicited, you must take special care in defining the problem, since you must convince the reader that the problem does exist and is worth solving.

You write *solicited* proposals in response to a Request for Proposal or *RFP*. Companies and governments at all levels regularly issue RFPs for problems they need solved. Many RFP announcements appear in the *Commerce Business Daily*, a publication of the U.S. Department of Commerce. Others appear in newspapers or are mailed to prospective bidders. An announcement in the *Commerce Business Daily* may be for either an RFP or an Information for Bid (*IFB*). An IFB is usually for physical objects like electric hospital bed parts or shaft and gear assemblies, and the requirements are very specific. An RFP is usually for services or studies, and the proposer has more latitude in designing a solution. Figure 20-2 shows two announcements of RFPs from the *Commerce Business Daily* (U.S. Dept. of Commerce, 1987). These announcements are summaries of the actual RFPs; note that the reader is instructed to write for the RFP.

Figure 20-3 shows a complete RFP that appeared as a newspaper ad. When you respond to an RFP, you must be sure to meet all the requirements that it lists, and often you must follow the method of organization that it specifies.

Internal and External Proposals

Internal proposals go to management within your company. You can write a short proposal in standard memo form (see Chapter 14). However, long internal proposals, or those that must go many levels up in the management hierarchy (perhaps to corporate headquarters) are usually written

---

Williams C. Roberts, Contracting Officer, National Institute of Dental Research, Westwood Bldg, Rm 521, Besthesda, MD 20892
A – CLINICAL EVALUATION OF THE OF THE RELATIVE EFFICACY AND SAFETY OF INTRAVENOUS PREMEDICATION IN DENTISTRY. RFP NIH-NIDR-3-87-1R, RFP avail o/a 27 Feb 87, with responses due 10 Apr 87. Williams Roberts, Contracting Officer, a collaborative clinical study of prototypical drug regimens administered intravenously to alleviate patient apprehension and pain during outpatient dental procedures. From two five contractors will be selected to carry out a collaborative protocol developed by the contractors and NIDR. Phase I of the study will include: design of the protocol; selection of prototype drugs and doses; development of a manual of procedures, subject data forms for a field test, and data quality control procedures; and development of a common data management and analysis system. Phase II will involve recruitment of subjects, data collection, and quality control monitoring. During Phase III, data analysis and collaborative publ of the study results occur. Ea contractor will be expected to provide a min of 100 male subjects or nonpregnant female subjects during ea of the projected two yrs of data collection. Ea subject must be in need of the surgical removal of 2-4 impacted third molars. Requests for a copy the RFP should be in writing. (029)

US Dept of Labor, Natl Capital Service Center, Office of Procurement Services, Rm S-5526, Attn D. Trippel, 200 Constitution Ave NW, Washington, DC 20210
M – A-76 PROCUREMENT FOR THE MANAGEMENT AND OPERATION OF THE US DEPT OF LABOR LIBRARY RFP L/A 87-2, due 1 Apr 87. Contact Dennis Trippel, 202/523-6445. This is third and final Advance Notice of RFP L/A 87-2 to be conducted as an OMB A-76 Cost Comparison Procurement for the management and operation of the US Dept of Labor Library. The contractor shall be required to (1) implement a collection development policy, (2) acquire info matls and info systems, (3) acquire info matls for other DOL offices, (4) catalog and classify matls for the library, (5) provide info, reference, and research assistance services, (6) provide circulation service for physical access to the DOL library for DOL library matls, (8) provide for receipt of incoming mail and for delivery of DOL library matls, (9) maintain a law library and provide associated services, and (10) provide physical security. As this procurement is an A-76 Study procurement no contr will be awarded if Government performance is determined to be more cost efficient. Only written requests will be honored. If a previous request has been submitted based on prior notice no addl request is necessary. (029)

Figure 20-2   Announcements of RFPs in the *Commerce Business Daily*

WE NEED OFFICE SPACE
REQUEST FOR PROPOSALS

DEPARTMENT OF
PRIVATE DEVELOPMENT

THE OFFERING

The City of San Jose is actively seeking to lease office space to house its Fire Department Administration and Traffic Operations Departments. Currently located at City Hall, these units are being required to relocate to provide space for other city functions.

REQUIRED PARAMETERS

Office space to be considered under this proposal must conform to the following parameters:

1. The City is seeking leased office space located in the vicinity of City Hall or in the downtown area.
2. Space offered must be integrated in two blocks containing 12,000 and 9,000 square feet and shall be located in one building.
3. There must be reasonable access to parking for employees and the public. Both on-site and off-site parking will be considered. Employee parking estimates are 70 spaces for Fire and 50 spaces for Traffic Operations.
4. Space offered must be reasonably accessible to the public both with respect to parking and location.
5. Space offered must be available by October 1, 1985.
6. Quotes offered shall be a monthly square foot cost for a full service lease (including parking) for a three year period and for two consecutive one year options with provisions and costs for tenant improvements provided separately.
7. Future rent increases and other costs shall be identified and included in the proposals.

EVALUATION AND AWARD CRITERIA

The basis for evaluation and award will be to identify the one proposal that best fulfills the needs of the City to provide office space for its Fire and Traffic Operations Departments. The City is seeking office space which can both best meet the needs of its employes and the public and be offered at a very competitive price. Items to be evaluated will include, but not be limited to the following:

1. Cost
2. Location
3. Easy public access
4. Availability of parking
5. Physical condition of space provided
6. Timely availability
7. Tenant improvements
8. Future rent increases and other costs.

SUBMISSION REQUIREMENTS

Each proposal shall contain information sufficient to permit the City to properly evaluate the offer. A written submission shall be provided covering each of the concerns expressed in this Request for Proposals.

DEADLINE FOR SUBMISSION AND DELIVERY OF PROPOSALS

The deadline for submission of proposals is 5:00 p.m., Wednesday, September 4, 1985. Proposals must be delivered to the attention of Edwin Louis, Room 200, City Hall, 801 N. 1st. Street, San Jose, CA 95110 before this deadline. Any proposal arriving after this deadline will be disqualified.

REQUESTS FOR SUPPLEMENTAL INFORMATION

The City reserves the right to require the submission of additional information which supplements or explains proposal material.

REJECTION OF PROPOSALS

The City also reserves the unqualified right to reject any or all proposals.

Figure 20-3 Request for proposal

as formal proposals, using a specific form like that explained in section 20.3.

External proposals go outside your organization to customers or granting agencies like the Ford Foundation. Thus, short external proposals can be written as standard business letters (see Chapter 15). Long external proposals, however, are written as formal proposals and accompanied by a letter of transmittal.

20.3 THE FORM AND ORGANIZATION OF PROPOSALS

Informal Proposals

Whether presented internally as a memo or externally as a letter, an informal proposal must answer these questions:

1. What is the problem or need?
2. What is the proposed solution and by what procedure will it be accomplished?
3. Who will solve the problem?
4. What is the timetable?
5. What is the cost?
6. Is this solution acceptable (request for approval)?

To answer these questions effectively, most informal proposals are divided into separate sections, with each section providing the answer to one question. However, section divisions are not rigid, and you can always adapt the form and organization to the needs of a particular reader.

In spite of their name, informal proposals are not written in informal style but in semiformal or formal style. "Informal" refers to the flexibility of the organization and the lack of front and back matter. Informal proposals generally include (1) an introduction, (2) a proposed plan, information on (3) staffing and (4) cost, and (5) an action request.

1. Introduction The introduction identifies the problem or need very specifically and states the proposed solution and the plan for achieving it. In addition, it may give background information, telling who has studied the problem and what they have found. It may list the goals or objectives of the project—both short term and long term. Finally, it may discuss the significance of the need and the importance of the proposed procedure. The most important function of the introduction, though, is to spell out clearly what the problem is. In an unsolicited proposal, you must convince the reader that a problem exists; in a solicited proposal, you must convince the reader that you understand the problem and all its ramifications. (Often you do this by careful repetition of the language used in the RFP.)

In the following introduction, for example, the student writer (Davis 1985) discusses both the need for the study and her specific objective.

> Introduction
> Purpose
> > The purpose of this memo is to introduce the proposed study of wood-burning stoves and request permission to proceed with the study.
> Assessment of Need
> > Davis Development Corporation will be building a group of luxury mountain cabins in 1988 and needs to plan for and later purchase

wood-burning stoves for the cabins. The choice of the stove is important for reasons of comfort, ease of operation, safety, and appearance;
therefore, primary considerations are quality and efficiency, with price
a secondary factor.

Objective

My objective in the proposed study is to evaluate several wood-burning stoves currently on the market and select the one that conforms
most closely to the criteria established by Davis Development Corp.

This student introduction (Cook 1985) defines the problem in order
to indicate the need for the study.

Introduction

Purpose

This memo describes a proposal to investigate eight surge suppressors
and to evaluate each model for optimum performance against a set of
criteria. The outcome of this report will be a concise listing that can
assist microcomputer owners in purchasing the proper surge suppressor for their model of microcomputer.

Problem Definition

In 1984 the microcomputer industry had combined sales in the bil-
lions of dollars. Several million computers were bought to fill needs in
offices and homes throughout the world. With that great number of
computers came a serious problem: the potential loss of data or the
damage of integrated circuits from power instabilities. Whether
through the whims of nature, accidents involving local utility companies, or a fault within a business or home's own electrical system, a
power surge, power spike, or power failure can wreak havoc with the
delicate circuitry of any microcomputer.

2. Proposed plan At the heart of any proposal is the proposed procedure
or plan. In this section you describe how you intend to solve the problem
or provide the service—detailing the procedure or methodology. You may
want to divide this section into

- a description of the approach (how the problem will be solved)
- a description of the evaluation (how you will judge success)
- a detailed schedule

Figure 20-4 shows an informal student proposal that details its proposed plan in a section called Proposed Procedure. Since the proposal is
written to the instructor, it is considered internal and uses memo form.
This proposal does not contain a separate section for staffing since the
writer himself is the only staff person; nevertheless, the writer's experience
as a music instructor is explained in the introduction. Also, since this is
a classroom assignment, the cost section is omitted.

3. Staffing The staffing section in an informal proposal may be very short,
especially if only one person is involved. In this section you discuss the

DATE: September 10, 1984

TO: Professor L. Rew

FROM: Jonathan Neher, English 100 WT   *J. Neher*

SUBJECT: Proposal to make a comparative study of 4—track cassette tape recorders for use in my home studio.

Purpose

    This memo describes a proposal to make a comparative study of currently available 4—track cassette recorders. I need a 4—track cassette recorder to make demonstration tapes for my music students. This study will determine which product would do the best job for the lowest price.

Introduction

    For the past two years I have been teaching music at my home studio, and I have found it necessary to repeatedly demonstrate musical techniques and sounds to students who then can't remember them between lessons. Since most of my students own cassette players, I have decided to make high—quality demonstration cassettes of the musical techniques they need to practice or have trouble remembering.

    These tapes will need to be revised frequently, so it is more practical to purchase the necessary recording equipment than to rent a recording studio for each revision. I need a 4—track cassette recorder so that I can synchronize multiple lines of music on the cassettes.

    This memo proposes that I make a comparative study of the currently available 4—track cassette recorders according to technical and financial criteria, and use this study to decide which recorder best meets my needs.

Proposed Procedure

    I have selected nine 4—track cassette recorders for evaluation:

Aria Studiotrack IIII	Clarion XD—5
Cutec MR402	Fostex X—15 and 250
Studiomaster Studio 4	Tascam 234 and 244
Yamaha MT—44	

    I propose to write to each of the manufacturers or distributors and ask them to send me technical and pricing information on the products. I will also call or visit local music stores and recording studios and gather information on the products' performance and service records. If needed, I will follow up my letters to the manufacturers with phone calls after two weeks.   *(Continued)*

Figure 20-4 Informal Proposal

Using the information gathered from these sources, I will evaluate each product according to criteria based on my personal needs.

Evaluation Criteria

1. Tape Speed—Faster recording speeds produce higher—quality recordings.

2. Noise Reduction—The type of noise reduction determines how much distracting background noise is added to a tape each time it is mixed down.

3. Is an equalizer included?—If no mixer is built into the recorder, an additional piece of expensive equipment may need to be purchased.

4. Rack Mount—Since most of the electronic equipment in my studio is rack mounted, I prefer a rack—mountable recorder.

5. Pitch control—A pitch control is necessary to compensate for out—of—tune instruments or voices.

6. Price—The most I am willing to spend on a 4—track cassette recorder is $1600.

After evaluating the recorders using these criteria, I will write a report that contains the results of the study and a recommendation on which product best meets my requirements for the lowest price.

Projected Schedule

Deadlines

Sept. 17, 1984	Write letters of inquiry
Sept. 21 "	Mail letters of inquiry
Oct. 5 "	Status report, follow—up phone calls
Oct. 15 "	Finish contacting music stores and studios
Oct. 17 "	Begin outline for formal report
Oct. 22 "	Outline for formal report (first draft)
Oct. 29 "	Outline for formal report (final)
Nov. 2 "	Status report
Nov. 9 "	Technical definitions
Nov. 14 "	Technical description
Nov. 16 "	Graphics completed; first draft of report
Nov. 30 "	Final edit; print out formal report
Dec. 3 "	Turn in report

Request

I request permission to embark on this project and would appreciate any comments or suggestions on how I could improve any facet of these plans. Thank you for taking the time to consider them.

Figure 20-4   Continued

credentials or qualifications of the people involved to verify their ability to do the job. Remember always that a proposal is a sales document, and one way to succeed is to sell your expertise or previous track record. Here is how a student might give credentials for evaluating skis.

> I am an avid skier and have been skiing for three years. During that time I have progressed rapidly in skill and am now rated an advanced intermediate skier. In addition, I have attended three sessions at Alpine Mountain Ski School, where ski selection was a topic of instruction.

Here is a brief statement of a project director's qualifications:

> The project director will be Dr. A. B. Cristos, whose resumé is attached. For 10 years Dr. Cristos has been involved in contamination control at Allied Laboratories and for the last 3 years he has directed the contamination control laboratory. In 1988, Dr. Cristos chaired the symposium Contamination Control in Semiconductor Manufacturing in Raleigh, NC.

See Personnel in the Formal Proposals section for more detailed considerations.

4. Cost The cost section details the budget for the project, including costs for staff, equipment, and material. This information is usually presented in a table for easy reference.

40 hours research @ $40 per hour	$1600
10 hours report preparation @ $15 per hour	150
.5 hours computer time @ $120 per hour	60
supplies	100
TOTAL	$1910

Remember that you must abide by your cost estimates, so make them accurate. If your proposal is accepted by a client, you are entering into a contract and have legal obligations to fulfill it, so any cost overruns will come out of your profits.

5. Request for approval Many informal proposals close with a request for approval. This section is similar to the action paragraph that closes a letter or memo of request; it should remind the reader of the advantages of accepting the proposal, and its tone should be courteous and enthusiastic. For example:

- The plan detailed above will improve test procedures and schedules by 200 percent. If you have any questions, please contact me at extension 434. I look forward to your favorable response.
- May I have your approval to pursue this study? Please notify me of your decision at your earliest convenience.

Formal Proposals

Like the recommendation reports explained in Chapter 21, a long formal proposal is divided into sections for the convenience of different types of readers. The sections correspond to those discussed for informal proposals, but each section is presented more formally and may be further divided into subsections. A formal proposal also includes additional material at the beginning and end called "front matter" and "back matter." Here are the typical parts of a formal proposal in the order they usually appear.

front matter	letter of transmittal
	abstract
	executive summary
	title page
	table of contents
	list of illustrations (or list of figures and list of tables)
the proposal itself	introduction
	procedure or methods
	personnel
	budget
back matter	appendixes

Each proposal will modify this organization to some extent to meet the reader's particular needs. Also, even though the sections appear in this order, they are usually not written in this order. You will probably want to begin writing with the introduction or procedures section. The following paragraphs explain each part.

Letter of transmittal   A letter of transmittal is a formal business letter directed to the person who will approve or accept the proposal. Like a job application letter, the transmittal letter should tell how you learned about the problem (for example, an RFP or a letter from the client), highlight the strengths of your solution or approach, and acknowledge any contact you have had or help you have received from the reader or the reader's organization. It should close by offering to answer questions and by requesting action (a response to the proposal). For correct letter form see Chapter 15. The letter should be a separate document from the proposal but attached to it.

Abstract   The abstract should be informational—that is, a brief summary of the proposal's key information—and aimed at specialist and technician readers. The abstract appears at the beginning of the proposal, but it should be written last—after you know exactly what the proposal includes. See Chapter 19 for details on writing abstracts.

Executive summary Most proposals will also contain an executive summary, again a summary of the key information but written for managers. An executive summary should be free of specialized language and should summarize whatever background the reader needs to have to understand the proposal. Chapter 19 explains executive summaries. They are placed either before the title page or before the introduction.

Title page The title page includes the proposal title, the name of the requesting organization, the number of the RFP or other announcement, the date of submission, and perhaps the name of the author or submitting division. Most of this information is repeated on the cover of the proposal when it is bound.

Table of contents This list shows the proposal's organization and provides entry points to the information. The table of contents can be constructed from the outline you create before writing the proposal. Since proposals often do not have an index, a good table of contents is essential. Figure 20-5 shows the table of contents for a formal proposal to provide a service— in this case "Performance Audits for Ambient Air Monitoring Stations at the West Hopkins Facility." The proposal also has front matter, including a letter of transmittal, an abstract, and an executive summary, but traditionally those items are not listed in the table of contents.

List of illustrations The list of illustrations (or separate list of figures and list of tables) is also vital because figures and tables often condense key information in an easily accessible form. An otherwise good proposal might be rejected if no good tables or figures were included to help readers find the details they need. Tables and figures are explained in detail in Chapter 12.

Introduction The introduction can be subdivided into

- *A statement of the proposal's purpose.* This statement discusses objectives or goals of the project, both short term and long term. It may detail the project's significance and its benefits to the company or to society in general. Proposals can be either for research projects (to discover information) or for technical applications (to build, analyze, apply, or administer projects). The purpose statement should identify which type of proposal you are making.
- *A definition and description of the problem or need.* If the proposal responds to an RFP, that fact is noted here with an identification of the RFP. If the proposal is unsolicited, the problem must be clearly presented to convince the reader that a need truly exists. The problem often needs to be presented in considerable detail and should be supported by objective data, such as monetary losses, numbers of defective parts, and low performance ratings.

Table of Contents

Figure 20-5 Table of contents for a proposal to provide a service: "Performance audits for ambient air monitoring stations at the West Hopkins facility"

- *An overview of the proposed project.* This overview allows the reader to see the whole picture and how the parts relate to the whole. It should tell what is included in the proposal and describe the order of the major points.

Procedure or methods The procedure section, sometimes called "Scope of Work," presents the work plan, or method of solving the problem. It describes, in chronological order, how you intend to achieve your purpose.

It is often written like the procedures discussed in Chapter 18; the only difference is that instead of using the present tense, it uses the future tense to refer to activities ("will study" instead of "studies"). The procedure section is usually written in the first or third person and the active voice. In this section you tell not only what will be done but how it will be accomplished; thus, this section often contains specifications, technical description, and functional description. When your proposal is evaluated, this section will be reviewed by specialists in the technical field, so before you submit the proposal, check the procedure section carefully for technical accuracy.

One portion of the procedure section may cover progress reports; some companies and funding agencies will not accept proposals unless regular intermediate evaluations are provided for as well as a final report. Evaluations can be done by (1) measurement, test, or audit, (2) questionnaire and expert review, or (3) progress and final reports. If you propose to evaluate by testing or measuring at regular intervals, specify in the procedure section the kinds of tests that you will administer or specify the measurements that you will take. If you propose questionnaire or expert review, provide a sample of a questionnaire or identify the expert reviewers by name, job title, and professional affiliation. If you propose progress reports, specify the points at which these reports will be submitted—such as at the conclusion of the design phase, modeling phase, and test phase. Most of the time, your proposal will be written to conform to specific requests made in the Statement of Work section of the RFP. However, readers will generally respond most favorably to proposals that have detailed and specific plans, so even if the RFP does not ask for details, the proposal should provide them.

A schedule can form part of the procedure section, or it can appear in a section by itself. Decide where to put it based on the length and amount of detail in the plan or proposed solution. One of the ways to sell a proposal is to show careful planning, and a detailed schedule provides such evidence. Schedules can be presented as a list of specific dates, as estimated hours or weeks to completion of each segment, or as a task-breakdown graph like that in Figure 1-4 (section 1.6).

Personnel   In a formal proposal, the personnel section must include (1) the credentials of project leaders, (2) the size and qualifications of the support staff, and (3) organizational support and resources, like laboratories, computer facilities, and instrumentation. Often the information on credentials is summarized in the personnel section and supported by resumés in an appendix. Support staff and organizational resources can be simply listed; you can judge by the RFP how detailed this section must be.

Budget   Since most formal proposals are bids to obtain contracts, the budget is obviously a key item—justifying item by item the total proposed cost. You need to prepare the budget carefully because if other factors are equal, cost is usually the deciding factor. Budgets include personnel costs (salaries, benefits, and overhead), material costs, and expenses for such things as computer time, travel, and consultants. The budget section should also include the length of time the quote is valid and terms for payment.

Appendixes   Tailor the number and content of appendixes to the proposal's needs. You might include resumés of project leaders, endorsement letters, examples (or a listing) of previous work, and technical material (including graphics) that supports the procedures section but may not interest all readers. A review of relevant literature can be included here, or it can appear as part of the introduction.

20.4 THE PROCESS OF WRITING A PROPOSAL

Planning and Gathering Information

In proposal writing, planning centers on defining the problem and determining the purpose of the proposal. Because you and your company need considerable background in the field in order to do these two tasks, information gathering is often part of the planning process. Remember that a successful proposal will

1. demonstrate your understanding of the problem
2. describe your proposed solution and the scope of your work
3. convince the client that you can carry out your proposed plan

You may want to divide information gathering into three parts:

1. technical information needed (what will be done)
2. management and personnel information needed (who will do the work and how they will do it)
3. schedule and cost information needed (how long it will take, what equipment is required, and what personnel costs will be)

As you plan, you must always keep in mind that *specific people* will approve your proposal. To be successful, then, you must understand your potential clients: how they perceive the problem and what they find important. One way to do this is to talk with them as you develop the proposal. Another way is to look at proposals they have approved. You always need to understand reader type, background, and purpose when you write; in proposal writing, however, such understanding is not just important—it's crucial. To see how a company works toward understanding its readers and develops a proposal from an RFP, read the following description of the process, written by a student intern after working in the proposal department of a large company that specializes in projects for the federal government (Mawk 1985).

The proposal process is a very complex, highly specialized field of writing. The marketing department of a company is responsible for obtaining information about potential customers and competitors and serves as the interface between the company and the customer. All government contracts are obtained via a proposal/bidding process; therefore, a company that secures business this way must develop strong proposal writing skills. The internship with this company provided an opportunity to observe the process of preparing proposals—from receipt of requests for information (usually before the customer begins preparation of the request for proposal) to shipment of the final proposal. I learned several things about the proposal process:

- In order for a company to receive information relative to upcoming government contracts, it must demonstrate its technical ability through presentations and demonstrations to potential customers. Such activity ensures that a company will be on mailing lists to receive requests for information (RFIs), requests for proposals (RFPs), and other such information.
- The proposal process is an interactive one, with much information exchanged between the customer and bidders. Prior to preparing an RFP, the government will frequently send out an RFI to obtain information about current technological capabilities. Also, the government often releases a draft RFP and solicits comments and questions from potential bidders before releasing the final RFP.
- Proposal writing is both an art and a science. The government is usually looking for specific things in a proposal. A successful bidder has learned the customer's idiosyncrasies and prepares the proposal with these in mind.
- The draft proposal goes through many reviews before it is released. There are technical reviews, reviews by those versed in government procedures, reviews by private consultants, and reviews by high-level management. Note that the reviewers don't necessarily always agree!

Learning to read and interpret an RFP is an awesome and sometimes overwhelming task. An RFP can consist of a few pages to several hundred pages; all instructions, specifications, and requirements need to be addressed. One of the tasks of the editor is to prepare a cross-reference of proposal paragraphs to requirement paragraphs, a very tedious, time-consuming job.

Once a proposal has been sent to the customer, the job has only begun. Then a series of customer inquiries follows, necessitating "change pages." This process continues until all questionable areas are clarified. After all of this, you can only hope that your company wins the contract!

Organizing and Reviewing

Many proposals follow a standard form and are, in fact, partly composed by a "cut and paste" procedure. In other words, certain standard information, like the background of the company and division, resumés, and related experience, can be written once and then used in more than one proposal with only minor changes. Having such standard segments (which are called *"boilerplate"*) is an advantage, because most proposals are written under severe time pressure and with inflexible deadlines. However, the danger of the cut and paste method is that the proposal will lack coherence—segments that seem unrelated weakening the impact of the document as a whole.

Students and beginning workers writing their first proposals have a related problem. It's very tempting to adopt an existing proposal model without evaluating whether its organization will work in a new situation and with new material. Before adopting any model, study its organization and the contents of each section. If they are not appropriate for your needs, change to a new organization or adapt what you have. Don't just copy the model. Use the various brainstorming techniques in section 2.1 to gain a fresh perspective on the material.

Many times you must, of course, follow your instructor's or your company's form and organization. Sometimes that method of organization is even specified in the RFP. But when you and your colleagues review the organization, make sure that the parts contribute to a unified whole. The proposal should present a logical, coherent, and clear argument for your plan.

Figure 20-6 shows an informal proposal for an evaluation of scanning electron microscopes. Since the proposal is internal, the writer uses memo form, but this proposal is longer and more detailed than the student proposal in Figure 20-4. Notice how the writer organizes the material into sections that make the proposal easy to understand.

Writing, Reviewing, and Editing

Writing a good proposal calls for nearly all the techniques covered in Chapters 9 through 12. Proposals often include definition, description, and analysis. The methods section is a procedure, and the introduction employs interpretation and persuasion. Longer and more detailed proposals usually include graphics. Examine the specific needs within sections of the proposal to see how graphics can help clarify information.

Once the proposal is written, it needs several kinds of review. Specialists in the technical field should review the introduction and the procedures sections (including graphics) to verify technical accuracy. Management representatives should review personnel and budget sections to ensure accuracy of time and cost estimates. Copy editors should review the writing itself for clarity and correctness of syntax, diction, and punctuation. As a student, you must sometimes take on these reviews yourself, or you may have access to peer review. Use outside reviewers whenever you can. You may be too involved to see the material objectively.

Some agencies or companies "grade" a proposal and comment on its strong and weak points as well as telling why they have rejected it (if they do). If your proposal is graded, value that information for helping you learn what it takes to succeed in this challenging form of writing.

To: Dr. Connie Martinez, Department Chair

From: Kim Rosetta   *K. R.*

Date: September 10, 1988

Subject: Comparative report for the Smith University Department
of Biological Sciences on scanning electron microscopes
(SEMs) compatible with image enhancement (IE).

Summary

I propose to write a comparative report on SEMs compatible with IE
as part of my research assistantship. I will submit the report by
December 15, 1988, to Dr. Bernard Colson, Chair, Subcommittee on
SEM Purchase, Equipment Committee, Department of Biological
Sciences, Smith University. My report will compare SEMs specified
by the subcommittee, with emphasis on features the subcommittee
selected for further study. The report will form a basis for
selection of a new SEM, to be purchased with funds from Federal
Grant # 1067A.

Introduction

The Department of Biological Sciences has determined the need for
an SEM with recently available features, including IE. On August
15, 1988, the Department was notified that the new microscope
would be funded through Federal Grant # 1067A. In light of funding
approval, the Department Equipment Committee, chaired by Dr.
Hilda Johnson, met with SEM users on September 1, 1988, to discuss
guidelines for purchase. The following guidelines were
established by the Committee:

GUIDELINES FOR SEM PURCHASE

Price Range, including IE	$120,000–$140,000
Electron Gun	tungsten; possibly also LaB_6
Saturation	auto
Focus	auto/manual option
Magnification	x250,000
Stigmator	auto/manual option
Brightness/Contrast	auto/manual option
Dynamic Focus	standard
Gamma	standard
Tilt Correction	standard

Further detailed study was delegated to a temporary Subcommittee
on SEM Purchase. Through preliminary inquiries, the subcommittee
identified eight SEMs, manufactured by five companies, for further
study. The committee felt that further study should focus on the
following features:

FEATURES FOR COMPARISON

SEM

    Lens system
    Vacuum systems   *(Continued)*

Figure 20-6 Informal proposal

Resolution
Specimen chamber size
Specimen stage—movement in x, y, and z axes; rotation; and
   tilt
CRTs—number, size, type, and number of slow—scan lines for
   direct record
Installation requirements
Maintenance requirements
Service agreements

IE

Number of pixels for digitized image record
Grey scale—number of bits
Pseudocolor—available?
Image processing—standard features available

Subcommittee members noted that the SEM is an expensive long—term purchase; also, they recognized that the SEM must fulfill the varied research requirements of over 30 professors and graduate students. In order to make an informed recommendation of a model for purchase, the subcommittee felt that a detailed comparative report delineating differences between the eight models should be prepared. The subcommittee decided that a careful review of these models by someone familiar with both SEMs and departmental research was necessary. Because such in—depth study will require considerably more time than subcommittee members can allocate, members felt that the study should be incorporated into a research assistantship. The graduate student assigned the study will present the report to the subcommittee at their December 19, 1988, meeting.

Proposal

I propose to perform the comparative study needed by the SEM Subcommittee and provide a detailed report by December 15, 1988.

I will request information on the eight models identified by the subcommittee from the following companies:

AMRAY
Cambridge Instruments
Carl Zeiss, Inc.
Hitachi
Jeol USA, Inc.

The requests will be by phone to initiate the study as quickly as possible, with a written follow—up to ensure receiving the correct information. I will clearly specify the model numbers and information needed, according to the SEM Subcommittee's decisions.

I will review the information received, compile it, and arrange it in table format. To obtain information not provided in company brochures, I will conduct discussions with company representatives by phone. I will also invite representatives to visit our department; I will notify all interested members of the Department of Biological Sciences of these visits.

Figure 20-6 Continued

In addition, I will send a one-page survey to department SEM users to clarify and confirm departmental needs. I will compile a list of preferred features and indicate how many members designated each as helpful, and how many designated each as essential.

My report will include the sections listed below:

(1) A table of information pertaining to features for comparison for the eight SEMs identified by the subcommittee
(2) Notes on special aspects of the SEMs that may be of interest to the subcommittee
(3) A list of SEM features preferred by departmental members, showing, for each feature, the number of SEM users who judged the feature as essential or helpful to their research
(4) A one-page summary of the advantages and disadvantages of the three models that, in my opinion, will best serve departmental members, according to priorities identified in the survey
(5) An appendix containing literature received, and notes from company presentations, organized by company and model number (in primary copy only)
(6) An appendix containing the names, addresses, and phone numbers of company representatives contacted
(7) An appendix containing evaluation comments on the report by Department members

Evaluation

I will provide a draft of the report to interested members of the department by December 1, 1988, for their review. Members will be asked to evaluate the report for completeness in terms of specifications given in this proposal. In addition, they will be asked to assess the value and usefulness of the report.

Schedule

The following dates are presented as a tentative schedule:

Sept. 19–27, 1988	Telephone companies
Sept. 26–Oct. 2	Write follow-up letters
Oct. 3–9	Write one-page survey for departmental SEM users
Oct. 10–14	Approve survey with SEM Subcommittee Chair
Oct. 17–21	Send out survey to departmental SEM users, to be returned Nov. 4
Oct. 17–Oct. 31	Review information; compile preliminary table "Features for Comparison"
Oct. 24–Nov. 11	Obtain information still needed for above table through discussions with company representatives
Nov. 7–13	Compile data from departmental SEM users survey into list form
Nov. 7–13	Complete table of information "Features for Comparison"
Nov. 14–20	Write report—first-draft stage
Nov. 21–25	Enter report into departmental word processor
Nov. 26–29	Edit report—second-draft stage
Nov. 29–30	Print and make 10 copies of second draft

Figure 20-6 Continued

Dec. 1   Give second draft to interested department members for evaluation, to be returned by Dec. 5
Dec. 5–11   Prepare final report
Dec. 5–11   Prepare appendices
Dec. 12–14   Print and copy final report
Dec. 15   Give final report to Chair, Subcommittee on SEM Purchase
Dec. 19   Present final report to subcommittee at their meeting

Time Line

1. Telephone companies
2. Write letters
3. Write survey
4. Approve survey
5. Send out survey
6. Review information
7. Obtain information still needed
8. Compile data from users survey
9. Compile table
10. Write first draft
11. Enter into word processor
12. Edit report to second draft stage
13. Print and copy second draft
14. Draft for evaluation
15. Prepare final report
16. Prepare appendices
17. Print and copy final report
18. Give final report to Chair
19. Present report to Subcommittee

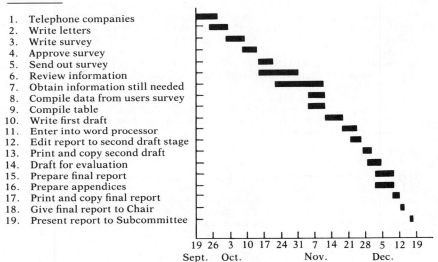

19 26   3 10 17 24 31   7 14 21 28   5 12 19
Sept.   Oct.   Nov.   Dec.

Staff/Qualifications

I plan to personally complete all work specified in this proposal.

I am a fourth–year Ph.D. student in microbiology, working under Dr. Pat Georges, Professor of Microbiology. My research work on the filamentous bacterium *Actinoplanes philippinensis* has demanded extensive SEM work. I have spent an average of fifteen hours/week in the SEM laboratory during the past two years. Thus, I have become familiar with Departmental SEM projects, and I have assisted many of the newer graduate students.

My formal training includes 15 quarter units of physics and 140 quarter units of biological sciences; courses in both SEM and transmission electron microscopy (TEM) were part of my graduate work. I also attended a one–week seminar in SEM–IE through Microscopy Seminars, Inc. I served as a teaching assistant for the Smith University SEM course during both my second and third years as a graduate student.

Figure 20-6 Continued

Cost

TIME

Obtain information from companies	10 hours
Write survey	2
Review and compile information from companies	10
Review and compile information from surveys	6
Write report	14
Type report	8
Edit report	4
Organize appendices	5
Printing and copying	2
Presentation	2

Total time	63 hours $\times$ \$10.00 / hour = (out of research assistantship hours)	\$630.00

COPYING

10 copies—second draft	15.00
15 copies—final report	23.00

TELEPHONE

Average of two 10—minute long—distance calls for each of five companies	50.00

TOTAL COST	\$718.00

Request for Approval

I request approval to conduct the research and prepare the comparative report as described in this proposal.

Figure 20-6   Continued

CHECKLIST FOR WRITING PROPOSALS

1. Have I fulfilled all the needs of a proposal? Does my proposal
_____ solve a problem or provide a service?
_____ follow a specified procedure?
_____ use identified people?

_____ adhere to an announced timetable?

_____ project a specific fee?

2. Is this proposal

_____ solicited?

_____ unsolicited?

_____ internal?

_____ external?

_____ informal?

_____ formal?

3. If the proposal is informal, does it include

_____ an introduction giving the need or the problem and background?

_____ a proposed plan for solution (procedure) that describes the approach?

_____ means of evaluation?

_____ time table?

_____ staffing?

_____ cost?

_____ a request for approval?

4. If the proposal is formal, does it include

_____ a letter of transmittal?

_____ an abstract?

_____ an executive summary?

_____ a title page?

_____ a table of contents?

_____ a list of figures?

_____ a list of tables?

_____ an introduction?

_____ a procedure or methods (solution)?

_____ personnel (staffing)?

_____ budget (cost)?

_____ appendixes?

EXERCISE 20-1   Analyzing a Proposal

Directions: The following proposal was written by students in the Division of Technology (DOTSA 1985) to establish an office for the student organization. Analyze it to see if it

1. includes all the parts of an informal proposal
2. makes its case effectively
3. is organized appropriately

Be prepared to discuss your findings.

SAN JOSE STATE UNIVERSITY•DIVISION OF TECHNOLOGY
1 Washington Square•San Jose, CA 95192-0061

Date: September 24, 1985

To: D. Betando, Director

From: DOTSA Executive Committee

Subject: Consideration of establishing
         DOTSA/EPT office

Purpose

This proposal presents the ideas of DOTSA to establish a common
office for DOTSA and EPT in the Division of Technology Resource
Center. This office would be the headquarters for these two
organizations, which would take on the initial reorganization and
continual maintenance.

Background

The Division of Technology Resource Center is currently closed to
students due to a lack of organization and security. Its use is
limited to DOTSA/EPT officers and graduate students under the
supervision of Dr. Bates. The estimated number of graduate
students using this room is three per semester. DOTSA/EPT
currently uses this room for planning events and meetings as well
as for records storage.

Upon investigation of other student organizations on the SJSU
campus, we noticed that, of the six organizations questioned,
each had an office of its own set aside for its use by the host
department. The organizations questioned are found in
engineering. Other departments will be questioned during the next
week. In general, these offices were observed to be disorganized,
bare, and visually unappealing. A list of the organizations and
their room/building numbers appears in Appendix A. By
comparison, we can make an outstanding center in our Division.

Proposal

Office Arrangement. The office equipment could be reorganized into
a more attractive arrangement. One such plan is shown in Appendix
B. By stacking the library shelves and anchoring them to the wall
the space can be utilized more efficiently. Purging the file
cabinets of unnecessary materials would eliminate the need for
all but three of the four-drawer file cabinets. The unwanted file
cabinets could be distributed to instructors who may need
additional file space. One desk and file cabinet would be specified
for each of the two organizations.   (Continued)

Figure 20-A

The two work stations are proposed to be used as computer stations. One computer has been offered to DOTSA by Dr. Mueller. The other work station might be used for faculty to develop curriculum with another PC. These two PC stations can share a common printer. Such a printer has been offered by Dr. W. White. Two bulletin boards have been offered to aid in communications to the student organization.

Library. The executive committee of DOTSA proposes to reorganize and add to the existing library contained in the DOT Resource Center. Additional library materials can be gained through donations from industry or national affiliations such as SME. The reorganization would be accomplished by separating the existing books into categories and discarding or selling those books which have no value to our students. By labeling the shelves into categories a student or faculty member could more easily find the valuable information sought.

Reorganization of the periodicals would be accomplished by labeling a specific place on the magazine rack for current issues of each periodical regularly received by DOT. All past issues of that year would be located in a labeled bin located near the display. Issues older than one year would be donated to an area school or faculty. Current periodical files would be sorted and purged. (See list of DOT periodicals in Appendix C.)

Student Use. The proposed office and library would not be open at all times, for security reasons. DOTSA suggests that a schedule of hours be set up for open student use. During these hours at least one officer of either student organization must be present. This would ensure security, and also increase the value of the library. No food would be allowed in the office. Social gatherings would be discouraged.

Summary

DOTSA seeks to establish an office which will be available for executive meetings of both organizations to increase camaraderie and professionalism within our division. This office would not be a social hall but a headquarters for the clubs to plan for future events. The responsibility for the appearance and maintenance of this room would be solely on the officers of both clubs under the supervision of Dr. Bates and Mr. Lichtenstein or other advisor.

We seek your input on this subject and wait eagerly for your reply.

APPENDIX A

Organization	Room Number
ASME	Eng 179
IEEE	Eng 234
ASM	Eng 123
ASCE	Eng 171
ACM	Eng 239
AIIE	Eng 315

Figure 20-A Continued

APPENDIX B

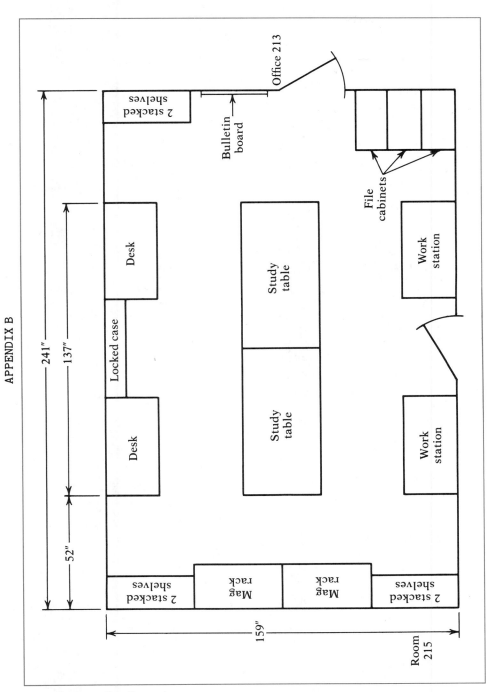

Figure 20-A Continued

APPENDIX C

Periodical List	Receiving Instructor
American Educational Research Journal	Betando
American Fabric & Fashions	Christensen
American Machinist	Chaplin
American Photographer	Helprin
Architectural Record	Centanni
Cal. Computer	Bates
Chronicle of Higher Ed.	Betando
Circuits Manufacturing	MacDonald/Chaplin
Communication Arts	Hazarian
Composites Tech Review	Bates
The Computer Journal	Mueller
Computers & Electronics	Mueller
Creative Computing	Mueller
Darkroom Photography	Helprin
Design News	MacDonald
Electronics	Mueller
Engineering News Record	Centanni
Fine Homebuilding	Centanni
Fine Woodworking	Betando
The Futurist	Markert
High Tech	Bates/MacDonald
Machine Design	Bates
Materials Eng.	Bates
M & C Measurements & Control	Chaplin
Metal Working News/American Metal	White
Modern Plastics	Bates
Photo Lab Index	Helprin
Plastics Design & Processing	Stivers/Bates
Power	MacDonald
Print	Hazarian
Progressive Arch.	Centanni
Radio Electronics	Mueller
Reprographics	Centanni/Moore
Science News	MacDonald
Scientific American	MacDonald
Solar Age	MacDonald/Centanni
Wood & Wood Products	White

Figure 20-A   Continued

EXERCISE 20-2   Writing a Proposal for a Change in Your
Major Department

Directions: Following the form given in this chapter, write an informal
proposal to the chairperson of your major department suggesting a change
in a course or a procedure. Justify the reasons for the suggested change.
Submit it to fellow students for review; then revise as needed.

WRITING ASSIGNMENT #2   Writing a Proposal for a Term
Project Report

Directions: Follow your instructor's RFP and write a proposal for a term
project. (See section 1.6 and Writing Assignment #2 in Chapter 1 for details). Submit it for review as instructed. If this is a team project, discuss
the proposal content in a group meeting, and then assign one or two people
to write the proposal. Review it in the group before submitting it to your
instructor.

REFERENCES

Cook, Ronald. 1985. *Proposal to investigate and evaluate surge suppressors for the
general microcomputer user.* Informal proposal, San Jose State University.
Davis, Mary. 1985. *Proposal for an evaluation and comparison of four wood-burning
stoves.* Informal proposal, San Jose State University.
Mawk, Kenna. 1985. *Internship report.* San Jose State University.
Rosetta, Kim. 1988. *Proposal for a comparative report on scanning electron micro-
scopes compatible with image enhancement.* Written by Anne Rosenthal, San Jose
State University.
U.S. Dept. of Commerce. 1987. *Commerce business daily.* Issue PSA-9268. 3 Feb-
ruary.

CHAPTER 21

Comparative, Feasibility, and Recommendation Reports

As a student you have been writing something called "a report" since the third or fourth grade, when you laboriously copied out paragraphs on whales from the encyclopedia. Since then you've written book reports, science reports, history reports, and committee reports. In college you've struggled with research papers and laboratory reports. By this time, you may wonder if there is anything new to learn about the process or the form of report writing. In fact, there probably is. Few students have ever written the kind of formal report that is the subject of this chapter: the comparative or feasibility report that can lead to a recommendation. Yet because these reports are widely used in technical and scientific fields, you need to know what they are, what form they take, and how they are written.

21.1 DEFINITIONS OF TERMS AND TYPES OF REPORTS

The word "report" comes from a Latin word meaning "to carry back." In a report, then, the technical writer carries back to the reader the information learned through research or analysis. The writer has become an expert on some subject or problem and shares that expertise through a report.

Some reports present information from original research (in the laboratory or field) as well as from a review of existing research (in the library or through interviews). These are called *research reports,* or scientific or investigative reports. They are often published in scientific and technical journals, in which case they are also called journal articles. The writer reports on "What it is, " "What happens," "How it works," or "What happens if . . ." Because scientists and technical specialists usually seek to publish the results of research or investigation, research reports are discussed in Chapter 22, along with other documents intended for publication.

Other reports analyze and interpret information, come to conclusions, and recommend action. These may be labeled "comparative," "feasibility," or "recommendation" reports. Those that recommend action are also sometimes called proposals. All these terms imply studying a subject carefully or matching one option against another to reach a conclusion. The writer asks "Will it work?" "Can we do it?" or "Which option is best?" and the report is structured to answer those questions. A *comparative report* examines two or more products or options and matches one against the other or against a set of criteria. A *feasibility report* looks at options, ideas, or products to see if they can be done or used and which one is preferable. Both may include recommendations and are, therefore, sometimes called *recommendation reports*. This chapter considers reports that examine alternatives in this way—reports that are usually not published but instead are written for internal company use or for external clients or customers. For simplicity, I will call them recommendation reports, though sometimes conclusions are presented without actual recommendations.

Recommendation reports are usually divided into two large classes based on style and form—what we call informal and formal reports. Informal reports are often written in conversational style and appear in various forms such as letters and memos. They are discussed in detail in Chapter 14. Formal reports are written in a more objective, impersonal style, and also have an established form—an arrangement of standard parts that work together to convey information efficiently to both specialists and nonspecialists. A sample formal report appears at the end of this chapter.

21.2 THE FORM AND ORGANIZATION OF A FORMAL REPORT

Like proposals, formal reports are divided into discrete parts that are identified by subheadings. The various parts meet the needs of different types of readers and provide a number of ways to get at the information.

Not everyone reads a report from beginning to end. For example, some managers read only the executive summary; others want to see the conclusions and recommendations. Specialists might be primarily interested in the abstract and in procedures or description. Even those who do read a whole report might not read it from beginning to end, but might move from abstract to conclusions to introduction to an appendix. Therefore, one of the writer's major tasks is to make the information in the report easily accessible.

Like most other forms of writing, formal reports are divided into three large sections: introduction, body, and conclusion. In addition, they have peripheral material at both the beginning *(front matter)* and the end *(back matter* or "end matter"). Listed below are typical parts of a formal report.

front matter	letter or memo of transmittal
	abstract (and often executive summary)
	title page
	table of contents
	list of illustrations (or separate list of figures and list of tables)
the report itself	introduction
	body
	conclusions (sometimes placed earlier)
	recommendations (sometimes placed earlier)
back matter	reference list or works cited
	glossary
	appendixes. Can include:
	technical data such as specifications, calculations, tables, and graphs
	supporting data such as letters, brochures, and sample questionnaires

Front matter provides an overview and a summary for the busy reader and appeals especially to managers, who want the broader view without all the details. For these readers, the writer may also want to move the conclusions and recommendations in front of the introduction.

The report itself, with its introduction, body, and conclusion, interests those readers who are specialists or technicians and who want to see how the conclusions and recommendations were reached. If these are the primary readers, the conclusions and recommendations are more likely to appear at the end of this section. They will also be summarized in the abstract.

Back matter can appeal to two types of readers. Most often the reader interested in technical back matter is the specialist, who wants detail, calculations, and supporting data. If the specialist is the only reader, this material can be in the body. Both specialists and nonspecialists consult back matter for the glossary and the documentation of sources.

Not all reports include every part, but those parts discussed in the next sections are important ones. After each explanation you will find an example of that part (or a reference to an example) from a student report on surgical lasers. Then, at the end of this chapter, you can study a complete student report to see how the various parts fit together. Be aware that a formal report is *presented* in the following order, but it is not *written* in this order. See section 21.3 and Chapter 5 for suggestions on what to write first.

Front Matter

Letter or memo of transmittal   The transmittal document follows standard letter or memo form, as explained in Chapters 14 and 15. It is called a "transmittal" document because the writer uses it to transmit, or transfer, the report to the intended reader. This is done by means of a personal communication: a letter for external communication and a memo for internal communication. The first paragraph should include the title and purpose of the report and the reason it was written. Middle paragraphs briefly mention the contents, highlighting important information or conclusions and discussing problems or limitations. As in a letter of request, the closing paragraph should suggest action, such as implementing the recommendations. In addition, it can convey thanks or note specific accomplishments. The tone should be confident, positive, and businesslike. The transmittal document is not numbered as part of the report; it may be bound at the front or clipped to the cover. Figure 21-1 shows a letter of transmittal from a student.

Abstract and executive summary   All reports should have an abstract to provide an overview and summary. Abstracts are frequently circulated separately to potential readers; if an abstract looks promising, the reader can order the report. Abstracts are also printed in some technical indexes in libraries. Abstracts can be either descriptive or informational, though informational abstracts are becoming more common. If the report is written for both specialists and managers, it often includes both an abstract and an executive summary—the abstract for technical readers and the summary for managerial readers. The abstract is not usually numbered as a part of the report; it may be bound at the front or left free so it can be detached. If bound, it usually precedes the title page, although in some companies it comes after. For details on writing abstracts and executive

14956 Rome Drive
San Jose, CA 95124

May 6, 1985

Professor L. Rew
English Department
San Jose State University
One Washington Square
San Jose, CA   95192

Dear Professor Rew:

As you requested on January 30, 1985, I have conducted an
evaluation of surgical lasers for Allied Hospital.

Since submitting the original proposal, I decided that a study
limited to Nd:YAG lasers would be most appropriate for Allied
Hospital's use. The four models I evaluated are

    Lasers for Medicine: FiberLase 100
    Cooper LaserSonics: Model 8000
    MBB–Angewandte Technologie: mediLas 2
    Laserscope: OMNIplus

I evaluated these models on versatility, cost, safety, mobility,
and ease of operation.

I believe Allied Hospital surgical personnel will find this report
very valuable in their purchasing decisions, and I am submitting
a copy directly to them. As an example of my ability to research,
organize, and present a comparative study, I submit this copy to
you. I hope you find it informative, complete, and interesting.

This project was very demanding and equally rewarding. The
technical, communicational, and organizational skills it taught
have direct applications in career fields. Additionally, as a
result of this project, I have become intrigued by the tremendous
value and relative youth of lasers in medicine. I am looking
forward to pursuing this topic as a career.

                        Sincerely,

                        David Leonard

                        David Leonard

Figure 21-1 Letter of transmittal, formal student report

summaries, see Chapter 19. The abstract of the student report on surgical lasers appears in Figure 21-2.

Title page   In addition to the title itself, the title page displays the author's name and the date. It can also include the name and title of the person for whom the report was written and, if appropriate, a document or contract number. Some or all of this information is repeated on the report cover, most reports being bound in a sturdy cover of some kind. The title page of the laser report appears in Figure 21-3.

Table of contents.   The table of contents provides both an overview of the contents and a way of finding information. It can be written from the report outline and may include three or four levels of headings as well as the page on which each is found. The headings in the table of contents should match the headings in the report exactly. The table of contents may be in outline form with numbered points, or it may simply use in-

ABSTRACT

An Evaluation of Nd:YAG Surgical Lasers
for Allied Hospital

by David Leonard

     This report evaluates four Nd:YAG surgical lasers for use
at Allied Hospital:

          o   Lasers for Medicine: FiberLase 100
          o   Cooper LaserSonics: Model 8000
          o   MBB—Angerwandte Technologie: mediLas 2
          o   Laserscope: OMNIplus

     Knowledge of surgical laser action, application, and
market competition is needed to choose the right laser. The
evaluation is based on five criteria: versatility, cost, safety,
mobility, and ease of operation. Each model is analyzed by these
criteria; comparison emphasizes key differences among the
models.

     The FiberLase 100 is recommended for Allied Hospital. It
costs about $20,000 less than its nearest competitor, the
LaserSonics Model 8000. Its audible fault indicators and self-
calibration system combine to give it the best overall safety
features of any model. Additionally, its broad delivery options
and push—button input panel give it creditable versatility and
easy operation. Following behind the FiberLase, and each with its
own advantages, are the LaserSonics Model 8000, mediLas 2, and
OMNIplus.

Figure 21-2   Abstract, formal student report

```
                    AN EVALUATION OF Nd:YAG SURGICAL LASERS

                           FOR ALLIED HOSPITAL


                        Prepared for the Surgical Staff




                              by David Leonard

                               May 6, 1985
```

Figure 21-3   Title page, formal student report

dentation to show subpoints. Since reports often do not have an index, the table of contents is an important aid to the reader. Entries should use key words and phrases that give information and tell the purpose of the entry. For example, "Body" is not a useful term in a table of contents, but "Analysis of the Four Calcining Processes" is. Front matter is not listed in the table of contents, which usually begins numbering the report with the first page of the introduction. Back matter, however, is listed; the pages may be numbered in sequence with the report, or each section of back matter may be numbered separately: A-1, A-2, B-1, B-2, and so on. Notice in Figure 21-4 how much information about the contents and organization you can get from a good table of contents.

TABLE OF CONTENTS

Figure 21-4   Table of contents, formal student report

List of figures and list of tables Graphics should be listed separately for the reader's convenience; that is, there should be both a list of tables and a list of figures. Both lists can be on one page, titled "Illustrations." Some readers are more interested in the graphics than in the text, and they should not have to search through the table of contents looking for either figures or tables. Tables are numbered consecutively and given titles that identify the purpose and content of each. All other graphics, such as drawings, graphs, and diagrams, are considered figures. They are also numbered consecutively but separately from tables and given explanatory titles. In the lists of figures and tables, indicate the pages on which they appear. See Chapter 12 for detailed information on tables and figures. The lists of tables and figures for the laser report appear in Figure 21-5.

The Report Itself

Introduction Each report has a slightly different kind of introduction, dictated by its purpose, intended readers, and topic. The introduction should launch the reader into the report by giving background and explaining

ILLUSTRATIONS

Figure		Page
1	Special qualities of laser radiation are not found in conventional sources	5
2	Three basic components of the laser mechanism are responsible for laser production	8
3	The LaserSonics Model 8000: A representative Nd:YAG surgical laser	11
4	Relative performance of the Fiberlase 100	41
5	Relative performance of the LaserSonics Model 8000	44
6	Relative performance of the mediLas 2	48
7	Relative performance of the OMNIplus	52
Table		
1	Key differences among surgical models	38
2	Relative performance of the models evaluated	55

Figure 21-5 List of illustrations, formal student report

the problems to be solved; it can also summarize what will follow. Introductions can include any of the following information, though most will not have all these parts. Also, while the order given here is standard, you can use other sequences.

- definition of the subject or problem to be solved
- background information
- purpose of the report and purpose of the study
- significance of the subject
- scope or focus: what is included, what is left out
- literature or historical review (sometimes included with background information)
- procedure or method of analysis
- definitions of key terms (can also be placed in an appendix)
- overview of the report itself

As you think about the whole report, keep the length of the introduction in proportion. It should not exceed one-fourth of the total report, so you may want to move parts you must treat at length into the body. However, all introductions should include:

- a statement of the problem
- the purpose
- the topic and scope
- a brief overview of the report

See section 5.4 for details on writing introductions and an example of the student introduction on surgical lasers.

Body   The central part of the report is the body. Divide it into sections that fit the report content, and give each section a meaningful title; do not call it "body." The body of a formal report can include:

- background
- theory
- description
- definitions
- procedure or methods
- discussion of criteria
- evaluations of products or options
- comparisons
- results
- analysis and evaluation of results

The body of the report on surgical lasers, for example, contains the following sections. Numbers refer to the position of sections in the master outline as presented in the table of contents (see Figure 21-4).

2 LASERS IN THE MODERN HOSPITAL
[This section is mostly background for the nonspecialist reader. It contains definitions, technical descriptions, a process description, an explanation of a complete hospital laser facility, and a brief discussion of competition among hospitals and the role of the government in laser regulation. You can find the definitions from the laser report in Chapter 10 of this book, the description in Chapter 11, and the process description in Chapter 18.]

3 PURCHASING A SURGICAL LASER
[This section explains how the hospital will justify the need for a surgical laser and how the decision was made to study Nd:YAG lasers. Most of the section, however, is devoted to a detailed explanation of the five criteria chosen for evaluation, why each criterion was chosen, and how each criterion can be be applied in evaluating the various brands.]

4 FOUR GOOD MODELS
[In this section the writer first describes the group as a whole and then provides a technical description of each model. Details of the key differences of each model are summarized in the following table (21-1), which appears in the report as Table 1.]

A recommendation report is a decision-making document, and the decisions must come from the analysis and evaluation that take place in the body of the report. Thus, as this student writer did, you must set up and discuss criteria or standards for judgment. How can you choose such criteria or judgment guideposts?

Sometimes the criteria will be given to you along with your assignment:

> Determine whether we should switch to a solar hot water system for the design laboratory. Look at installation cost and maintenance cost vs. our current cost for the gas-fired heaters. Also find out how long the system would be inoperable if we made a change.

This is a feasibility assignment: the reader wants to know if it's feasible to change to solar hot water. As the writer, you know that your primary criteria will be the two kinds of cost and the time lapse for installation.

Often, as the person who has (or will have) the most technical knowledge, you must choose the criteria yourself. One way to do that is by brainstorming at the planning stage for a list of all the criteria you can think of. Then you can refine the list by considering the purpose of your study and the intended reader. As you gather information, you will be able to refine the list further.

For example, the student writing on surgical lasers brainstormed a list of the following criteria—only some of which he used in his final report.

adaptability	cost
ability to be upgraded	height
safety	warranty
weight	ease of use
accessibility of parts	repair record

Table 21-1   KEY DIFFERENCES AMONG SURGICAL MODELS

Category	FiberLase 100	LaserSonics 8000	mediLas 2	OMNIplus
versatility power (W)	15-100	1-110	15-100	0-15
approval	GI, ENT	GI, neuro	GI	investig.
delivery systems	nonfocusing, focusing, and malleable pieces with range of focal lengths	curved and straight pieces, interchange-able lenses	extensive range of pieces, endoscopes, guides with liquid or gas circ.	nine specific delivery systems
cost (list)	$79,000	$99,000	$105,000	$117,000
safety	audible fault self-calibrate short focal lengths	audible fault and delivery no kill switch	audible delivery	calibrate delivery devices no kill switch
mobility	660 lb.	800 lb.	400 lb. laser head projects	600 lb.
ease of operation	push-button controls	He-Ne and white aiming light	no high flow gas circ. needs ext. gas	remote operation, self-diagnostic increase/ decrease touch sense controls
availability	U.S.	U.S. California based	W. German	U.S. California based

training availability   technical support
projected life span   quality of written instructions

When you have criteria, you use them to compare alternative methods or products. You must objectively analyze the advantages and disadvantages of each option, for only then can you reliably conclude and recommend. The comparison section can be done in two ways: the point-by-point method or the block method.

In the point-by-point method, the criteria are the major organizing factors. Each criterion becomes a subsection in which the various options or products are compared. For example:

Cost Criterion
    Option 1
    Option 2
    Option 3

Reliability Criterion
    Option 1
    Option 2
    Option 3

In the block method, the options or products are the major organizing factor. Each option or product becomes a subsection in which the various criteria are applied. For example:

Option 1
    Cost
    Reliability
    Safety

Option 2
    Cost
    Reliability
    Safety

You also need to think about the order in which you present criteria. You may want to put the most important criterion first and complete the list in descending order. This is probably most common, but if you have a reason for doing it, you can follow a different order. See Chapter 3 for more on methods of organization.

Conclusions and recommendations   The concluding section should briefly summarize what has been said and done and then discuss and interpret the results. Conclusions and recommendations are the reason for writing this kind of report, so they are often moved in front of the body, where they can be found and read easily. They can be combined in one section or treated separately. Both conclusions and recommendations should be stated confidently, for the whole report exists to back them up. *Conclusions* are the results of the analysis and interpretation—telling what the data means. For example:

> When the results are tabulated (see Table 1), the Hearthstone I emerges as the clear winner of the four wood-burning stoves evaluated. Its soapstone exterior is beautiful, durable . . . (Davis 1985).

A *recommendation* is like a call to action—telling what should be done about the results. For example:

> Metal Technology Corp. should purchase Anvil-4000 software and H-P 9000 hardware for the new CAD/CAM system. These systems are clearly superior to the others evaluated (Bautista 1984).

If your presentation is thorough, your data complete, and your analysis sound, the reader will accept your conclusions and recommendations as valid.

Don't feel, though, that you must always support one action or one option. For example:

> I recommend the SL-1 office telephone system because it provides all the required features and has a better service record than the other systems. However, the SL-1 system is also the most expensive. If the cost is restrictive, I would recommend the NEAX 2400 system as a second choice (Edwards 1985).

Your conclusion might also be that the data is insufficient and more study is needed; it might be that no product or option meets the criteria and that other products or options need study. It might even be that the criteria need revision. Remember that through your study you have become an expert on this subject. You have an obligation to the reader to share whatever you have learned as clearly and objectively as possible.

In the student report you have been studying, conclusions and recommendations are combined in section 5.

5 A RECOMMENDATION FOR ALLIED HOSPITAL
[This section summarizes what the writer has learned, recommends one brand of surgical laser, and presents the reasons for that choice. It also summarizes the strengths and weaknesses of the three other models. The writer uses a table to summarize the relative performance of each laser, shown below as Table 21-2.]

Back matter in a formal report consists of information that is important but peripheral to the central content. Even though the report is written in sections, it should form a unified whole; thus, you place in the back matter those details that would clutter the main report, sidetrack the reader, or interest only some types of readers. Back matter can all be put

Table 21-2   RELATIVE PERFORMANCE OF THE MODELS EVALUATED

	Versatility	Cost	Safety	Mobility	Ease of Operation	Total
FiberLase 100	3	4	4	2	3	16
LaserSonics 8000	3	2	3	2	4	14
mediLas 2	4	2	2	2	2	12
OMNIplus	2	1	2	2	2	9

Key: 4 = best, 3 = good, 2 = fair, 1 = poor.

in one section (called an appendix) or in separate sections (or appendixes), and the order of the sections can vary. If you choose to label each section as a separate appendix, identify each one with a letter of the alphabet and a title. Typically, back matter will include the following.

Reference list or works cited   This may also be called References Consulted. It can list interviews, tapes, and speeches as well as company reports and library information. If you annotate the references, you will help the reader see which sources contributed what information. Usually the reference list is the first section in the back matter. See Chapter 13 for details on constructing this list and for excerpts from the annotated reference list used in the laser report.

Glossary   Definitions of terms may appear in the back matter or the front matter, but the longer the glossary, the more likely it is to appear at the end. In the glossary, you should alphabetize terms for easy reference and define them with phrases. You may also include symbols in the glossary, or you may have a separate section for symbols and abbreviations. A glossary is especially important if the report will have several types of readers and if the readers are generalists or managers. Terms should also be defined the first time they are used in the text, or the words should be highlighted (by caps, italics, or boldface, for example) and the reader referred to the glossary. See Chapter 10 for more on definition. The first five entries in the laser report glossary look like this:

> AIMING LIGHT is an accessory component of a surgical laser system that indicates the target spot with its own beam.
>
> BRONCHOSCOPE is a delivery component of a surgical laser system used for examining the air passage to a lung.
>
> COLLIMATION is the process by which divergent rays are converted into parallel rays.

CONTINUOUS WAVE is a mode of laser operation in which the laser beam is emitted continuously as opposed to pulsed mode.

DOUBLED Nd:YAG is a Nd:YAG laser beam whose frequency has been doubled and its wavelength halved by passing it through a special crystal.

Technical data Specialists and technicians often want technical details, including specifications, calculations, tables of data, copies of lab reports, and physical or procedural descriptions. This material should go in the body if it is crucial to understanding the text, in the appendix if it provides only support or added information. The laser report, for example, includes a list of hospitals using each of the four kinds of surgical lasers and the names of physicians who can be contacted by interested professionals.

Supporting data Material in this section might include copies of relevant letters, proposals, and brochures. Such material helps validate the report's information, but it does not directly relate to the content. The laser report includes a copy of the request letter written to laser manufacturers and a copy of the original proposal for the study.

The student report on four-track cassette recorders at the end of this chapter will give you an idea of how all these major parts fit together.

21.3 THE PROCESS OF WRITING A FORMAL REPORT

Like proposals and research reports, recommendation reports are long, complex, and composed of many parts. One challenge in writing such a report is the sheer volume of the material; another is keeping the parts intelligible by themselves but also smoothly integrated into the whole document. Before you undertake a feasibility, comparative, or recommendation report, it will pay you to review Chapters 1 through 8. You will want to refresh your memory about details that will be mentioned only briefly here.

Planning

If you are asked to write a recommendation report in industry, your primary purpose is already set, for the purpose dictates the choice of this form. The primary purpose is to interpret, evaluate, judge, or recommend—and often all of these. Secondary purposes might be to define, describe objects or procedures, or explain. If you are assigned to write a recommendation report in the college classroom, you also have a general purpose and a form, but you may not have a topic. To refine your purpose and your topic, you must think through what you intend to analyze and how your conclusions and recommendations can come from that analysis.

Recommendation reports can be written for specialists, managers, or generalists, so you must also determine your primary reader and all secondary readers. Often you will write these reports for some level of management, using your time and expertise to help managers make decisions. You could also write for generalists like consumers. As section 1.2 points out, you need to evaluate potential readers carefully so that you can meet their needs with useful information, understandable vocabulary, and logical report organization. Knowing your readers will also help you decide what material to place in front matter, in the report itself, and in back matter.

For a major project like this, you need to do both long-term and short-term scheduling. Use a Planning Sheet like that in section 1.6 and schedule the *whole* project first, allowing as many days or weeks for each step as seems necessary. Then schedule tasks *within* each step so that you can work on the analysis and the report itself a little at a time. My students almost always underestimate the time they need for gathering information, for writing, and for typing; you should add contingency time to these categories. Your instructor can give you advice about time allotments based on the length of the term and your specific writing requirements.

Gathering Information

Before you start gathering information, review section 2.1 and spend some time brainstorming to see what you already know, where the gaps are, and how the material groups itself in your mind. Then plan your search strategy. Establish tentative criteria to help guide your research. Library research can help you understand basic descriptions and procedures and is a good place to begin. But in gathering data for analysis, you are more likely to be doing field research of some kind:

- writing to or interviewing specialists
- gathering specifications and descriptions from brochures and other current data not likely to be in a library
- conducting surveys
- observing processes
- examining places or objects
- experimenting with options, methods, or products

It's important to keep careful notes as you proceed in order to have all the needed information when you are ready to write. You may even find that data can't be recovered after an experiment is over. As you proceed, you may need to revise some criteria or eliminate some options. Do make changes as you learn more about the subject, because those changes will move you toward the organizing step. At some point you will be schedule-driven to stop gathering information and begin organizing, but if necessary, you can go back later to fill in gaps.

Organizing and Reviewing

As you begin to organize your data, don't worry about the front and back matter of the final report. Those sections can be tackled after the body is written. In fact, you may want to ignore the introduction as well—just don't lose sight of the report's purpose (which should be stated in the introduction). Concentrate your first efforts on organizing the body itself— that central section where you present data and judge it by established criteria.

Read through all your information in order to settle on a main point. Then sort the data and shuffle the bits and pieces into a tentative arrangement. Choose a method of organization that will help you achieve your purpose. Probably in a recommendation report your organization will involve some kind of hierarchy, such as comparison-contrast or problem-cause-solution. These methods of organization and others are explained in Chapter 3. Remember that data can almost always be organized in more than one way, so try two or three possible organizations, checking each one to see if it fulfills your purpose and meets your readers' needs. Once you have the body organized, choose material for the introduction and create an outline for that. Determine what the conclusion and recommendations will cover and note the main points. (Or write the conclusions and recommendations first and then the introduction.) Put all the parts together and see if they fit. Adjust sections as necessary.

When you have a rough outline, review it by applying the questions in Chapter 4. Then, if possible, have three or four other people review your organization. You will benefit from their perspective, which can be more objective than yours.

If you are writing this report as part of a team, meet with other team members to examine your organization in relation to theirs. Remember that while each of you may be writing separate segments, the segments must interlock to produce a coherent whole.

Writing

Before you begin writing, review Chapter 5 on organizing the writing process. That first plunge into writing is scary, so begin with easy material that you know well. That might be definitions, a description, or a procedure—all concrete sections for which you probably have plenty of information. Once you've written that material, you might turn to an explanation of your criteria. Then you can apply the criteria to your options, and so move on to conclusions and recommendations. Next you might want to write the introduction (though you could also try it earlier). Note places in the manuscript where graphics would be helpful. Later you can find or create the best graphics for those spots.

Only after you've written the main report should you worry about front and back matter. By then much of the work will be partly done. You might have created most of the glossary with the definitions you wrote to warm up; now you only need to finish and alphabetize it. If you kept track of source materials as you researched and wrote, creating your reference list will be a simple matter. The outline will convert to a table of contents with only a bit of polishing (though the page numbers can't be added until the report is typed). That leaves only the abstract, executive summary, and letter of transmittal, which must be written last since they are based on the main report. So you see, if you take one small step at a time, you'll have the draft of your report written before you know it.

Reviewing Between Editing, Typing, and Assembling

A long report needs to be reviewed several times:

- before it is typed, to check for accuracy of content, organization, grammar, syntax, and punctuation. This is called copyediting.
- after it is typed, to check for correctness and agreement with the draft. This is called proofreading.
- before the graphics are put in final form to ensure accuracy of content and details.
- after the graphics are incorporated, to make sure they're where they're supposed to be, properly titled, and referenced in the text.
- after the pages are punched and assembled between the covers. (Nothing is more unprofessional or distracting to a reader than a page inserted upside down or backwards.)

All those reviews should ensure a report that is presented as well as it is written. *Presentation* is important. To create a professional-looking document, follow these suggestions:

1. Allow an extra inch for the left margin so that the pages can be bound without covering up any text.

2. Follow the suggestions in Chapter 6 for use of white space, print sizes, and varied levels of section headings.

3. Allow plenty of time to type the final copy. Even numbering pages and finishing the table of contents with page numbers can take more than an hour.

4. Use a word processor if possible. It will speed up changes and revisions in the text.

5. Schedule time for all the levels of review discussed in Chapter 7.

Study the complete student report at the end of this chapter to see how one student tackled a recommendation report. Remember, though, your report should not be organized, written, or presented exactly like this one. The material, your purpose, and your intended reader should motivate you to create your own unique report.

Note: Because this report was done as a classroom assignment on writing skills, the definitions, description, and procedure are included in the main document. In a report for industry, that material would either be omitted, or it would appear in an appendix. To save space, the sample report is single-spaced. Your report, however, should be double-spaced for ease of reading.

CHECKLIST FOR WRITING A FORMAL REPORT

1. What is my purpose? my topic? Who is my primary reader?
2. In the front matter of my report, have I included the
   _____ letter or memo of transmittal?
   _____ abstract and/or executive summary?
   _____ title page?
   _____ table of contents?
   _____ list of figures?
   _____ list of tables?
3. In the report itself, do I have
   _____ an introduction?
   _____ a body?
   _____ conclusions?
   _____ recommendations?
4. Should I combine the conclusions and recommendations or place them in separate sections?
5. In the back matter, do I have
   _____ a reference list?
   _____ a glossary?
   _____ appendixes?

EXERCISE 21-1   Analyzing Tables of Contents for Completeness

Directions: Study the following student tables of contents and determine if they contain the major elements for an effective report and if the subheadings give enough information to help the reader. Be prepared to discuss in class.

TABLE OF CONTENTS

TABLES AND FIGURES

Example 1

TABLE OF CONTENTS

LIST OF TABLES AND ILLUSTRATIONS

Example 2

TABLE OF CONTENTS

Example 3

WRITING ASSIGNMENT #13   Completing a Formal Report

Directions: Following your instructor's specific suggestions, write a comparative or feasibility report leading to a recommendation. Related writing assignments earlier in this book will help you prepare. Consult Writing Assignments #1 Choosing a Topic for a Report, #2 Writing a Proposal, #3 Writing Letters that Ask for Information, #4 Writing an Outline for a Term Project, and #6 Draft of a Term Project Report.

If this is a collaborative project, set deadlines for individual sections well in advance of the due date. Individual assignments might be the

> introduction
> definitions
> technical description
> procedures
> analysis of options
> graphics
> conclusions and recommendations
> front matter (abstract, table of contents, list of illustrations, title
>   page)
> back matter (list of references and appendixes)

Appoint one or two people to edit the individual sections and coordinate them so that the document reads as smoothly as though one person had written it.

REFERENCES

Bautista, Carl. 1984. *Evaluation of CAD/CAM computer systems for the purpose of purchase.* Formal report, San Jose State University.

Davis, Mary. 1985. *An evaluation and comparison of four wood-burning stoves.* Formal report, San Jose State University.

Edwards, Kellie. 1985. *Office telephone systems: a comparative report.* Formal report, San Jose State University.

Sample Student Recommendation Report

675 Bell Avenue
Palo Alto, CA   94306

November 30, 1984

Prof. L. Rew
English Department
San Jose State University
One Washington Square
San Jose, CA   95192

Dear Prof. Rew:

Enclosed is the formal report you requested on September 3, 1984.
The report, "A Comparative Study of Four-Track Cassette
Recorders," makes a comparison of available four-track cassette
recorders and recommends one for use in my home recording studio.

The research and evaluation process involved in this project
helped me to decide that the Tascam 244 is the best four-track
cassette recorder for use in making demonstration tapes for my
music students. Although I could have arrived at this conclusion
by merely reading the technical and price specifications on each
product, this project taught me how to clearly and persuasively
present the comparative process that my mind went through in
arriving at this conclusion. I can see how such a presentation is
mandatory in a more complicated project involving large amounts
of money and resources.

The practice in research, organization, and writing provided by
this assignment was valuable experience for me, and I feel that
the report itself is a strong addition to my writing portfolio.

I request that you accept this report as fulfillment of the
proposal I submitted at the beginning of the semester.

Sincerely,

Jonathan Neher

Jonathan Neher

A COMPARATIVE STUDY OF FOUR-TRACK CASSETTE RECORDERS

Prepared for

Professor Lois Rew, English 100WT

by

Jonathan Neher

November 30, 1984

INFORMATIVE ABSTRACT OF

A COMPARATIVE STUDY OF FOUR—TRACK CASSETTE RECORDERS

This study, using a set of six evaluation criteria, compares five four—track cassette recorders in order to recommend one or more of them for use in a home recording studio.

The products compared are Aria Studiotrack IIII, Clarion XD—5, Cutec MR402, Fostex 250, and Tascam 244. The evaluation criteria used are tape speed, pitch control, noise reduction, equalization, rack mount, and price.

The study includes (1) a technical definition of a four—track cassette recorder, (2) a description of the principal components of a multitrack recorder, (3) a description of the multitrack sound recording process, (4) an explanation of the evaluation criteria and the point system, (5) individual product evaluations, (6) product comparisons, and (7) conclusions and recommendations.

The study recommends the Tascam 244 for use in a home recording studio because it best satisfies the criteria set up for the study, with a 3 3/4 inch—per—second tape speed, flexible pitch control, an excellent noise reduction system, and built—in equalization. The Tascam 244 combines all the features necessary for recording, processing, and mixing a high—quality four—track cassette tape.

Key words: four—track recorders, cassette recorders, multitrack
               recording, recording studios

TABLE OF CONTENTS

ii   LIST OF ILLUSTRATIONS

1

I. INTRODUCTION

The purpose of this study is to make an objective comparison
among five high-quality four-track cassette tape recorders in
order to recommend one or more of them for use in a home recording
studio. The recorders in the study are the Aria Studiotrack IIII,
Clarion XD-5, Cutec MR402, Fostex 150, and Tascam 244. The
comparison is based on a set of six criteria: Tape Speed, Pitch
Control, Noise Reduction, Equalization, Rack Mount, and Price.
Information for the study comes from the product manufacturers,
magazine articles, and a book on multichannel recording.
Technical terms are defined in a glossary in Appendix B.

The products are evaluated and points are assigned according
to how well each product measures up against the criteria. The
point totals serve as objective indicators of each recorder's
quality in relation to the criteria and the other products in the
study.

The conclusions and recommendations that follow are based on
the product evaluations, point totals, and particular needs of a
home recording studio.

II. CONCLUSIONS AND RECOMMENDATIONS

A. Conclusions

Table 1 is a breakdown of the points awarded to the various
products in the study during the evaluation process. It also
shows the point total for each product.

2

Table 1: EVALUATION POINT TOTALS

EVALUATION CRITERIA	ARIA STUDIO‑TRACK IIII	CLARION XD‑5	CUTEC MR402	FOSTEX 250	TASCAM 244
TAPE SPEED	4	4	4	1	4
PITCH CONTROL	1	1	1	1	2
NOISE REDUCTION	1	2	0	3	4
EQUALIZATION	0	0	0	4	4
RACK MOUNT	2	0	0	0	0
PRICE	5	2	4	0	0
POINT TOTALS	13	9	9	9	14

Interestingly enough, the two products that have the highest point totals are at the opposite ends of the price spectrum. The Tascam 244 lists for $1300 and the Aria Studiotrack IIII for $800.

The Tascam 244 is one of the most expensive recorders in this study, but it includes everything that is needed to produce high‑quality four‑track casette recordings: built‑in equalization, the best noise reduction system, flexible pitch control, and a 3 3/4‑inch‑per‑second tape speed.

On the other hand, the Studiotrack IIII has many of the features of the Tascam 244 and costs $500 less. It lacks built‑in equalization and the highest quality noise reduction system, but it is possible to purchase these units separately if the need for them arises. I was told by Phil Delancy, a recording engineer at the Music Annex recording studio in Menlo Park, CA, that an external equalizer and a dbx noise reduction system would cost approximately $300 each. If both of these units were purchased along with the Studiotrack IIII, the total expenditure would be approximately $1400.

The other recorders in the study did not receive high point totals because they are overpriced or completely lacking in some area that can't be compensated for, such as tape speed.

B. Recommendations

The Tascam 244, which received the highest number of evaluation points in this study, is clearly the best four‑track

3

cassette recorder to buy for use in a home recording studio. For
$1300 it combines into one cabinet all the features necessary for
recording, processing, and mixing down a high—quality four—track
cassette tape.

The Aria Studiotrack IIII, which received the second—highest
number of evaluation points, is a good lower—priced alternative
to the Tascam 244. It includes most of the features necessary for
getting started with four—track recording, and additional
equipment, such as an equalizer and a higher—quality noise
reduction system, can be purchased later.

III. TECHNICAL DESCRIPTION OF THE FOUR—TRACK CASSETTE RECORDER

A. Technical definition

A four—track cassette recorder is an audio recording device
used to record audio signals (sounds converted into electronic
signals) from four separate sources onto four separate tracks on
magnetic cassette tapes.

1. <u>Physical Description</u>

Most four—track cassette recorders are compact because the
main advantage of recording on cassette tapes is their small size
and portability. The recorder shown in Figure 1 measures
473 × 110 × 300 mm and weighs 10 kg.

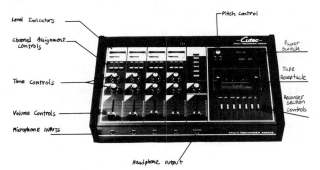

Figure 1   Exterior view of a typical four-track cassette recorder. Source:
Cutec Sound Reinforcement and Home Recording Equipment

4

The knob, switch, and slider controls located on the left half of the top panel control the tone, channel assignment, and volume of each audio signal. The audio signals are input from microphones, which are attached to the phono jacks on the front panel during recording. The meters above the controls for each channel are used to monitor the power level of each audio signal.

The switches on the right half of the top panel are used to control the record, playback, fast-forward, rewind, stop, pause, and eject functions. The tape cassette resides in the receptacle located just above these switches on the top panel. The pitch (tape speed) control, tape counter, and power switch are located just above the cassette receptacle.

The main components inside the recorder are the record/playback head and erase/bias head, which transfer audio signals to and from the tape; the tape drive system, which moves the tape back and forth across the heads at a constant speed; the amplifier, which boosts or reduces the audio signals to appropriate power levels; and the noise-reduction system, which helps eliminate distortion or background noise from the recorded signals.

2. Functional Description

During the recording process, sounds are converted into audio signals using microphones. These audio signals are then converted into magnetic variations on the tape's magnetic coating as the tape is passed over the recorder's record/playback head.

In order to reproduce the sounds, the tape is again passed over the record/playback head; here the magnetic variations previously recorded on the tape reproduce the original audio signal.

A one-track (mono) recorder stores only one audio signal on a tape, while a four-track recorder divides the width of a tape into four separate channels where it records four separate audio signals. These four signals are synchronized during playback because they are passed over the record/playback head simultaneously.

5

B. Technical Description of the Principal Components

1. Recorder Section

   The recorder section is the section of the tape recorder
where electronic audio signals are recorded on, or played back
from magnetic tape. The tape, driven by a tape drive system, moves
from a supply reel to a takeup reel, as shown in Figure 2. On the
way, the tape passes over an erase head, a record head, and a
playback head (the record head and playback head are combined
into a single record/playback head in four-track cassette
recorders).

   The erase head erases any audio signals previously recorded
on the tape. The signal to be recorded is fed to the record head,
which records it onto the magnetic surface of the tape. During
playback the tape is again passed over the record/playback head,
which picks up the previously recorded signal from the surface of
the tape and sends it to the mixer section for amplification and
adjustment.

a. Record/Playback head. A record/playback head is a small coil
of wire capable of producing or sensing variations in a magnetic

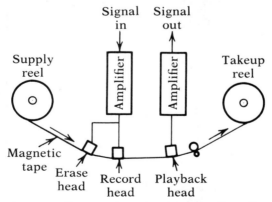

Figure 2 Basic magnetic recorder/reproducer configuration. Source:
Handbook of Multichannel Recording

6

field. During the recording process, the audio signal is sent through the coil causing small variations in the magnetic field, which are recorded on the magnetic surface of the tape as it passes over the coil.

During playback, the record/playback head reproduces the original audio signal by picking up the variations in the magnetic field on the tape's surface.

The record/playback head in a four-track recorder consists of four separate coils stacked on top of one another so that each coil comes in contact with 1/4 of the tape's width. Each coil can record or play back a separate audio signal.

b. Erase/Bias head. An erase/bias head is a small coil of wire that, like a recording head, produces a magnetic field that affects the recording tape. It records a high-frequency bias signal on the tape, which erases previously recorded material and helps eliminate background noise. The tape passes over the erase/bias head before a fresh signal is recorded on it by the record/playback head. The erase/bias head is not used during playback.

c. Tape Cassette. A tape cassette is a small plastic box about the size of a cigarette case. It contains magnetic recording tape wound onto a supply reel and a takeup reel. An opening in one side of the cassette makes it possible for the tape to come in contact with the record/playback and erase/bias heads when the cassette is inserted into the tape receptacle in the recorder.

d. Tape Drive. There are also openings in the top and bottom of the tape cassette at the center of the supply and takeup reels. Drive shafts fit into these openings when the cassette is inserted into the tape receptacle. During record, playback, and fast forward functions, the shaft in the takeup reel rotates counterclockwise and winds the tape to the right. During the rewind function, the shaft in the supply reel rotates clockwise and winds the tape to the left. Both drive shafts are driven by a precision electric motor.

e. Record Section Controls. The recorder section has a variety of external controls for affecting the speed, motion, and direction of the tape. They are located next to each other on the top panel of the recorder, just below the tape receptacle (see Figure 1).

7

The RECORD BUTTON starts the tape moving forward, at normal recording speed, over the record/playback and erase/bias heads. It also engages the electronics involved in the recording process.

The PLAYBACK BUTTON also starts the tape moving forward at normal speed, but it only engages the record/playback head.

The FAST—FORWARD BUTTON causes the tape to move forward faster than normal. It is used to get to a specific spot on the tape quickly.

The REWIND BUTTON moves the tape ahead quickly, and is used to get to the beginning of the recording or locate something on the tape quickly.

The PAUSE BUTTON is used to temporarily interrupt the motion of the tape without disengaging the tape drive or the process being performed at the time.

The STOP BUTTON is used to completely stop the tape drive and disengage all other functions.

The PITCH CONTROL is used to adjust the normal record/ playback speed slightly.

2. Mixer Section

The mixer section allows the recording engineer to control the relative volume, tone, and channel assignment of each signal being recorded or played back on the recorder.

a. Amplifier. The amplifier is a signal—processing device used to boost the power level of audio signals being recorded or played back on the tape recorder. These signals need to be boosted (amplified) during recording so they will produce a magnetic variation in the record/playback head that is strong enough to leave a solid impression on the tape. In addition, the audio signals need to be amplified in order to drive headphones or loudspeakers for monitoring purposes.

b. Monitors. During recording and playback, the recording engineer needs to continuously monitor the audio signals to make

8

sure that the mixer controls are set properly. The engineer uses both aural and visual monitoring techniques.

Aural monitoring is a process in which the engineer listens to the audio signals on headphones or on loudspeakers and adjusts the mixer controls accordingly.

Visual monitoring is a process in which the engineer watches the level indicators to make sure the audio signals are not too weak or strong. Level indicators are meters located above the volume and tone controls of each channel (see Figure 1). Each meter shows the strength of the audio signal being processed on its channel.

c. <u>Controls.</u> Each of the four channels on the recorder has an identical set of controls for the volume, tone, and channel assignment of its audio signal (see Figure 1).

The VOLUME CONTROL (FADER) is used to continuously adjust the level of amplification given to the audio signal during recording or playback.

The TONE CONTROLS are used to compensate for deficiencies in the audio signals and to discriminate selectively against unwanted noise.

The CHANNEL—ASSIGNMENT SWITCHES are used to individually assign each channel to record, playback, or mixdown functions. Using these controls, one channel can be played back to provide a cue for a performer who is simultaneously recording on another channel.

d. <u>Noise Reduction Systems.</u> Almost all four—track recorders have a noise reduction system. This system is used to reduce the level of background noise and distortion that accumulate on a tape during recording and mixdown. The two types of common noise reduction systems used in four—track cassette recorders are "Dolby" and "dbx."

The Dolby system divides the recorded sound spectrum into four separate bands and operates on each band separately to filter out unwanted noise. The dbx system operates on the entire sound spectrum as a unit. The dbx system provides better noise

9

reduction than the Dolby system, but it is also more expensive and
is less commonly used.

IV. DESCRIPTION OF THE MULTITRACK SOUND RECORDING PROCESS

A. Overview

Multitrack recording is the process of recording two or more
audio signals onto separate tracks on magnetic recording tape.

The main purpose of multitrack recording is to
simultaneously or consecutively record multiple sound sources so
that they are synchronized on the recording tape, and so it is
possible to individually adjust the level, equalization, and
special effects on each sound source as it is being recorded or
mixed down.

Multitrack recordings are generally made in a soundproof
environment, such as a recording studio, so that no extraneous
noises are recorded on the tape.

In making a multitrack recording, a recording engineer
usually operates the recording equipment, while at least one
performer is needed to produce the sounds. The sounds can be
produced with musical instruments, voices, other recordings, or
sound effect devices.

The equipment needed to make a multitrack recording is a
multitrack tape recorder with a mixing section, and at least one
microphone for each sound source. During recording, each sound
source is picked up by a microphone, which converts the sound into
an electronic audio signal. Each signal is fed into the mixing
section of the recorder where the input level and tone are
adjusted. The processed signals are then recorded onto their own
separate channels or tracks on the recording tape. The resulting
multitrack recording can then be mixed down onto a standard
stereo tape for distribution.

The principal steps in making a multitrack recording are

1. Setting up and adjusting the recording equipment
2. Making a master recording
3. Mixing down the master recording

10

B. Principal Steps in the Process
1. Setting Up and Adjusting the Recording Equipment

    The recording engineer and the performers work together to
set up the appropriate microphones for each sound source. The
engineer decides the type and placement of these microphones.
While the performers produce sample sounds, the engineer adjusts
the input level of each signal as it is fed into the mixing section
of the recorder. Distortion results if the input level is set too
high, and background noise results if the level is set too low.
The engineer also adjusts the equalization or tone of each
signal.

2. Making a Master Recording

    Once the recording equipment is set up and properly adjusted,
the master recording is made. The engineer turns on the recorder
and signals the performers to begin. During the performance, the
engineer monitors the record levels and tone adjustments to make
sure they stay within acceptable bounds. When the performance is
over, the engineer stops the tape and rewinds it. If, after
playing back the tape, the engineer and performers are satisfied
with its quality, the tape is ready to be mixed down. Otherwise,
the recording process may need to be repeated, or individual
performers may do their parts over (overdubbing).

3. Mixing Down the Master Recording

    Since most tape players can only play one or two tracks,
master tapes having more than two tracks need to be mixed down
onto more common two-track tape for distribution. The engineer
uses the multitrack recorder and a second, mixdown recorder to
accomplish this task. The master recording is played on the
multitrack recorder, and the audio signals that are produced are
routed through the mixing section. The signals are combined into
a composite two-track signal, which is then recorded by the
mixdown recorder. During this mixdown process, the engineer can
again adjust the volume, equalization, and relative balance of
all the tracks being mixed down from the master tape.

C. Conclusion

    Multitrack sound recording can produce high-quality
demonstration tapes of specific musical techniques and sounds. The

11

advantage of multitrack recording is the degree of control the
engineer has over the sound quality.

V. EXPLANATION OF EVALUATION CRITERIA

A. The Point System

   This study compares five available four-track cassette tape
recorders by rating each of them on a point system. The products
are assigned points according to six criteria, which are scaled
in value according to their importance. The point totals for each
product are used as objective indicators for the overall
evaluation.

B. The Criteria

1. Tape Speed

   Tape speed is one of the most important factors in
determining sound quality in a recording machine. The faster the
tape is passed over the record head, the higher the sound quality
will be for the recording. The two tape speeds used on four-track
cassette recorders are 3 3/4 ips (inches per second) and 1 7/8
ips.

   For the purpose of this study, machines using the slower tape
speed (1 7/8 ips) get one point for the tape-speed evaluation.
Machines using the faster tape speed (3 3/4 ips) get four points.

2. Pitch Control

   Pitch control is essential on a recorder that is going to be
used for recording music. It is used to speed up or slow down the
tape speed slightly to compensate for out-of-tune instruments or
voices. All of the products in this study have pitch controls, but
some allow for greater flexibility than others. Recorders with
pitch controls that allow for + / - 10 percent pitch adjustment
get one point for pitch control. Recorders that allow for + / - 15
percent adjustment get two points.

3. Noise Reduction

   Noise reduction systems help control unwanted distortion or
background noise on a recording. The best type of noise reduction

12

system is dbx. Recorders using the dbx system get four points for noise reduction. The next best noise reduction system is the Dolby system. Of the Dolby systems, Dolby C is the best and Dolby B is next. Recorders using the Dolby C system get three points, and recorders using Dolby B get two points. Some recorder manufacturers come up with their own noise reduction systems, but these systems are not widely accepted or tested. A recorder using an unknown noise reduction system gets one point. If no noise reduction system is built into a recorder, it gets zero points for noise reduction.

4. Equalization

    Equalization controls allow the engineer to individually adjust the tone of each audio signal. This allows for more accurate reproduction of the original sound quality and helps filter out background noise. The equalizer doesn't have to be built into the recorder, but external equalizers are expensive and inconvenient. A recorder with an equalizer built in gets four points for equalization, and a recorder without one gets zero points.

5. Rack Mount

    Audio equipment is commonly mounted on a 19-inch rack. Since I already own a 19-inch rack cabinet, I would prefer to own a recorder that could be mounted on it. However, this is not one of the most important features to be considered. A recorder that can be rack mounted gets two points, and one that can't be rack mounted gets zero points.

6. Price

    The prices for the four-track cassette recorders in this study range from $800 to $1300. A recorder costing $800 gets five points. Each increase of $100 in price lowers the score by one point. Therefore, a recorder costing $900 gets four points, and a recorder costing $1300 gets zero points.

13

VI. INDIVIDUAL PRODUCT EVALUATIONS

A. ARIA STUDIOTRACK IIII

    Aria Music USA, Inc.
    1201 John Reed Court
    City of Industry, CA   91745

    The Aria Studiotrack IIII is a rack-mountable four-track cassette recorder that measures 482(W) × 132(H) × 235(D) mm and weighs 7.5 kg. Its volume controls are knobs, unlike the convenient slider controls found on most other recorders.

1. <u>Tape Speed</u>

    The tape drive in the Studiotrack IIII drives the tape at 3 3/4 ips. This tape speed produces higher-quality recordings than the slower 1 7/8 ips tape speed used on standard cassette recorders.

    RATING: 4

2. <u>Pitch Control</u>

    The pitch control, mounted on the front panel of the Studiotrack IIII, allows for + / − 10 percent adjustment of the tape speed.

    RATING: 1

3. <u>Noise Reduction</u>

    The Studiotrack IIII doesn't use any of the industry-standard noise reduction systems; the manufacturers designed their own system on which no objective data is available.

    RATING: 1

4. <u>Equalization</u>

    The Studiotrack IIII has no built-in equalizer. Because of this recorder's low price, it might be worth purchasing an external equalizer to improve the recording quality.

    RATING: 0

14

5. <u>Rack Mount</u>

This product is the only one of the five recorders in the study that is designed to be mounted in a 19-inch rack cabinet.

RATING: 2

6. <u>Price</u>

Listing for $800, the Studiotrack IIII is the least expensive of the five products in this study.

RATING: 5

B. CLARION XD-5

Kamen Music Distributors
P.O. Box 507
Bloomfield, CT   06002

The XD-5 is a four-track cassette recorder with built-in noise reduction. This product doesn't include an equalizer, and the meters for monitoring the signal levels are hard to read. It measures 528(W) × 254(H)× 167(D) mm and weighs 9 kg.

1. <u>Tape Speed</u>

Like the Studiotrack IIII, the XD-5 uses a 3 3/4 ips tape speed resulting in higher-quality recordings than standard stereo recorders.

RATING: 4

2. <u>Pitch Control</u>

The XD-5's pitch control allows for + / − 10 percent adjustment of tape speed.

RATING: 1

3. <u>Noise Reduction</u>

The Dolby B noise reduction system built into the XD-5 is adequate for most recording purposes.

RATING: 2

15

4. Equalization

The XD—5 has no built—in equalizer, and it wouldn't be practical to buy an external equalizer because of the recorder's high list price.

RATING: 0

5. Rack Mount

This product will not fit in a 19—inch rack.

RATING: 0

6. Price

The list price for the Clarion XD—5 is $1100.

RATING: 2

C. Cutec MR402

Dauphin Company
P.O. Box 5137
Springfield, IL   62705

The Cutec MR402 is a basic four—track cassette recorder without built—in noise reduction or equalization. Unlike the Studiotrack IIII, the MR402 has convenient slider volume controls. It measures 473(W) × 110(H) ×   300(D) mm and weighs 10 kg.

1. Tape Speed

Like most of the other products, the MR402 uses the 3 3/4 ips tape speed, which improves the sound quality of recordings.

RATING: 4

2. Pitch Control

The pitch control on the MR402 allows for + / — 10 percent adjustment of tape speed.

RATING: 1

16

3. Noise Reduction

   The MR402 is the only recorder that doesn't provide some sort
of noise reduction system. It would be necessary to buy an
external noise reduction system to make high—quality recordings
with this product.

   RATING: 0

4. Equalization

   Like the Studiotrack IIII and Clarion XD—10, the MR402
doesn't have a built—in equalizer.

   RATING: 0

5. Rack Mount

   This product is not designed to be mounted on a 19—inch rack.

   RATING: 0

6. Price

   The Cutec MR402 lists for $895.

   RATING: 4

D. FOSTEX 250

   Fostex Corporation of America
   15431 Blackburn Avenue
   Norwalk, CA   90650

   The unique feature of the Fostex 250 is its flexible channel
assignment controls, which allow the mixer to be used
independently of the recorder, or vice versa. In addition, all
four microphone inputs can be assigned to record on a single
track, leaving the other three tracks free for later use. This
product measures 430(W) × 80(H) × 355(D) mm and weighs 8.5 kg.

1. Tape Speed

   The main drawback to the Fostex 250 is its slow tape speed
(1 7/8 ips), which adversely affects the recording quality in
comparison to recorders with a tape speed of 3 3/4 ips.

   RATING: 1

17

3. Noise Reduction

The Fostex 250 uses the Dolby C noise reduction system, which is only surpassed by the dbx system used in the Tascam 244.

RATING: 3

4. Equalization

Fostex has built an excellent equalizer into its four-track recorder. Each channel has separate high- and low-frequency tone controls, which can be used to reduce background noise and approximate the sound qualities of the original sound source.

RATING: 4

5. Rack Mount

The Fostex 250 has no provisions for 19-inch rack mounting.

RATING: 0

6. Price

The Fostex 250 lists for $1300. This price may seem high, but this includes the high-quality noise reduction and equalization systems that are missing from some of the other recorders.

RATING: 0

E. TASCAM 244

TEAC Corporation of America
7733 Telegraph Road
Montebello, CA   90640

The Tascam 244 includes a built-in dbx noise reduction system and a high-quality equalizer. It measures 455(W) × 120(H) × 370(D) mm and weighs 9 kg.

1. Tape Speed

The tape speed on the Tascam 244 is 3 3/4 ips.

RATING: 4

18

2. Pitch Control

   The pitch control on the Tascam 244 is the most flexible of the controls being studied. It allows for $+/-$ 15 percent adjustment of the tape speed.

   RATING: 2

3. Noise Reduction

   The Tascam 244 has a built-in dbx noise reduction system. In the recording industry, dbx is acknowledged to be the best noise reduction system available for small multitrack recorders.

   RATING: 4

4. Equalization

   The Tascam 244, like the Fostex 250, has a built-in equalizer for eliminating background noise and improving tone quality.

   RATING: 4

5. Rack Mount

   There are no provisions for rack mounting the Tascam 244.

   RATING: 0

6. Price

   The list price for the Tascam 244 is $1300.

   RATING: 0

VII. PRODUCT COMPARISONS

   Table 2 summarizes the differences among the five four-track cassette recorders according to evaluation criteria and points awarded on the basis of those criteria.

19

Table 2 PRODUCT COMPARISONS

	Tape Speed (ips)	Pitch Control (%)	Noise Reduction	Equali- zation	Rack Mount	Price ($)	Point Totals
Aria Studio– track IIII	3¾	±10	Aria NR	No	Yes	800	
Pts.	4	1	1	0	2	5	13
Clarion XD-5	3¾	±10	Dolby B	No	No	1100	
Pts.	4	1	2	0	0	2	9
Cutec MR402	3¾	±10	No	No	No	900	
Pts.	4	1	0	0	0	4	9
Fostex 250	1⅞	±10	Dolby C	Yes	No	1300	
Pts.	1	1	3	4	0	0	9
Tascam 244	3¾	±15	dbx	Yes	No	1300	
Pts.	4	2	4	4	0	0	14

20

A. Tape Speed

The Fostex 250 is the only recorder in the study that uses the
1 7/8 ips tape speed rather than the more efficient 3 3/4 ips speed
used by the other products.

B. Pitch Control

All of the recorders in the study have a pitch control that
allows the user to adjust the tape speed to compensate for minor
variations in pitch and tuning. However, the Tascam 244 allows
for a + / − 15 percent adjustment, which is 5 percent better pitch
control than any of the other products offer.

C. Noise Reduction

The Cutec MR402 is the only recorder in the study that doesn't
have any type of built−in noise reduction system. This is a major
drawback because a noise reduction system is necessary for making
high−quality recordings, and an external system to use with the
MR402 would be expensive. In addition, the Tascam and Fostex
recorders have the two highest−quality noise reduction systems
(dbx and Dolby C respectively).

D. Equalization

The two highest priced recorders, the Fostex 250 and Tascam
244, are the only products with built−in equalizers.

E. Rack Mount

The Aria Studiotrack IIII is the only one of the products that
can be mounted in a 19−inch rack case. Being rack mountable adds
to its desirability, but this is not an overwhelming advantage.

F. Price

The Aria Studiotrack IIII and Cutec MR402 both list for under
$1000 and define the low end price range for four−track cassette
recorders, the Clarion XD−5 defines the middle price range at
$1100, and the Fostex 250 and Tascam 244 define the high end price
range at $1300.

21

APPENDIX A: REFERENCE LIST

Aria sound equipment. 1983. City of Industry, CA: Aria Music USA. This data sheet includes technical and operational information on the Tascam 244 multitrack recorder.

The Clarion XD–5: A complete recording studio. 1983. Lawndale, CA: Clarion Corp. of America. This brochure includes technical and operational information on the Clarion XD–5 multitrack recorder.

A complete portable 4–channel multitrack recording/mixing system. 1982. Montebello, CA: Teac Corp. of America. This brochure includes technical and operational information on the Tascam 244 multitrack recorder.

Cutec sound reinforcement and home recording equipment. 1984. Springfield, IL: Dauphin Co. This data sheet includes technical and operational information on the Cutec MR402 multitrack recorder as well as two other Cutec products.

Everest, Alton. 1981. Handbook of multichannel recording. Blue Ridge Summit, PA: TAB Books, Inc. This book provides an overview of the internal functioning and general use of multitrack sound recording equipment. It also describes the processes involved in making a multitrack recording.

Fostex personal multitrack. 1983. Norwalk, CA: Fostex Corp. of America. This brochure includes technical and operational information on the Fostex 250 multitrack recorder.

Frets looks at Cutec mixer/recorder. 1983. Frets Magazine. December, 20. This article critiques the Cutex MR402 multitrack recorder in terms of tape speed, pitch control, ease of use, and sound quality.

Peterson, George. 1984. Multitrack cassette recorders. Mix. July, 15. This article provides basic technical and price information on 12 multitrack cassette recorders including the Aria Studiotrack IIII, Clarion XD–5, Cutec MR402, Fostex 250, and Tascam 244.

Riggs, Michael. 1983. How tape recording works. High Fidelity. February, 16. This article explains the way sound is recorded on magnetic tape and discusses the relative merits of Dolby B, Dolby C, and dbx noise reduction systems.

22

APPENDIX B: GLOSSARY

bias
>In magnetic tape recording, the signal to be recorded is
>mixed with a high-frequency bias current to reduce
>distortion in the magnetization process.

equalization
>Adjustment of frequency response of a channel to achieve a
>flat or other desired response.

erase
>Magnetic tape having previously recorded signals on it must
>be erased before reuse. This is accomplished by a high-
>frequency current in an erase head over which the tape passes
>before it reaches the record head.

fader
>An adjustable resistance element in an electrical circuit
>that controls the sound level (volume control).

flux
>The lines of force of a magnetic field.

frequency
>The number of signal cycles occurring during each one-second
>period. Frequency is the physical measurement corresponding
>to the sensation of pitch.

level indicators
>Any visual device for indicating signal levels in various
>channels of a console or tracks on the recorder.

master
>A master tape is the original tape on which is recorded the
>signals mixed down from the multitrack tape.

mixdown
>The process of combining several signals on a multitrack
>magnetic tape with appropriate levels.

mixer
>An electronic device capable of mixing (summing) two or more
>signals into a composite signal.

23

monitor
>   The means for hearing and checking signals for control and
>   evaluation.

monophonic
>   Recording and reproduction of sound by a single channel.

multitrack recording
>   Recording 2 to 40 audio tracks on magnetic tape.

overdub
>   Individual tracks can be recorded or redone at a later time
>   by overdubbing one track at a time.

potentiometer
>   An adjustable resistance element in an electrical circuit.

Research Reports and Other Articles for Publication

Being published—with your name in print and your words and ideas officially set forth in a journal or magazine—may seem like a far-off and impossible dream while you are a student. But as you gain knowledge and expertise in your field, you may soon feel the urge to share what you have learned with others, and you can often do that best through the printed word. Indeed, in the technical and scientific community, a research project is not considered complete until the information has been published in a book or a journal article. Furthermore, publication will give you prestige and will keep your employer's name before the public—both real incentives to writing for publication.

The purpose of a research report and most other published articles is usually not to evaluate, recommend, or choose from alternatives, but to present information, refute earlier theories, or suggest new interpretations of data. Thus, factual and verifiable information forms the central content and makes this kind of report different in purpose and form from the comparative, feasibility, and recommendation reports discussed in Chapter 21. This chapter will help you write research reports and articles for publication. It contains three major sections.

22.1 DEFINITIONS OF TERMS AND TYPES OF PUBLISHED ARTICLES

As noted in Chapter 21, the word "report" comes from a word meaning "to carry back." In a report, then, you carry back or present information to the reader, and to do that, you must become an expert in the subject. Who your readers are and how much they know will determine the type of article you write and where you seek publication. The three basic kinds of published articles are

1. research reports
2. technical or trade articles
3. popular articles

Research Reports

A *research report* is a document that reviews existing research and describes original research results for publication in a professional journal, such as the *International Journal of Environmental Studies* or the *Journal of Bacteriology*. Research reports are also called journal articles or scientific, investigative, and technical papers. These reports are written for your professional peers—that is, by a specialist for other specialists—and they usually follow a prescribed form. Section 22.2 explains this form and contains an example of such a report.

Technical or Trade Articles

A *technical* or *trade article* is a report that describes applied research or development, typically giving information about new techniques or products. These reports appear in the semitechnical journals published by professional societies or, more often, in trade journals supported by advertising. Two examples of trade journals are *Machine Design* and *EDN* (for engineers and engineering managers). Again you are usually writing for your professional peers—specialists, technicians, or managers. The reader group is likely to be broader than for a research report, and some readers will be specialists in related areas. These reports also contain standard elements but do not all follow the same rigid form, varying by the style of the journal in which they appear.

Popular Articles

A *popular article* is a report that simplifies and explains a scientific or technical subject so that it can be understood by generalists. Popular reports appear in newspapers and general commercial magazines. However, readers of these periodicals can range from those with considerable technical background (who might read *Scientific American*) to those reading primarily for interest (who might read *Popular Science.*) You write as a specialist, and you must determine who your readers will be and design your report accordingly. Popular articles do not follow a prescribed form, though each periodical has its own preferred style.

Table 22-1 summarizes the features of each kind of published article.

Table 22-1   TYPES OF ARTICLES FOR PUBLICATION

	Purpose	Published in	Primary reader
Research reports	to review existing data and describe original research results	professional journals	specialists
Technical or trade articles	to describe applied research or development	semitechnical or trade journals	specialists, technicians, managers
Popular articles	to simplify and explain technical subjects	commercial magazines and newspapers	generalists, varying in degree of technical background

22.2 THE FORM AND ORGANIZATION OF PUBLISHED ARTICLES

Research Reports

A research report is the most formal of the published articles in technical and scientific fields. It is based on research in the laboratory or the field, and it is written to share information—usually including a review of published literature on the subject. The basic parts of most research reports are these:

- title
- abstract
- introduction, including literature review
- materials and methods
- results
- discussion
- references

These parts can be grouped into preliminary matter (the title and abstract), the report itself, and back matter (the references). Since research reports are often brief, these divisions may not be set off by subheadings; each journal has a slightly different style. Note that even though the parts are *presented* in this order, they need not be *written* in this order. For example, in writing the draft, you might begin with the materials and methods section, which deals with specifics and might be easier to write than the introduction.

To supplement the following brief discussion of the organization of research reports, you can see either Robert A. Day, *How to Write and Publish a Scientific Paper* (Philadelphia: ISI Press, 1983) or Herbert Michaelson, *How to Write and Publish Engineering Papers and Reports* (Philadelphia: ISI Press, 1986). Also, you might obtain several recent issues of your target journals and their guidelines for authors and study them carefully. Look to see how each of the following elements is handled.

Title   The title should contain both key content words and function words so that it accurately reflects the contents and can be usefully indexed. See section 6.1 for more on writing good titles.

Abstract   The abstract can range from 50 to 300 words but should be proportional in length to the article. You may want to write an abstract to help you in planning the purpose, content, and organization of your report, but you should rewrite it when the report is finished and the form and content are fixed. An abstract serves several functions: (1) it can be sent as part of a query to see if journal editors are interested in your projected report; (2) it can be printed in a volume of abstracts to provide condensed information about the main report to potential readers; (3) it can precede the article, allowing readers to preview the article and see if they want to read it. Abstracts may be either descriptive or informational; determine which to write by consulting the journal's guidelines or by matching the kind used in your target journal. See Chapter 19 for details on writing abstracts.

Introduction   The introduction states the purpose and the main point or thesis of the report. It gives background information, including a review or summary of what has already been written on the subject. In order to preview for the reader what will follow, the introduction should clearly define the problem and indicate the method of investigation. It should also define any unusual terms or symbols. Finally, it should briefly summarize the results. See section 6.5 for more on introductions.

Materials and methods   The central section of a report, usually called Materials and Methods, is much like a procedure: it discusses the tools, equipment, and materials used and then presents the procedure itself in a sequential method of organization. Make this section specific enough so that another researcher can duplicate the procedure, and analyze the target publication to see how its writers ensure that the same procedure can be duplicated. The ability of others to duplicate results and verify conclusions will validate your research. See Chapter 18 for more on writing procedures.

Results In the results section you present the data gathered when you followed the method explained in the previous section. Often these results can be presented in a table or graph. Although they may be presented in a brief form, results actually provide the most important information in the report.

Discussion Since the discussion interprets the results, sometimes you can combine these two sections. The discussion section is like the conclusions section of a feasibility or comparative report; you should tell what the results mean and why they are important. Often the results suggest the need for further testing or experiments, and in the discussion section you should explain and support that need.

References In the reference section you list the published sources you have cited in your literature review and elsewhere in the article. The documentation method depends on the journal for which you are writing, so obtain a copy of the journal's guidelines for authors and study previous issues. Also see Chapter 13 for information on documenting sources.

Figure 22-1 shows a brief research report published in the Notes section of the *Review of Scientific Instruments*. Notice how the various parts of the article blend together. Notice too that this article is written for fellow specialists, so it assumes common knowledge of terms and procedures. Therefore, unless you have experience in this field, the report will be difficult to understand—even though the writing itself is clear.

Technical and Trade Articles

Technical and trade articles also convey information, but they are more likely than research reports to show how the information can be applied in development or manufacturing. Because form and organization vary from one publication to another, you must study recent issues and the target journal's guidelines in order to write an article with a good chance of acceptance.

Technical and trade articles tend to be less formal in structure and language than research reports. Some use a chronological narrative structure, and others present information through case studies. The introduction of a technical article may be written to catch the reader's eye and interest, and these articles often use many graphics—both tables and figures—to supplement and condense information. Typical parts of a technical article are

A Low Energy Electron Source for Mobility Experiments*

James A. Jahnke and M. Silver

Department of Physics, University of North Carolina, Chapel Hill, North Carolina 27514
(Received 11 December 1972; and in final form, 16 January 1973)

Thin film Al-Al$_2$O$_3$-Au diodes were used as sources of low energy electrons ($\sim$ 1 eV) in a drift velocity spectrometer. These diodes were found to offer several advantages over other sources used for electron mobility studies in dense gases and liquids.

The electronic states of liquids and dense gases have increasingly been studied by the injection of excess electrons. The measurement of the drift velocities of excess electrons by mobility experiments has proven to be one of the simplest and most useful methods for investigating the conducting states of gases and liquids. In our studies in dense helium gas,[1] we have used a cold-cathode emitter as a source of low energy electrons and have found that it provides several advantages over more conventional methods.

Radioactive sources and field emission tips have been most commonly used,[2-4] particularly at low temperatures. They have specific disadvantages however, since the emitted electrons have a high initial energy of several kilovolts or more. The possibility that excited species or impurity ions may be produced by these high energy electrons has continually added a dimension of uncertainty in the interpretation of drift velocity data throughout the literature. A radioactive source also has several practical experimental problems associated with it, namely, low current levels, the inability to easily turn the source off, and the procedural problems of dealing with radioactive materials themselves. Field emission tips are quite useful at liquid densities,[4] but they tend to have a short lifetime at lower gas densities where they are easily blunted. Also, at high current levels, a bubble of ionized gas tends to form around the tip, leading to increased noise and measuring problems in the experiment.[5] Photoelectric sources have proven useful in drift and other experiments.[6,7] They are, however, most easily constructed in glass systems, which makes it difficult to study pressure and temperature effects over a range of fluid densities.

The cold-cathode emitter used in the present experiments has previously been used to study the current–voltage characteristics of gases and liquids.[8] These emitters inject electrons at energies of about 1 to 2 eV[9] and easily can be operated to provide currents over the range of 10^{-10} to 10^{-13} A. The Al-Al$_2$O$_3$-Au diode has been found to be one of the easiest emitters to construct and one of the most reliable.[10] An aluminum strip about 2000 Å thick is first evaporated unto a clean glass substrate and then oxidized to a depth of about 100 Å by wet electrolysis. Transverse gold strips are then evaporated onto the oxide layer to a thickness of approximately 100 Å. It is important that the gold layer be not too thick since higher voltages are then needed for emission. In the past few years, techniques in fabricating thin film diodes have become more clearly defined[10] and by carefully following these techniques we have been able to obtain satisfactory results from better than 50% of those prepared.

For the drift velocity spectrometer, two diodes were constructed on a 2.5 cm square glass substrate which was then clamped between two Plexiglas plates as shown in Fig. 1. An accelerator grid was separated from the diode by a 0.4 mm Teflon spacer. In the normal operation of the spectrometer, the system was cooled to liquid nitrogen temperature since these diodes tend to have a short lifetime at temperatures above 200 K.[10] Emission occurred when a potential difference of about 8 to 9 V was applied between the aluminum and gold layers. It has been found necessary to approach this voltage slowly and then to slowly "age" the diode at incremental voltage increases after the first emission. In the spectrometer system, the diode was first aged by floating it at −100 V above the accelerator grid which acted as the collector and biasing the diode with a 45 V battery subdivided with a 500 kΩ Helipot. A Keithley 602 electrometer was used to monitor the current on the accelerator. After a stable current was obtained, the electrometer was then connected to the collector of the spectrometer which was held at ground potential. Measurements were performed at fields of from 50 to 200 V/cm over the 2 cm length of the spectrometer. Since the diode was kept at −100 V above the accelerator, it was therefore floated at from −300 to −400 V above the collector.

In vacuum, currents on the order of 10^{-9} to 10^{-10} A were obtained between the diode and accelerator grid at diode voltages of from 10 to 11 V. Current levels decreased less than an order of magnitude when the current was passed through the grid system to the collector. The addition of gas or liquid to the system further decreased the current level by approximately an order of magnitude, so that

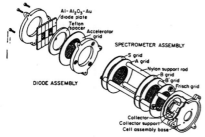

Al-Al$_2$O$_3$-Au
/diode plate
Teflon
spacer   Accelerator
grid

SPECTROMETER ASSEMBLY

S grid
A grid
Nylon support rod
B grid
B' grid
Frisch grid

DIODE ASSEMBLY

Collector
Collector support
Cell assembly base

Fɪɢ. 1. Diode holder and drift velocity spectrometer grid system.

(Continued)

Figure 22-1   Journal article for specialists

currents on the order of 10^{-12} A were normally measured during an experiment. The collected current is of course dependent on both the gas density and the accelerator voltage. Previous papers on injection experiments consider this in greater detail.[8,9] We have found it more practical here to maintain the accelerator voltage constant and vary the current when necessary by changing the diode voltage. In this manner, greater detail over the drift velocity spectrum may be obtained by increasing the current level to 10^{-11} or even 10^{-10} A. Driving a diode to higher current levels, however, greatly reduces its working lifetime, so a balance must often be found between problems of spectral analysis and experimental lifetime. At moderate current levels, we have had diodes operate for more than 24 h. Since the diodes do operate well over long periods at low gas densities, as well as at higher densities, they offer advantages over field emission tips in addition to avoiding some of the problems associated with radioactive sources.

*Research supported by the Advanced Research Projects agency of the Department of Defense and monitored by the U. S. Army Research Office, Durham, under Grant no. DA-AROD-31-124-72-G80, and by the Materials Research Center, UNC under contract DAHC 15-67-C-0223 with the Advanced Research Projects Agency.
[1]J. A. Jahnke and M. Silver, Chem. Phys. Lett. (to be published).
[2]F. Reif and L. Meyer, Phys. Rev. 119, 1164 (1960).
[3]H. Schnyders, S. A. Rice, and L. Meyer, Phys. Rev. 150, 127 (1966).
[4]J. A. Jahnke, L. Meyer, and S. A. Rice, Phys. Rev. A 3, 734 (1971).
[5]B. Halpern, and R. Gomer, J. Chem. Phys. 51, 1031 (1969).
[6]B. Halpern, J. Lekner, S. A. Rice, and R. Gomer, Phys. Rev. 156, 351 (1967).
[7]J. L. Levine and T. M. Sanders, Jr., Phys. Rev. 154, 138 (1967).
[8]M. Silver, P. Kumbhare, P. Smejtek, J. Chem. Phys. 52, 5195 (1970); D. G. Onn and M. Silver, Phys. Rev. 183, 295 (1969); and M. Silver, D. G. Onn, and P. Smejtek, J. Appl. Phys. 40, 2222 (1969).
[9]D. G. Onn and M. Silver, Phys. Rev. 3, 1773 (1971).
[10]D. G. Onn, P. Smejtek, and M. Silver, Proceedings of the Low Temperature Physics Conference, Boulder (1972).

Figure 22-1 Continued

- title
- introduction
- body
- conclusion
- references (sometimes these are omitted)

Title The title should include both key content words and function words. Titles should be interesting but not frivolous; since technical and trade articles may not have abstracts, the title should fully describe the contents. See section 6.1 for more on writing titles.

Introduction The introduction should state the purpose, main point or thesis, and the way this information may meet the reader's needs. In technical articles, the introduction gives background information but usually does not include a literature review, although the writer certainly must know what has been published on this subject. If the subject is a new product or process, the introduction may include a technical description, and it should define new terms, but it should emphasize what's new and how it might be useful to the reader. If you plan to write a technical article, carefully identify your potential readers so you know how technical you can be. See section 6.5 for more on introductions.

Body The central section of a technical article may contain the following information:

- results and analysis
- detailed description of a product, procedure, or design
- application to a variety of circumstances

In writing this section—which is the meat of your article—you should keep in mind the three questions editors ask in reviewing potential manuscripts:

- What does the manuscript tell readers that they don't already know?
- Why would readers want to know the information?
- How can readers use the information?

As you plan and write the body of your article, follow the general organization you find in past issues of the journal, adapting that organization to fit your own purpose. Include graphics, but try to balance the number of graphics with the length of the article. Graphics should not overwhelm the written material.

Conclusion In most trade journal articles, the section called Conclusion summarizes the main points and interprets the results—sometimes recommending actions or applications. The Conclusion section of a trade article, unlike that of a recommendation report in industry, usually appears at the end of the article.

References If you include references, give complete bibliographical information so readers can find the information easily, and follow the documentation style of the journal to which you submit the article. See Chapter 13 for more on documentation.

The article in Figure 22-2 appeared in 1986 in *Machine Design*, a publication primarily for mechanical engineers and technicians. Notice how the information is specifically aimed for use by the readers. The writer assumes technical knowledge and terminology but supplies excellent drawings to aid in understanding.

Popular Articles

Popular articles—those written for generalists—range from simple explanations for children to fairly high-level information in such magazines as *Technology Review* and *Psychology Today*. Specialists in science and technology can provide an important service to the community by sharing their knowledge with interested generalists. However, no standard organization exists for such reports, so you must study recent copies of target publications to determine an acceptable form. You can also consult Barbara Gastel's useful book *Presenting Science to the Public* (Philadelphia: ISI Press, 1983). Here are some considerations in writing the following key sections:

- title
- introduction
- body
- conclusion

ROBOTIC HAND APPROACHES HUMAN DEXTERITY

A touch of reality.

Most industrial robots have end effectors or hands that are designed to carry out a few very specific functions. However, for a robot to be able to take on diverse tasks, the hand has to have capabilities approaching that of a human hand. Researchers in the U.S. and Japan have made several attempts to build a fingered hand with varying degrees of success. The main problem is control, coordinating the motion and force in several multi-joint fingers plus a wrist. Then, there are the difficulties of making a low-impedance actuation system. Finally, these elements must be put together in a practical, lightweight package.

One of the more advanced hand designs of this type is the Utah-MIT dextrous hand. This hand was designed as a research tool for the investigation of machine dexterity. (Its present configuration is not for industrial application.) It was designed to minimize inertial, viscous and frictional impedances. This means that intrinsic resistance to forces caused by external objects is small. This allows a greater range of controllable impedances. An algorithm in the digital control computer changes the impedance to the task. For example, the force needed to grasp an egg is set differently from that needed to ring a bell.

The Utah hand has three four-degree-of-freedom fingers, one four-degree-of-freedom thumb, and a three-degree-of-freedom wrist. The three fingers are arranged in an almost planar sequence to form a surface for the thumb to act against.

In operation, a digital computer generates signals that trigger pneumatic actuators. The actuators, in turn, move lines or "tendons" that move the fingers and wrist.

Tendon-operated

The entire hand is tendon-operated so that the finger actuation systems can be remotely located. This allows space for internal instrumentation that feeds back joint and tendon strains and hand structure that includes individual finger joints. Remote actuation also keeps the hand's weight to a minimum.

The tendons are intricately routed over a series of pulleys to move the hand's 16 joints. The tendon system is arranged in an "antagonist-pair" configuration. Each joint requires two actuators, that work against each other because the tendons must be pulled, not pushed.

Tendons consist of a multifiber Kevlar and Dacron weave. Kevlar fibers run the length of the tendon to support tension loads while Dacron fibers weave in and out of the Kevlar ribbon of fibers at a 45° angle to distribute the load over the pulley evenly.

Multifiber, sheathed steel cables were avoided in the tendon design because they possess inherent problems with strength versus fatigue life, especially if cables are routed over small pulleys that produce high internal bending stresses. Metallic strap systems usually have sudden modes of failure because they are a one-piece structure and their sides tend to curl up.

Pneumatic actuation

The actuation system is pneumatic which allows the construction of a low weight, compact actuator that can generate required speeds and forces. This selection has far less intrinsic impedance than hydraulic or electrical actuation systems.

However, even pneumatic actuators exhibit an undesirable springy actuation if standard flow control valves are used. This is the result of the compressibility of the working gas. Consequently, a pressure control valve was developed. It negates the effects of compressibility provided its speed dominates the natural system dynamics generated by the structural masses and operating fluid compressibility.

An electropneumatic valve selected must be carefully controlled to eliminate the problems of nonlinearity due to complex dynamic and fluid mechanical interactions. The valve is controlled by an analog electronics package that accepts inputs related to joint position and stiffness, joint torques, or individual tendon forces.

The pneumatic actuation from each electropneumatic two-stage valve pressurizes a glass cylinder

Figure 22-2   Article in a trade journal

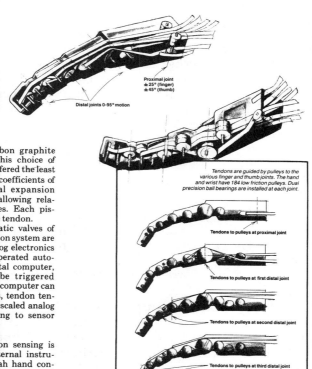

Proximal joint
±25° (finger)
±45° (thumb)

Distal joints 0-95° motion

Tendons are guided by pulleys to the various finger and thumb joints. The hand and wrist have 184 low friction pulleys. Dual precision ball bearings are installed at each joint.

Tendons to pulleys at proximal joint

Tendons to pulleys at first distal joint

Tendons to pulleys at second distal joint

Tendons to pulleys at third distal joint

which houses a carbon graphite composite piston. This choice of cylinder and piston offered the least drag due to very low coefficients of friction and thermal expansion compatibility while allowing relatively close tolerances. Each piston/cylinder pulls one tendon.

The electropneumatic valves of the pneumatic actuation system are controlled by an analog electronics package. They are operated automatically with a digital computer, or the signals can be triggered manually. The digital computer can supply desired angles, tendon tensions, and can access scaled analog voltages corresponding to sensor signals.

Instrumentation

Torque and position sensing is achieved through internal instrumentation in the Utah hand consists of 16 position sensors located in each of the joints and 32 tendon tension sensors mounted immediately behind the wrist section. The position sensors are Hall effect devices that give an analog voltage corresponding to the angular position of each joint. Each of the tendons passes over a pulley mounted on a cantilever whose deflection is measured through a pair of thermally compensated strain gages. The sensor is calibrated to give an analog voltage corresponding to the tendon tension.

Space has been allowed for tactile sensors that may be added to the fingers later. These sensors would give the force of contact position.

Also under development are joint angle sensors, tendon force transducers, compliant covering materials to provide a closed glove for the dextrous hand, and various electronic components for implementation of control functions.

The Utah/MIT hand is a joint project of researchers at the University of Utah Center for Engineering Design and the Artificial Intelligence Laboratory at the Massachusetts Institute of Technology. Under development since the early 1980's, 20 hands will ultimately be built to serve as test beds for artificial intelligence and machine-based manipulation in laboratories throughout the United States. ■

Figure 22-2 Continued

Title   Titles can be straightforward and serious with an emphasis on key words, like "The Value of Fundamental Science." They can also be eye-catching rather than informative, like "Horseplay and Monkeyshines" for an article on play among animals and what it means. The type of title depends on the selected publication and the intended reader.

Introduction   Like titles, introductions of popular articles are frequently designed as "hooks" to pull the reader into the content. Thus, they use rhetorical devices like anecdotes, analogies, questions, humor, quotations, provocative statements, statistics, and personal application—anything that will attract the reader's interest. The variety is endless; for example, look at these opening sentences from articles in the December 1983 issue of *Science 83* magazine.

- Like little spring-clips, hemoglobin molecules snap up molecules of oxygen in the lungs, carry them through the bloodstream, and release them to the cells (Preuss, 81).
- "It's nice," says Bruce Beasley, "when big companies say something is impossible. If they just said, 'Yeah, it can be done, but it's real, real hard'—then it might be more discouraging." (Rogers, 39).
- About 600 cubic miles of snow cover the Northern Hemisphere, and in every cubic foot of the fluffy stuff there are about 18 million snow crystals (Allman, 24).
- Pictures, like people, deteriorate with age (Matthews, 97).

Body   After the hook (and sometimes instead of it), introductions to popular articles present the purpose and the thesis, or main point. The body paragraphs then support these general statements. In supporting the thesis, popular articles frequently quote authorities, mixing their comments with description, definition, and case studies. Popular articles follow no standard organization, but each one must have a clear organizational pattern. Many popular articles use a general-to-specific method of organization, so the reader gets an overview before details. Popular articles also use many examples and analogies with familiar objects or procedures to help clarify complex ideas. Articles for generalists should avoid jargon whenever possible; thus, when new terms need to be introduced, they should be defined using common synonyms and simple language. Before you try your hand at writing a popular article, read current issues of your target publication to see how organization and vocabulary are handled.

Conclusion   Popular articles often conclude by using standard rhetorical devices like quotations, predictions for the future, summaries, questions, and implications. Key words or phrases should link the conclusion to the introduction in order to give the article continuity and coherence.

The article in Figure 22-3 illustrates some of these principles. Notice how the title plays on words for interest and the introduction takes a

A Well–Seasoned Tasting Machine

Scientists try to unravel the mystery of how humans distinguish tastes

By Glennda Chui
Mercury News Science Writer

Bite into a ripe strawberry. Revel in the sweet, tangy flavor and the crunch of tiny seeds between your teeth.

You are operating the most versatile tasting machine known to science. It kicked into action even before the berry reached your lips, as your nose began to analyze some of the hundreds of chemicals that give the fruit its flavor.

It is incredibly sensitive. The nose, which contributes most of what we perceive as the taste of food, can pick up the whiff of a few molecules of musk floating in the air. Taste buds can tell whether water has been distilled once or twice.

Scientists use the human taste machine as a laboratory tool to analyze the flavor of food and the sensual appeal of such consumer products as carpet deodorizer and kitty litter.

But don't ask them what makes it tick.

"For the most part, we don't understand how it works," said Richard Mattes, a specialist in human

nutrition at the Monell Chemical Senses Center in Philadelphia. "We're light years behind knowledge on hearing and sight."

The basics are easy: Taste originates in the taste buds, which send signals to the brain. The particulars remain a mystery.

Scientists theorize that molecules of food attach to receptors in the taste buds. They recently found that salt, in the form of sodium ions, actually enters the sensitive cells of the taste buds through tiny channels.

But they know "practically nothing" about how other receptors work, Mattes said, and it is not clear how the receptors trigger the nerve impulses that send taste signals to the brain.

The confusion extends to the brain itself, said Monell biochemist Joseph Brand. Three major nerves carry taste sensations from the mouth to the brain stem—but where they go from there is unknown.

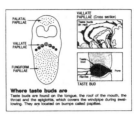

Where taste buds are
Taste buds are found on the tongue, the roof of the mouth, the throat and the epiglottis, which covers the windpipe during swallowing. They are located on bumps called papillae.

Further, much of what people think they know about taste is wrong, the experts say.

✔ Most textbooks insist that there are four basic tastes—sweet, sour, salty and bitter. But some Japanese researchers think there is a fifth basic taste, monosodium glutamate, which they call "umami," or "delicious." Experiments show that there are receptors for glutamate on the tongue, Brand said. And some researchers argue for a sixth taste, metallic.

✔ Not all taste is in the tongue. There are taste buds scattered in the roof of the mouth, the throat and the epiglottus, a flap that covers the windpipe during swallowing.

Beyond pure pleasure, flavor has a serious purpose.

(Continued)

Figure 22-3   Popular article

"It looks like taste has real global effects," said Israel Ramirez, a physiological psychologist at Monell. "It affects digestion at nearly every stage," fine-tuning the system to prepare for particular foods.

Flavor alone can stimulate the body to increase its flow of saliva, dump gastric juices into the stomach, absorb nutrients from the intestine and release digestive enzymes, Ramirez said. It also influences the level of insulin in the blood.

One study at Monell suggests that dogs digest food more efficiently when they like its taste. If this proves to be true, it could have implications for human health.

Ramirez speculated, "Maybe the people who spend their lives thinking up ways to make TV dinners taste better . . . are doing something beneficial for mankind's health."

Mattes said, "It would open up whole new avenues of dietary therapy." For instance, dieters could eat certain flavors that inhibit the absorption of fat in the intestines, he said. Conversely, people could get more nutrition out of their food by eating flavors that increase absorption.

Mattes plans to test the role of taste in the nutrition of premature babies who are being fed by stomach tube. He will have the babies suck flavored nipples while they are being fed to see whether this makes them grow faster or secrete more digestive hormones and enzymes.

While some scientists are unraveling the mechanisms of taste, others are putting the sense to work. In the field of sensor evaluation, human tasters—and sniffers—are trained to produce hard scientific data.

"You can't predict from laboratory instruments what the flavor of a food is going to be," said Michael O'Mahony, a professor of food science and technology at the University of California at Davis. "You have to add the human senses to your set of laboratory instruments."

People, he said, are more versatile than any machine because they automatically take in all the important aspects of a food—its taste, its smell, the way it crunches between the teeth and the way it looks and feels.

Human testers often save money for the food industry by analyzing the flavor of new products before expensive market tests are conducted.

People also evaluate a wide range of consumer products—soaps, bleaches, cosmetics, anything that appeals to the senses, O'Mahony said.

"When your cat has peed all over the kitty litter," he said, "does the kitty litter smell or not? How does it kill the odor?"

One place that investigates such things is Tragon Corp. of Redwood City.

Company founders Herbert Stone and Joel L. Sidel developed a scientific method for evaluating products while working at SRI International in the early 1970s.

They recruit testers by calling people at random. Volunteers—who are paid for their services—are tested to see if their taste buds are keen. Some people can't tell whether there are six teaspoons of sugar or two teaspoons of vinegar in a cup of coffee, Stone said. Out of 30 people, about 20 qualify.

Then the testers, as a group, draw up a

Figure 22-3 Continued

list of the food's qualities.

For chocolate chip cookies, the distribution of chips, shape, salty taste or baked flavor might be listed. The group must agree on what each of these terms means.

Finally, the testers go into individual booths to ogle, sniff and taste the cookies and fill out score sheets. There's one drawback—they don't get to swallow. They spit each chewed mouthful into a cup.

A computer scans the score sheets and translates the markings into numbers, which are compiled to create a precise description of the food.

These numbers are useless if the samples offered to the tasters are not identical, Sidel said. So the company takes precautions, such as hiring experts to pour samples of beer so the amount of head is the same.

Tragon has tested steak sauce, mustard and salad dressing, and it brings in children as young as 7 to evaluate cereal and candy.

But, "it's not all candies, pizza, beer and ice cream," Sidel said. "We have a lot of people smelling other people's underarms. You won't find testers admitting to that."

In addition to testing underarm deodorants, the company asks mothers to evaluate the before-and-after performance of disposable diapers and coops up cats in small rooms to see if the odor of their waste is neutralized by kitty litter. Some volunteers smell each other's breath to see how well mouthwash is working.

Tragon finds volunteers for even the unpleasant tests, Sidel said.

"When we call to recruit people, their curiosity often takes over, and they're willing to participate," he said.

At Davis, O'Mahony goes further by isolating one sense at a time.

In a typical test, he wanted to see if a new canning process made peaches more sour. He put tasters in a booth flooded with red light so they could not see color differences between the peaches, and he put clips on their noses to block out odors.

The tasters also drank an herb tea from India called Gymnema sylvestre, which deadens the taste receptors for sweetness.

"We've blocked out as many sensations as possible so they can concentrate on sourness," O'Mahony said.

Recently, his laboratory has been testing the effects of irradiation on oranges, cherries and peaches under a grant from the state Department of Food and Agriculture. The technique, which kills insect pests, slightly changes the flavor of the fruit, O'Mahony said, but it remains to be seen if consumers will notice the difference.

He also is working on new ways of describing taste sensations. One of the nagging problems in the field is that two people don't mean the same thing when they use a term like "salty," he said. O'Mahony hopes to solve this by studying how the mind forms concepts.

"With luck," he said, "this should allow us to communicate more precisely about taste and smell."

San Jose Mercury News ∎ Tuesday, June 17, 1986

Figure 22-3 Continued

personal approach. Simple line drawings help introduce the subject, and quotations from experts support the article's thesis: we don't understand how the tongue works.

22.3 THE PROCESS OF WRITING FOR PUBLICATION

In writing for publication, you follow the eight steps detailed in Part 1: planning, gathering information, organizing, reviewing, writing, revealing your organization, reviewing and revising, and editing. You also add a ninth step: submitting the document for publication. This intent to publish forces you to consider a second group of readers. Not only must you write for the intended reader of the magazine or journal but also for the editors of that publication—often the toughest readers that exist.

Planning

As you plan any article for publication, you should (1) choose your topic and summarize what you want to say about it in a brief abstract and (2) research publications that might be interested in a report on this subject. Both tasks must be done early, because you need to know your topic to help choose a target publication, and you need to choose a publication to help refine your topic.

To help choose a target publication, decide whether you want to report research, explain a new product or development, present information, or simplify technical matters for generalists. Spend some time in your company or college library examining major publications in your field. Study at least six recent copies of each, reading the articles and analyzing their organization. Check to see that your topic has not been treated recently, and be sure that you have something new to say. When you choose a target publication, write to ask for its guidelines for authors, then follow those suggestions. In the journal itself, you can usually find information about submitting manuscripts near the masthead (where publication information is given) or immediately following the table of contents.

Follow the guidelines in this book to create a complete abstract or description of the article you plan to write. Then secure your company's approval before you send a query. Most companies support writing for publication because it helps keep the company name before potential customers. However, your company may give you access to confidential and proprietary information, and you must avoid breaching that confidentiality.

Once you've secured company approval, follow the guidelines for authors to draft a query. Many publications do not want to see finished articles; they prefer a query letter that summarizes the article, tells why it will be of interest to readers, and explains your approach. Add an outline to the query letter showing how you plan to organize the material. Remember that a query letter is a sales document. It must be written as carefully as a job application letter, which it resembles in many ways. Send the query and outline to your target publication, but send it to only *one* at a time; most technical journals will not accept simultaneous submissions. If you want your material returned, enclose a self-addressed stamped envelope (SASE).

Some professional journals *do* want to see a completed manuscript. This information will be given in the guidelines for authors. If you must submit a finished manuscript, read the rest of this chapter before you write it and send it off. While you are a student, make use of your professors as sources of information and advice about your ideas for publication and about journal practices.

While you wait for a response to a query letter, you can start the scheduling that will enable you to proceed efficiently. Schedule the whole writing project first; then break it into segments and schedule each segment. Use a Planning Sheet like that in section 1.6. Often you'll write for publication on your own time—in addition to your regular work—so you will need to schedule carefully.

If your proposed article is not accepted, don't become discouraged. Try again at a different publication, after using whatever feedback you have gotten to revise. It's not at all unusual for an article to be accepted by the fourth or fifth publication that examines it. If your proposal is accepted (as perhaps 5 percent are), you can begin gathering information. Remember that a go-ahead based on a query does not guarantee acceptance of the article; when the article is written, it must also be approved.

Gathering Information

Research reports require a literature review, which means that you must search carefully for *all* relevant articles, keeping notes on their content and main contributions. You may cite only a few of them in the review, but you must know about them all. Other articles may not require a literature search, but to ensure accurate and up-to-date writing, you should still search thoroughly and take good notes. In addition, you must often gather nonlibrary information: laboratory or field data, surveys, interviews, calculations, and analyses. As you proceed, document your sources, because it will be much harder to reconstruct a source once you are away from it. Even if you are writing for generalists, sources and information must be accurate and complete. Generalist readers may not spot errors, but editors probably will, and they will quickly reject your article.

Organizing and Reviewing

If you created a good outline as part of your query, you can simply follow the steps in Chapter 3 to flesh out the outline and prepare for writing. If your article was accepted subject to revisions, follow the editors' suggestions to create a new organization. Those suggestions may require a totally different approach, and you may have trouble rethinking your organization. Remember, though, that all material can be organized in at least two or three different ways, and you need to be flexible to meet editors' demands.

Many technical and scientific papers report on a project or experiment that was conducted in an organized sequence: first we did X, which resulted in Y, and then we did Z. Even though the project may have been well designed, in most cases the reader does not want to follow you through that sequence. That means you must choose a single important point to emphasize—the results, the conclusions, a significant failure—and organize your report to support that point.

When you review your organization, try to ask the kinds of questions editors will ask. Does your method of organization follow the pattern used in this particular publication? Does the organization emphasize and support your single main point? Does the article clearly move from the introduction, through the body, to a genuine conclusion?

If you're writing with a team, examine your outline to see how it meshes with outlines from other writers. You may have to shift your emphasis or change the method of organization to achieve a coherent document.

Writing and Communicating Your Organization

Scientific papers, which have a prescribed form, can often be written almost in the order they are presented—from introduction, to materials and methods, results, and discussion, to abstract. This procedure works because these reports do not attempt to hook the readers with an interesting introduction or explain unfamiliar material by analogy. The introduction sets up the situation, giving background, stating the purpose, describing the method of investigation, and telling the results. The body follows a procedural organization, first telling what materials were used and what methods followed, then closing with the results and explaining what they mean. Once written, the whole report can be summarized in an abstract and the abstract placed in front of the article.

Other papers, however, might be easier to write out of order. Instead of struggling with a difficult introduction, you might write the body first, since you probably know this material best. Next write the conclusion and recommendations (if any), and then try the introduction. The conclusion may even suggest an introduction. Since many methods of organization can be followed in a paper for publication, you should review section 3.2 to investigate possibilities. Rough out graphics as you go, but don't stop to finish them until you have completed writing the draft.

Once the paper is written, write your abstract to reflect the actual contents of the paper, add any documentation needed, and rewrite the title if necessary.

Reviewing, Revising, and Editing

Because you are competing for space in a journal or magazine, your articles and reports for publication may be reviewed more stringently than company reports; nevertheless, the criteria for editing are the same. You should review the manuscript yourself for technical accuracy, for clarity of content and organization, and for mechanics like spelling and punctuation. Focus on accuracy of graphics, labels, and titles. Double-check all calculations and entries in tables to verify numbers and placement of decimal points.

Both editors and subject specialists will usually review the content for accuracy, but as author you have final responsibility. Likewise, copy editors will review for syntax, punctuation, and spelling, but you are ultimately responsible, and once your article is published, it will be nearly impossible to correct any inaccuracy.

Once your article is set in type, you will probably be sent proof copy to ensure that the version in type agrees with the manuscript. Review with that in mind—constantly comparing the proof with the original manuscript. If possible, work with a partner, one person reading the manuscript aloud (including marks of punctuation and spelling of uncommon words), while the other person follows on the proof copy. It's a good idea to go through all tables together, all figures together, and all text together. If you can't work with a partner, you must read very slowly and carefully— word by word. Check any word or number that looks suspicious. Mark changes with standard proofreading marks like those on the back cover of this book. On the proof, correct the error where it occurs, with a note in the margin at that point so the typesetter will see the change.

Do not make major changes in the manuscript at this time; correct errors only. You may wish to change the wording of a sentence or add material, but this is not the time to do it. Substantial changes in proofs are very expensive and may also introduce new errors.

All these regulations may seem to make the process of writing for publication very intimidating. But remember, every month thousands of articles by ordinary people working in scientific and technical fields *are* published. You too will soon have the expertise to write about what you are learning in your own field. Each time you write a report for your class, make it the very best report you can—and think about the possibility of submitting it to a journal or magazine. The competition is keen, and you must have something new to say, but there is great satisfaction (for you personally) and great value (for your career and company) in having your writing published.

CHECKLIST FOR WRITING FOR PUBLICATION

1. Is my potential report or article a research report?
   _____ Does it describe original research or findings?
   _____ Is it written for fellow specialists?
   _____ Am I aiming for publication in a professional journal?
   _____ If so, which one?
2. Have I written a trade or technical article?
   _____ Do I describe applied research or development?
   _____ Is it written for specialists, technicians, and managers?
   _____ Am I aiming for publication in a semitechnical or trade journal?
   _____ If so, which one?
3. Have I written a popular article?
   _____ Do I simplify and explain a technical subject?
   _____ Am I writing for generalists?
   _____ Do I want this to appear in a general magazine or newspaper?
   _____ If so, which one?
4. Have I followed this general procedure?
   _____ choose the topic and write a summary of the information
   _____ research potential publications
   _____ secure company approval
   _____ send a query (or write the article) following the steps below
   _____ review relevant information in the field and gather facts
   _____ organize the information to fulfill the purpose and meet reader needs
   _____ plan graphics and write a draft
   _____ review for myself and then solicit peer review
   _____ submit to the target publication
   _____ revise as needed
   _____ proofread printed copy

EXERCISE 22-1   **Revising an Article for Different Types of Readers**

Directions: The article in Figure 22-2 was written for specialists and technicians, but the subject itself could be of interest to generalists. Based on this article, write a draft of a popular article about robotic hands. Use a technical dictionary if necessary to understand the technical terms, and explain whatever terms you think are needed, omitting others. Add an explanation of how you would change the graphics.

WRITING ASSIGNMENT #14   Writing a Popular Article

Directions: Choose a subject from your major field that interests you and that you know well. Write a two- to three-page article explaining that subject to generalists. Follow the style of a specific publication like *Popular Science* or *Scientific American*. At the top of your paper, write the name of your target publication. Exchange papers with a classmate (preferably one whose major is different from yours). After reading the draft of the popular article, write a memo to your classmate evaluating the article as a generalist reader. Point out any terms or concepts you did not understand and suggest ways to explain them more clearly.

WRITING ASSIGNMENT #15   Writing for a Young Audience

Directions: Choose a subject from your major field that would interest 12- to 14-year-olds and write a two- to three-page article aimed at publication in a magazine for that age group. Use examples, analogies, and graphics to convey your ideas.

EXERCISE 22-2   Reviewing Publications in Your Major Field

Directions: Find out the names of four journals in your field—two published by professional societies and two trade journals. Read through one issue of each journal that has been published within the last two years. Write a short report in which you (1) name each of the journals and summarize in a paragraph the kind of article contained in each, and (2) explain who the intended reader group of each journal is and how you can tell.

EXERCISE 22-3   Planning an Article for a Professional Journal

Directions: Examine all the issues for the last two years of *one* professional journal in your field. Then choose a topic of current interest that has not been written about and write a detailed outline of a proposed article. Submit it for instructor review.

REFERENCES

Allman, William F. 1983. Are no two snowflakes alike? *Science 83* 4 (December) 24.
Matthews, Anne. 1983. The aging face of modern art. *Science 83* 4 (December) 97-8.
Preuss, Paul. 1983. The shape of things to come. *Science 83* 4 (December) 80-87.
Rogers, Michael. 1983. The sculpture transparent. *Science 83* 4 (December) 38-45.

CHAPTER 23

Speeches and Oral Presentations

Why does a book on technical writing include a chapter on speeches and oral presentations? There are two important reasons.

First, good speeches are much like good written documents. They share a basic structure of introduction, body, and conclusion. Within the body, a good speech uses a method of organization appropriate to the purpose, listeners, and content. Effective speakers use graphics (or visuals) to enhance major points. And finally, in preparing a speech, you follow the same eight steps as in writing, except that you replace the final review with rehearsals.

Second, as a professional in a technical or scientific occupation, speaking well is important to you, whether you speak informally to a few colleagues or give a planned, practiced presentation to a large group. In fact, you may spend as much as 25 percent of your time communicating orally to clients, to managers, and to colleagues, and often you will succeed on the job as much or more by your ability to communicate effectively as by your technical knowledge. As one engineer put it (quoted in Almazol 1982 PC-1), "The most successful people around here tend to be the ones who are able to give good oral presentations."

Speaking and writing are different, of course, primarily because readers can go back to review what you write, while listeners must understand what you say the first time. So, because you have listeners instead of readers, a speech must be simpler, clearer, and more repetitive than a written document. You keep it simpler with short sentences, straightforward language, strong transitions from one idea to the next, good overviews, and frequent repetition of main ideas. Even the graphics for a speech must be simplified to present only one idea at a time.

In this chapter you will learn five things about communicating orally.

23.1 DEFINITIONS AND TYPES OF ORAL PRESENTATIONS

A *speech* or *oral presentation* is any kind of planned spoken communication to listeners for a given purpose. "Speech" and "oral communication" are general terms that in themselves say nothing about the content, the length, or the amount of planning involved. This chapter uses the two terms interchangeably. The focus is on planned speeches, though of course you will have to give unplanned *(impromptu)* speeches from time to time. The best preparation for an impromptu speech is thorough knowledge of your subject and a quick outline for a guide.

Planned speeches are of two types: (1) those written out in full and either memorized or read from the manuscript, and (2) those presented extemporaneously—that is, using planned and outlined main points in the form of notes but allowing for flexible wording at the time of presentation.

Written Speeches

A *written speech* has both advantages and disadvantages. One advantage is the careful polishing you can give each word and phrase. A second advantage is that when you are dealing with a highly technical subject, writing down all the details ensures that you will present them accurately. A third advantage is that your speech may be published after you present it, and for publication you will need a written version.

From the listener's point of view, however, the disadvantages of a written-out speech may outweigh the advantages. The primary disadvantage is lack of listener contact: if you read a text, you must concentrate on it, and you cannot look listeners in the eye and talk directly to them.

In addition, people can seldom read aloud as well as they can speak, and—in fact—most audiences dislike being read to. Memorizing a written speech is not the solution. Few speakers try to memorize a speech because of the time involved and the potential for forgetting a section. Besides, most memorized speeches sound rather mechanical because the speaker is concentrating more on remembering the next sentence than on clarifying the next point. Thus, if you are asked to "read a paper," at a conference for example, the best approach is to write out the speech but then to speak from an outline of that document. This is called an *extemporaneous* speech.

Extemporaneous Speeches

For most occasions, an extemporaneous speech is best. It sounds interesting because it seems spontaneous, and it enables the speaker to have eye contact with listeners, getting feedback from them and responding to the feedback. However, an extemporaneous speech is definitely a planned and practiced presentation, even though the words "impromptu," "ad lib," and "extemporaneous" are sometimes used interchangeably. An extemporaneous speech is planned to fit a designated time, organized (with main points and supporting facts written in an outline or on notecards), and rehearsed several times before it is presented. As Mark Twain said (quoted in Plimpton 1982), "It takes three weeks to prepare a good ad-lib speech."

This chapter concentrates on the form and the procedure for writing an extemporaneous speech. Before taking up form, though, you need to know some of the special considerations involved in speech writing.

23.2 SPECIAL CONSIDERATIONS IN SPEECH WRITING

Four factors make speech writing especially challenging:

- the distraction potential
- comprehension loss
- time limitations
- the group dynamic

If you understand these four factors, you can use them to your advantage, but if you neglect them, you may have a problem communicating effectively.

The Distraction Potential

When you give a speech, you rarely have an audience of only one person at a time as you do in writing. Whether your listeners number 3 or 300, speech-making is a group activity, and a group means distraction. Individual listeners can be distracted by outside noises like fire sirens or passing jets or by audience noises like whispers and coughs. Their thoughts can be diverted by things they see: a fly persistently settling on a neighbor's forehead, the glare from an open window, the speaker's sway or nervous clutch at the microphone stand. And even if the room is quiet and enclosed, listeners can be led astray by their own thoughts. Thought is four times faster than speech—a fact that causes listeners to fill in the gaps between your words with all kinds of extraneous material: "That's a good point . . . I wonder what Bill would think about it . . . Bill sure was mad yesterday when he heard about the contract revision . . . Well, it makes me mad too . . . Oops—what was that she said?"

You probably can't eliminate all external distractions, and you certainly can't control your listeners' every thought. But you do need to choose your place carefully (if you can). Arrange the room to minimize distraction and allow listeners maximum comfort, thus encouraging concentration on what you're saying. You can corral your listeners' attention by making eye contact with individuals, by responding to the signals you perceive, by organizing, and by repeating your main points.

Comprehension Loss

Add to the distraction potential the problems of limited memory and comprehension loss, and it's a wonder that speeches make an impact at all. I've frequently mentioned the important fact that people can only hold seven plus or minus two items in short-term memory at one time. That finding has enormous implications in speech writing: in fact, if you present seven or nine major points, you are probably overloading your listeners' short-term memory. In speeches you must simplify, group, and underscore each major point with graphics and with repetition. Three or four major points are probably a speech maximum.

Memory is also related to comprehension, and again the facts are disturbing. According to financial writer Sylvia Porter (1979), ". . . Immediately after listening to a 10-minute oral presentation, the average consumer or employee has heard, understood, properly evaluated, and retained only half of what was said. Within 48 hours, that sinks another 50 percent. The final level of effectiveness—comprehension and retention—is only 25 percent!"

If you accept those percentages as given, your task may seem impossible. Yet with careful planning and practice, you can reach your listeners with solid information and your own point of view. One reason for comprehension loss is that readers can't take breaks to mentally chew over what has been said before they go on to the next point. What's more, they can't backtrack to check a fact or statistic or to review the heading of the previous subsection—all actions that careful readers take when they seek to understand. As a speaker, you must take breaks for them—pausing momentarily to let a fact sink in, backtracking with summaries and repetition. The great comedians are masters of timing and the pregnant pause: Bob Hope and Bill Cosby know when not to talk. They pause, wait, let the audience catch up.

Time Limitations

As a speaker, you almost always have a time limit. Whether it's 5 minutes or 45 minutes, your listeners expect you to fit your speech into that time box. You may have page limits in writing, but your readers can usually spend as much time as they need to read—and reread—whatever you've written. One reader can skim 10 pages (2500 words) in 6 or 7 minutes, while another reader might study, ponder, and review the same 10 pages for 30 minutes. As listeners, however, both individuals have the same time exposure to your ideas. You can help your listeners understand by intensifying their exposure to the main ideas. You do that with repetition of key ideas, graphics for main points, handouts, even a question-and-answer period.

A second time factor is psychological: listeners expect you to respect the time limit. If they're anticipating a 15-minute speech, they will begin listening for the final point after 13 minutes. After 14 minutes they'll start to shift in their chairs and check their watches. If you pass the 15-minute mark, they'll be distracted—wondering how much longer you'll go on. What's the solution? Respect the time limit: plan your speech to fit that time box, rehearse it and time it, and then when you speak, keep track of the time.

The Group Dynamic

Listeners in a group behave differently than solo listeners, and you can use that fact to your advantage. So far the special factors in oral presentations seem to put a burden on you, but if you use the group dynamic, speaking can be easier than writing. What happens when a speaker is communicating—really reaching a listener? That listener becomes an ally, sending positive signals both to the speaker and to other listeners. Signals include sitting up and leaning forward, looking directly at the speaker, becoming still and concentrated, and even nodding, smiling, or frowning (sometimes clapping and cheering) in response to what's being said. Even though they may not realize it, other listeners are affected by those signals, and they are likely to respond in the same way. Be aware of these signals. Meet your listeners where they are and speak directly to the good ones, knowing that they can influence others.

23.3 THE FORM AND ORGANIZATION
OF A SPEECH

Like nearly every written document, a speech has an introduction, a body, and a conclusion. Each segment should be shaped by the time allotment for the whole speech. For example, if you are assigned 20 minutes to speak, the breakdown might look like this:

Introduction—3–4 minutes.
Body—about 12 minutes. Perhaps 3 minutes for each major point or argument and time to summarize at midpoints. Allow about 2 minutes to present each graphic.
Conclusion—3–4 minutes.

For an extemporaneous speech, write an outline of these three sections, including major points and supporting details. The outline can be the same as one you'd compose in preparing to write a document, except that it may have fewer main points. If you're writing out the whole speech (perhaps for later publication), work from the same outline—estimating delivery time at approximately 100 words a minute. Your final outline can be transferred to index cards that you'll hold in your hand, or you can use standard sheets of paper, perhaps fastened in a binder so they won't get mixed up. Whichever method you use, your notes should be unobtrusive. Use them for reference, not for reading, and organize them well to avoid shuffling paper.

A speech must begin by creating a friendly atmosphere and breaking the ice. You can do this by acknowledging your introduction (or perhaps introducing yourself), and by saying something complimentary to the audience. Be wary of using canned jokes to begin; they often do not fit the audience or the occasion. As soon as possible, launch into the real introduction of the speech, which should hook the listener's attention. In the following introduction to a technical speech, notice how the speaker (Jahnke 1985) allies himself with the listeners by talking about their common ground.

> This conference marks an anniversary. Ten years ago, on October 6, 1975, Performance Specification Test Procedures for Continuous Emission Monitoring Systems (CEMS) were promulgated by the Environmental Protection Agency. These performance specifications, in conjunction with new regulations requiring the installation of CEMS on industrial sources, encouraged the development and application of these systems. As users, vendors, regulators, or consultants, we have all been part of the significant growth of this industry over the past ten years.

Since speeches are generally less formal than reports, you might want to use the techniques discussed in Chapter 22 for hooking a generalist's attention in a popular article. Your introduction should also define key terms and tell the listener where the speech is going by giving an overview. The only time you won't bluntly set forth your major point in the introduction is when you're delivering bad news. Then you should establish rapport first and give the facts that lead up to the major point—which will come at the end. This technique is explained in Chapter 15.

The body of a speech is where you present and support your major points. Here you should apply another known fact about listeners: they pay the most attention at the beginning and again at the end of a speech. That means that whatever method of organization you choose, you should put your most important points at either the beginning or the end. The following methods of organization work well in speeches:

- increasing order of importance. You start with an overview and then build your case, ending with the most important point.
- general to specific. Again you begin with the larger picture and move to details. At the end you must return to the larger picture.
- problem-solution. You can make problems dramatic and therefore attention-getting, which will start you off strongly. Then you supply details in the middle and again end strongly with the solution.
- cause-effect. The links between causes and effects (or effects and causes) can be used as glue to bind your points together. Often cause and effect will develop into a narrative or chronological sequence. This sequence is effective with listeners because of its storylike approach, and you can plan a dramatic climax of the narrative near the end of your speech.
- specific to general. Interesting examples or illustrations will attract your listeners' attention at the beginning. Then you can move into more general statements in the body, culminating with your main points and so ending on a strong note.

Chapter 3 gives details on methods of organization that you can use in planning speeches.

The conclusion of a speech almost always includes a summary, designed to reinforce the main points and increase listener comprehension. Conclusions can also propose or recommend action. Many effective speakers close with an anecdote or vivid example; some use a quotation. The kind of conclusion depends, of course, on the occasion and the type of audience. The important thing is that you *plan* for an effective conclusion. Don't just let your words dribble off with a vague "That's about it." Listeners also expect you to signal the conclusion with your voice and body. Drop your voice or slow your pace and use terms like "finally" and "in conclusion" to tie up the last ends neatly.

23.4 GRAPHICS FOR ORAL PRESENTATIONS

Important as the sense of hearing is, we receive only about 5 percent of our information through our ears. Knowing this, good speakers supplement and underscore what they say with graphics, which in speeches are often called visuals. Graphics also help focus listeners' attention, and they provide variety. Figure 23-1, for example, shows the key points of a speech printed on a transparency that will be projected while the speaker discusses and elaborates on them (Rew 1986). Keeping the key points in front of the listeners helps them concentrate on the topic.

All the forms of graphics discussed in Chapter 12 can be used in a

PURPOSES OF AN INDEX

1. To help the reader find an item easily

2. To arrange the information by another sequence

3. To disclose relationships among items

4. To indicate omissions

Figure 23-1 A transparency summarizes key points in a speech

speech, but they need to be simplified. Drawings, simple tables, a list of main points, graphs, and even photographs can be effective. You can present these graphics in several different ways:

- *Chalk Board*. Because it is permanent and reusable, the chalk board will probably always be with us. If you write as you speak, your actions will draw the listeners' attention. But elaborate drawings can take too much time, and you must turn your back on your listeners to do them. If left on the board when you go on to other topics, the drawings may be a distraction.
- *Posters and Flip Charts*. Both posters and flip charts can be prepared in advance, using bright colors to highlight key points. Posters (on card stock) should be placed on an easel for better viewing and should be large—typically 2 × 3 feet. Flip charts are prepared on very large sheets of paper fastened together at the top and placed on an easel. Leave the top sheet blank to prevent distraction. Then uncover each sheet as you reach that point in your speech, flipping the used sheet over the top as you finish. You can also create the graphics as you go with felt-tipped pens.
- *Slides*. Because they can be seen most easily, slides are often the most effective vehicle for graphics if you're speaking to a large group. Their disadvantage is that the room must be darkened, and you may lose some listeners if they can't see you.
- *Handouts*. You can put graphics and other key information directly

into your listeners' hands. However, passing them out is a distraction in itself, and you'll probably lose many listeners as they switch their attention to what's directly in front of them. For this reason, limit handouts to what is needed as illustrative material during your presentation and pass them out ahead of time. Never spend your speech time reading aloud what's on the handout, but highlight points as the readers follow along. If the handout material summarizes your speech, hand it out at the end.

• *Transparencies.* For small and medium-sized audiences, a transparency used with an overhead projector is the most effective and versatile way to present graphics. The advantages of transparencies are that you can face the listeners, leave the room lights on, and still project the graphics for easy viewing. In addition, you can prepare transparencies in advance from either typed or hand-drawn materials, because most copy machines will also produce transparencies. You can also write or draw on the surface as you talk, or you can overlay transparencies to show a process or change. Perhaps most important, you can remove the transparency or turn off the projector when you are finished, thus removing potential distraction when you go on to another point.

All graphics used in speeches must be big enough to be seen easily. Even on transparencies, you should use oversized type and limit the number of words. Keep drawings and graphs simpler than those you'd use in a written report, and split complex tables into sections, presenting only one section at a time. In a graph or drawing, eliminate extra lines or focus on only one portion. Whether you do them in advance or as you speak, spend enough time on your graphics to make them neat and well organized, remembering that they should show you at your professional best.

Finally, don't overburden your speech with graphics; they should enhance, not dominate what you say. Use graphics for the main points and let your spoken words fill in the details.

23.5 THE PROCESS OF WRITING A SPEECH

Planning

As in writing, your first planning task must be to determine your purpose in giving this speech. You need to ask: "What do I want my listeners to go away with?" "Why was I asked to speak?" Presumably, you speak for the same reasons you write—primarily to share information, perhaps to recommend or persuade. But sometimes your primary purpose in a speech will be to entertain, even if you're also providing information. Therefore, you need to evaluate the social context and the time of day of the speech. A speech during working hours to a group of professionals can be very different in purpose from a speech after a luncheon or dinner meeting—even though the general topic and the audience may be the same.

You must try to learn as much as you can about your listeners, both as a group and as individuals. How many will there be? What are their interests? How can most of them be characterized? Are they specialists, technicians, operators, managers, or generalists? Will the audience be mostly of one type or a mix of two or three types? What do they need—and want—to know? What is their anticipated attitude; that is, will they be supportive and enthusiastic or perhaps hostile and antagonistic? What is your relationship to your listeners? You may be speaking down to subordinates, horizontally to peers, or up to your bosses. In each case, your preparation and approach will differ, for you must meet your listeners' needs.

Listener type and background will also influence your topic, or at least its level of difficulty and the amount of detail that you include. No matter what type of audience you have, though, the content of a speech must be simpler than the comparable content of a written report for the same audience.

Finally, learn about the location for your speech. Determine the size of the room, its physical makeup (and how many distractions come with it), and whether a lectern, lights, microphones, overhead projectors, and slide projectors will be available. Many speakers carry their own slide projectors with them to ensure that the equipment is there and working. If your speech is one of a series, find out how many other people will be speaking and where your speech will fall in the sequence. If you are at the beginning, the middle, or the end of a sequence of speakers, you will face different kinds of expectations (and perhaps meet with different levels of interest) from your listeners.

Gathering Information

In writing a speech, you will follow the same information-seeking route as in writing a report. Many times, of course, your speech will result from work you have been doing and will even be accompanied by a written document. Two hints may help you in searching out information:

1. Plan to know more than you will put in the speech itself. Have a broad base of facts, figures, comparisons, and expert opinion to undergird the three or four points around which you'll build your speech. This extra information will help you select the information to present, and you can draw on it later if there's a question-and-answer period.

2. Seek interesting information and dramatic ways of presenting facts that will make them interesting. Remember that to be effective, your points must be heard and understood. That means you must keep your listeners' attention. One way to do that is to have a theme—with variations—that you return to throughout. Another is to focus on the human side of facts and figures and especially to present facts in terms of how they will affect the listeners personally.

SAY: You can't depend on high tech for substantial gains in employment openings. According to the U.S. Bureau of Labor Statistics, only 4 of every 100 new jobs will come from what we call high technology.

NOT: Employment trends as reported by the U.S. Bureau of Labor Statistics do not substantiate the belief that high technology will provide vast numbers of new jobs, the latest estimate being that not more than 4 percent of new opportunities will be in high technology.

A third way is to include anecdotes and human interest narratives. In planning, though, don't spend so much time looking for the human interest angle that you neglect your purpose and your primary content.

Organizing, Writing, and Reviewing

Since an extemporaneous speech is given from an outline, the outline provides both the backbone and the substance of your speech. One scientist who frequently gives speeches on technical subjects (Jahnke 1987) says, "I never give prepared speeches because they come out too stilted. I outline the basic points I want to talk about on one sheet of paper. Then I highlight the key words I wish to emphasize and weave my talk around them."

As you sort your information, consolidate the bits and pieces into three or four major points, each of which can be strongly stated and frequently reviewed. Organize the body portion first to ensure that the topic is covered and the method of organization is dynamic—going somewhere and bringing the listeners along to the desired destination. To personalize the speech, plan to speak directly to your listeners with second-person pronouns, contractions, and relatively short sentences.

SAY: We've recently revised the long-range projections of employment and population in Madison County. Here's how they will affect you . . .

NOT: The agency has completed within the last month a revision of the long-range projections involving both employment and population in Madison County. The principal effect of these changes will be . . .

As a speaker you can emphasize words by using greater volume, higher (or lower) pitch, changes of pace, facial expressions, and physical gestures like pointing or clenching your fist. Albert Mehrebian, a communication expert at UCLA, says that up to 55 percent of a person's impact is from nonverbal signals (Almazol 1982). Since none of these features is available in writing, you have a real advantage as a speaker. But because of listeners' limited attention span and the distraction potential, your sentences do have to be less complicated than in writing. To remind yourself where you want to use transparencies, slides, or other graphics, mark those spots in your speech.

Organize the conclusion next to see where you will end. Make sure that somewhere in the conclusion you summarize the main points that you've raised. Then go back to the introduction. Perhaps now you'll have found a good way to attract your listeners' attention and focus it on your content. Don't try to be clever in the introduction; instead, let the material shape how you begin. Be sure that the introduction provides an overview of the whole speech—telling your audience where they're going. Review the organization to make sure that you've included all the key information. Then you are ready for the first rehearsal.

Rehearsing

Most people feel silly talking to themselves, so you may be tempted to skip this step. Don't. You may have the best organized speech in your county, but until you've actually used that outline to speak from, you don't know if it will work. What's more, until you actually say all the words you intend to say, you don't know how long it will take to say them. Some speakers go for long walks and practice the speech as they walk; some lock themselves in the bathroom with a shower running so no one will hear them. These may be extreme measures, but do what works best for you. Follow these hints to make the practice sessions easier.

- Look up all the tricky words before you start and mark the pronunciation on your cards in big letters. Practice the pronunciation until you can say the words without stumbling.
- Stand up when you practice and try to duplicate the conditions you'll have when you speak (lectern or not, overhead projector, and so on). If possible, practice in the room where you'll be making your speech, or at least visit the room beforehand so you know what it's like.
- Use your cards or outline, but make sure you're not reading from them. Practice in front of a mirror to make sure you're looking up.
- Time your presentation with a stopwatch. Listen to yourself when you speak and make sure you are not rushing.
- After the first rehearsal, adjust for timing and smooth out any places where you stumbled. Be aware that the actual presentation usually takes longer than rehearsals, so allow for that. Write out transitions if necessary. Practice two or three times more, but don't overpractice.
- Videotape or audiotape one rehearsal and review it to see how you look and sound to others.

Speaking

On the day of your speech, dress professionally for the occasion, avoiding loud or clashing colors without dressing like a mouse. For men, unless the occasion is informal, a jacket and tie are usually expected. For women, a tailored suit or dress is a good bet. Avoid jangling jewelry, which can distract your listeners. Arrive at least 15 to 20 minutes before your speech to check on the room and equipment and to prepare yourself mentally and emotionally. Don't worry if you're nervous—even professional speakers usually are, and you can use your extra energy to good advantage. Instead of thinking about potential problems, think positively about what you can accomplish. Meet the person who will introduce you, or learn if you must introduce yourself. (It's even better to know this in advance.)

Start slowly when you begin to speak, deliberately keeping the pace down at the beginning. Look your listeners right in the eye; in fact, find three or four supportive people in different parts of the room and talk directly to those individuals. That will ensure that your eyes move through the audience. Don't be afraid to move and to gesture, point to your graphics, and even underline key points. Keep your watch in front of you so you can stay within your time box, adjusting your pace as you go. Don't glance at the watch constantly; instead, look at your audience. End on time and with a real ending, not a weak gasp.

If a question-and-answer session follows, repeat or summarize each question to ensure that everyone in the audience has heard it and that you understand it. This also gives you time to plan your response. If you don't catch the gist of a question, feel free to ask that it be repeated. Don't let yourself be drawn into an argument with a listener; instead, offer to speak with the person later. If a question stumps you, admit you don't know the answer—perhaps offering to find out or giving an alternate source of information. Usually you can field questions for 10 to 15 minutes before they begin to repeat, but don't be afraid to end the session if it's not productive.

Finally, try to get some feedback from your listeners. Sometimes you can chat with individuals later, and sometimes you can ask for written comments. Responses, of course, will depend on the purpose and nature of the speech as well as the type of audience, but the more you can evaluate what you did, the better you'll be prepared for the next speech you're asked to give.

In this chapter I've concentrated on rather formal speeches, but more often you will give informal oral presentations to your colleagues or managers. Take these presentations seriously, because they are an important part of your professional career. Adapt the procedure and the form to your needs, but always spend time carefully preparing and rehearsing whenever you must give a short—or long—speech or oral presentation.

CHECKLIST FOR WRITING A SPEECH

1. In preparing my speech, have I considered
   _____ the distraction potential?
   _____ comprehension loss?

_____ time limitations?

_____ the group dynamic?

2. Does the introduction of my speech

_____ create a friendly atmosphere?

_____ hook the listener's attention?

_____ present an overview of what's coming?

3. Does the body present and support the major points, using the most appropriate method of organization for the purpose and the type of listener?

4. Does the conclusion summarize and, if appropriate, propose action?

5. Have I included anecdotes, vivid examples, or quotations to provide interest?

6. Have I planned graphics (visuals) to highlight key points?

7. Should I use

_____ a chalk board?

_____ posters or flip charts?

_____ slides?

_____ handouts?

_____ transparencies?

EXERCISE 23-1   Giving a Speech That Explains a Procedure

Directions: Choose a procedure from your major field that interests you and that you understand well. Plan a 5-minute extemporaneous speech in which you explain this procedure to the members of your technical writing class. In your planning, determine your classmates' level of technical knowledge so you know how technical you can be. Use at least one kind of visual in your speech. As a member of the audience, write a brief critique of each classmate's speech, commenting on the content as well as the presentation and noting areas that were well explained to a generalist audience.

EXERCISE 23-2   Planning and Giving a Speech on Your Term Project

Directions: Near the end of your term project, plan and present a 10-minute speech to your technical writing class giving the results and conclusion of your study. Prepare an outline of your speech to give to your instructor before you begin, and include at least two visuals.

REFERENCES

Almazol, Susan. 1982. Professionals getting the word on speaking skills. _San Jose Mercury News_, 31 January: PC-1.

Jahnke, James A. 1985. Conference introduction. Continuous Emission Monitoring: Advances and Issues. Air Pollution Control Assn. Baltimore.
———. 1987. Letter to author 8 September.
Plimpton, George. 1982. How to make a speech. Advertisement for International Paper Co. New York: International Paper Co.
Porter, Sylvia. 1979. Are you listening? Really hearing? *San Francisco Chronicle*, 14 November.
Rew, Lois. 1986. Lecture on indexing for Hewlett-Packard. 9 November.

HANDBOOK

PUNCTUATION

NUMBERS

CAPITALIZATION AND SPELLING

This Handbook contains useful information and practice exercises. It does not pretend to explain every detail of punctuation, numbers, or spelling, but it will answer many questions that you may have about those three subjects and will give you guidelines to follow in writing technical papers.

1.0 Punctuation

As a student you've studied the rules of punctuation for years, and you already know many of the topics covered in this Handbook. However, you may be unsure how to use some punctuation marks, and this course may well be your last chance to polish your punctuation skills before you enter the working world. Punctuation is important. When it is done well, it provides the reader with subtle guidelines in interpreting meaning, saying "Stop here," or "Pause here and then go on," or "These two ideas are related." Readers accept those signals subconsciously, not noticing the marks themselves but concentrating on the content. But if a sentence is either unpunctuated or badly punctuated, readers do notice the punctuation because they must slow down to puzzle out the sentence's meaning.

This Handbook covers those marks of punctuation you are most likely to use in your writing, and it provides explanations, examples, and exercises to help you practice using particular punctuation marks. You can use the Handbook both as a reference and as a workbook.

1.1 GENERAL PUNCTUATION

This section includes

- apostrophes
- commas in a series

- periods in abbreviations, acronyms, and initialisms
- commas with introductory and interrupter words and phrases
- quotation marks

Apostrophes

The apostrophe is used for three reasons:

- to show possession
- to show the omission of letters or numbers
- to prevent misreading of numbers or figures used as words

To Show Possession.

- Add 's to singular nouns and pronouns to show possession.

a product's reliability
John Parker's letter
somebody's badge

- Add 's to singular nouns that end in s. This is never wrong. If the pronunciation is awkward, however, you may use the apostrophe only.

Charles's report
this class's schedule
Keats' study

Note: The government printing office (GPO) *Style Manual* requires only the apostrophe; follow your company style guide on this point.

- Add only the apostrophe to plural words ending in s.

three hours' worth
all managers' reports

- Add 's to plurals that do not end in s.

men's contracts
women's rights

Note: Personal pronouns are already possessive and do not need apostrophes.

hers	its	ours
his	theirs	whose

Remember: Write "its" (no apostrophe) to show possession.

The snake shed its skin.

To show omission. Use the apostrophe to replace the missing letters or numbers in a contraction. Put the apostrophe where the letters or numbers would otherwise be.

she'll	she will
he's	he is
can't	cannot
I'm	I am
'87	1987
it's	it is

Remember: Write "it's" (with apostrophe) for the contraction.

It's time for work.

To prevent misreading. Use the apostrophe for clarity when you are using numbers, letters, or symbols as words.

The chart omits all 0's to the left of the decimal point.
Technical people use a crossbar on 7's to avoid mistaking them for 1's.
In my handwritten memo he misread my *a's* for *e's*.

If there is no chance of confusion, the apostrophe can be eliminated.

EXERCISE A Apostrophes

Directions: Use copyediting marks to correct any apostrophe errors in the following sentences.

Insert apostrophe _____ Delete apostrophe _____

1. In 1942 American LaFrance built a 125-foot aerial ladder, but in 43 the truck supporting it was crushed by a walls collapse.
2. The Germans intensive bombing on December 29, 1940, caused more than 1500 fires in London.
3. Due to equipment shortages, some fire brigades engines were made from cut-down cars or trucks.
4. The Coventry-Climax engines adaptability made it widely used in portable "wheelbarrow pumps."
5. New York Citys highest fire of the 40s occurred when a Mitchell bomber crashed into the 78th and 79th floors of the Empire State Building.
6. Due to the elevators failure, firefighters had to carry equipment from the 67th floor.
7. The firefighters job was made more difficult by the blazing aviation fuel cascading down the stairs to the 75th floor.
8. Modern Super Pumpers have been developed to fight urban centers large fires.

9. The combined force of the Super Pumper and it's satellites can hurl up to 37 tons of water on a fire each minute.
10. Super Pumpers high-pressure hose's use all synthetic materials that can withstand more than 700 psi.
11. Ships fires, even at dockside, cant always be fought by simply pouring on water.
12. The liner *Normandie* wasnt destroyed by fire but by the weight of the water poured on the fire.
13. It's said that an aircraft fire must be dealt with within 60 seconds of the crash.
14. All major airports fire trucks are capable of laying down foam blankets to deter fires from wheels-up landings.

Commas in a Series

Commas are used to separate three or more individual items in a series. Series items include the following:

- dates

   The order was shipped on June 12, 1989, by U.S. Mail.

- places

   The central manufacturing plant is located at 1066 S. Main Street, Rockford, Illinois.

- sentence elements

   The manufacturing plant requires bulk, container, and pallet facilities.

Always use the comma before the conjunction that precedes the last item. This comma is never wrong, and it may avoid confusion.

   NOT: The cafeteria stocks strawberry, mocha nut, vanilla and chocolate swirl
        ice cream.

Can you tell how many flavors the writer intended to list?

   INSTEAD: The cafeteria stocks strawberry, mocha nut, vanilla, and chocolate
            swirl ice cream.

EXERCISE B   Commas in a Series

Directions: Use copyediting marks to correct any comma errors in the following sentences.

   Insert comma ⟋⟍   Delete comma ⟋

1. The four most abundant elements in seawater are oxygen, hydrogen, chlorine and sodium.

2. Seawater is always slightly alkaline because it contains earth minerals such as sodium, calcium magnesium, and potassium.
3. Divers need to know about all five kinds of pressure: atmospheric hydrostatic absolute gauge and partial.
4. The second edition of *The NOAA Diving Manual* was published on December 1 1979 and is for sale by the U.S. Government Printing Office Washington DC 20402.
5. Advantages of open-circuit scuba are mobility, portability adaptability to small-boat operations and availability of training.
6. In many cases the instruments, tools and techniques were made or modified by individual scientists to meet the needs of the project.
7. A base-line study was made by Turner, Ebert, and Given at Point Loma San Diego County California.
8. The Pratt Macrosnooper is an underwater magnifying system consisting of three lenses, spacers, and plastic housing.
9. Soap mineral or fungus deposits on the Macrosnooper lenses may be removed by overnight soaking in bleach vinegar or laundry detergent.
10. Shellfish studies by divers have been directed toward the ecology of the organism its behavior relative to sampling gear and artificially implanted tags, and the efficiency of the sampling gear for capturing the organism.

Periods in Abbreviations, Acronyms, and Initialisms

An abbreviation is a shortened form of a word; like a contraction, an abbreviation is formed by omitting some of the letters of the word. As explained below, in many abbreviations a period replaces the omitted letters, but in special types of abbreviations, no periods are used. A good rule of thumb: If the abbreviation includes lowercase letters, periods are usually used. If the abbreviation is in capital letters, usually no periods are used.

Common abbreviations that usually require a period   Usually a space follows the period.

- A person's initials

  E. T. Jones
  Robert P. Waring
  BUT: FDR and JFK

- Designation of streets, avenues, and buildings in addresses

  4320 Mission Blvd.
  Bldg. 12

- A person's title

  Ms. Shirley Agnew
  Maj. Alexis Prudhomme
  Atty. Roger Hewlett
  Lawrence Tegren, Jr.

- An academic degree

  B.S.   Ph.D.
  M.D.   M.B.A.

- A company name. Always use the abbreviation if it is part of the official company name; otherwise, use the abbreviated form only in addresses, lists, and bibliographies.

  Smith Bros.
  Chrysler Corp.
  A. B. Price, Inc.

- A date or a time

  Jan. 3, 1989
  3:00 P.M.
  346 B.C.

- A part of a book in a cross-reference, table, or figure

  Chap. 14
  p. 372

- A country name, especially one that is long or unwieldy

  U.S.A.
  U.S.S.R.
  But also becoming common: USA, USSR

- A state name in ordinary writing. The two-letter postal service abbreviations for states are beginning to replace the traditional abbreviations.

  Calif. or CA
  Fla. or FL

- Measurement notations that could be confused with words

  in.   = inch
  tan.   = tangent
  at. wt. = atomic weight

Abbreviations that do not require a period. These also do not need a space between letters.

- Measurements. Use the abbreviation with a unit of measure only, and only when it cannot be confused with an existing word (see above). Otherwise, write out the word.

3000 kc	175 Hz	450 mm

- Postal abbreviations for states.

VA	WI	WA

- SI (Système International) units

m	=	meter
kg	=	kilogram
s	=	second
A	=	ampere
K	=	kelvin
mol	=	mole
cd	=	candela
min	=	minute
d	=	day
h	=	hour
L	=	liter

Acronyms and Initialisms   An acronym is formed from the first letter or letters of several words. Acronyms are pronounced as words. Initialisms are also formed from the first letter of each word, but the letters are pronounced individually.

radar	=	radio detecting and ranging
laser	=	light amplification by stimulated emission of radiation
scuba	=	self-contained underwater breathing apparatus
NATO	=	North Atlantic Treaty Organization
RAM	=	random access memory
IRS	=	Internal Revenue Service
UFW	=	United Farm Workers

The first time you use an acronym or initialism, put the identifying words first, then write the acronym in parentheses. Always do this unless you are sure that *all* your readers will recognize the shortened term.

EXERCISE C   Periods in Abbreviations, Acronyms, and Initialisms

Directions: Using standard copyediting marks, edit the following sentences for errors in abbreviations, acronyms, and initialisms.

Insert period ⊙
Insert space #

Delete period ⌐
Delete space ⌒

1. Cubic Corp developed a mine detector for the US Army that uses radio waves to find buried explosives of any kind.

2. The vehicle-mounted road mine detector (V.M.R.M.D.) can operate up to 8 mph to clear roads and communication lines of metal and plastic mines.
3. The device uses a Texas Instrument Co microprocessor.
4. Radio signals are transmitted and received by antennas on a search head placed 16 ft in front of a tank or other vehicle.
5. Signals are analyzed by the T.I. microprocessor and displayed on a dash-mounted control unit.
6. The unit contains a 4 in square T.V. displaying vertical red lines that correspond to each antenna.
7. Antennas ride 2 in above the surface for maximum sensitivity.
8. Older detectors could only find metal mines, according to Sgt GF. Frannis.
9. The VMRMD was announced in New York on Sept 1, 1979; a hand-held unit with similar technology is also being developed by Cubic Corp.
10. R Keller, product manager at Cubic, says that the VMRMD also has potential nonmilitary applications.

Commas with Introductory and Interrupter Words and Phrases

Commas are used to set off introductory and interrupter words or phrases from the rest of the sentence. A phrase is a group of related words.

Introductory words

- Nouns of address

  Paul Smith, will you please give your analysis first?

- Transitional words that break the continuity of the writing

  The manufacturing plant will be closed the first week in September. However, the development lab will operate as usual.

- Conversational openers like "yes," "no," "now," "well," or "why"

  No, the shipment has not arrived.

Introductory phrases

- Prepositional phrases of three words or more

  At each level of design, the review committee meets to solve problems.

- Verbal phrases (made from verbs, but not acting as verbs)

  To advance the spark, change the ignition timing.
  Printing to screen, he checked the format of the document.

- Short prepositional phrases that could cause misreading

  In time, all bearing surfaces will show wear.

- Long modifiers out of usual order

  Safe for humans, the spray is toxic to insects.

Interrupter words and phrases

- Direct address

  Call me, Dr. Howard, if you have any questions.

- Appositives (a second noun that amplifies the meaning of the first noun)

  Alan South, manager of publications, will be the principal speaker.

- Parenthetical elements (extra or added information)

  The temperature varied, he explained, because of the inversion.

- Contrasting statements

  The mechanical components, not the electrical, were causing the problem.

- Transitional expressions like "for example" and "especially"

  So-called stone fruits—for example, peaches and nectarines—are technically the soft tree fruits with pits.

EXERCISE D Commas with Introductory and Interrupter Words and Phrases

Directions: Using standard copyediting marks, edit the following sentences for comma errors.

Insert comma ___⋀___   Delete comma ___⨍___

1. Pyrotechnics the craft of making fireworks is beginning to use scientific chemical applications.
2. However the industry in the U.S. has been dominated for decades by small family firms.
3. To fuel an aerial firework pyrotechnists traditionally use charcoal and sulfur.
4. For the lifting charge and the bursting charge the pyrotechnist uses a mixture of gunpowder and fuel.
5. To create the colorful streaks across the sky fireworks manufacturers add "stars," for example strontium compounds for red and barium compounds for green.
6. New fuels especially magnesium and aluminum alloys can create brighter and deeper colors than were previously attainable.
7. Bill Page, manager of Celebrity Fireworks in California has per-

fected a new oxidizer potassium perchlorate that is more difficult to ignite.

8. Less hazardous than other powders, the new oxidizer is however harder to use in aerial shells.

9. Some fireworks accidents according to Page, are blamed on discharge of static electricity.

10. To avoid static discharge workers at the fireworks plant wear plug-in wristbands to keep themselves grounded.

11. Even so the primary safety rule is to keep the amount of exposed explosives to a minimum by dividing operations among several widely scattered buildings.

12. With inexpensive and high-quality products Chinese pyrotechnists have captured much of the American market.

Quotation Marks

Double quotation marks   Use double quotation marks as follows.

To set off someone's exact words.

Thomas Gold said, "Oil and gas come from methane gas deep in the earth."

- Use a comma or (in formal writing) a colon after the identifier, the word or words telling who made the statement.

The engineer concluded by saying: "The fog horns sound whenever visibility drops below three miles."

- Use commas before and after the identifier if it divides a complete sentence.

"I am convinced," Gold reported, "that oil and gas can be found anywhere if you dig down far enough."

- Place commas and periods inside end quotation marks; place semicolons and colons outside. Question marks go inside or outside depending on whether the quotation is a question or is part of a question.

A new theory uses a three-dimensional map to show that the universe is composed of gigantic "bubbles."

"If we are right, these bubbles fill the universe just like suds filling the kitchen sink," said John P. Huchra of the Harvard-Smithsonian Center for Astrophysics.

The observations "pose serious challenges for current models for the formation of large-scale structure"; conventional explanations say that gravity played a dominant role.

Are the stars and galaxies gathered on the surface of these "gigantic bubbles"?

To identify new technical terms or words discussed as words. Sometimes such words are printed in italics instead.

Gold was an originator of the "steady state" theory, which held that the universe was unchanging.

To identify parts of a larger published work.

Winters reported on Chapter 7, "Direct Redistributional Issues."

Note: Do not use quotation marks around indirect quotations.

He reported that a telescope was not required.

Single quotation marks   Use single quotes to enclose a quotation within a quotation.

Cruikshank reported last week: "We think the asteroid-meteorite link has been strengthened considerably by the recent discovery of '1982 RA,' which has an orbit that crosses earth's orbit."

EXERCISE E   Quotation Marks

Directions: Using standard copyediting marks, edit the following sentences for the punctuation associated with quotation marks.

Insert comma ⌃	Delete comma ✗	Transpose ∿
Insert period ⊙	Delete period ✗	Insert quotation mark ∀

1. "The biological theory of petroleum arose when all assumed that complex compounds of carbon could be formed only biologically" a science writer reported.
2. "Now it is clear he continued, that hydrocarbons are present everywhere in the solar system".
3. "If the earth started with a quantity of carbon compounds comparable to that on other planets, the question must be 'Where is it all?' the science writer said.
4. Geologist Gold contended that most of the methane is deeper than people have looked. "I expect the region between 15,000 and 30,000 feet below the surface to be very rich in methane, he said.
5. The Swedish government announced plans to drill a deep test hole in granite bedrock. Geologist Michael Halbouty wrote: The Siljan Ring area represents one of the best prospects in the world to test the abiogenic gas theory"
6. Gerard Demaison, however, recently said "There are people that are overreacting to Dr. Gold and think he's wasting everybody's time, and some who believe everything he says. I think both are extreme positions."

7. Gold countered that several kinds of evidence already established provided as firm a proof as is ever possible in science.
8. "Analysis of petroleum and gas from adjacent fields in the Middle East Gold contended, "shows that the oil is very similar chemically".
9. "However," he continued, "the three fields span totally different geological regions: folded mountains in Iran, the Tigris River valley in Iraq, and flat-bedded sediments in Saudi Arabia.
10. Gold concluded Detailed studies of the chemistry of oil and gas fields in Oklahoma and Texas that occur in geologically different formations provide additional insurmountable evidence for the theory."

1.2 PROBLEM PUNCTUATION

This section is called Problem Punctuation because most college students are not sure how to use commas and semicolons in clauses. If you sprinkle commas in your sentences like pepper, or if you follow a vague rule that says, "If I need a breath, I'll throw in a comma," this material is for you. If you are willing to learn two simple rules, you can solve 80 to 90 percent of your punctuation problems. What's more, those two rules will help you make choices as you write—choices that will allow you to punctuate for meaning in your sentences.

The problem punctuation section includes:

- subjects and verbs
- compound sentences
- complex sentences with subordinators
- complex sentences with relative pronouns

Subjects and Verbs

The key to problem punctuation is a clear understanding of what makes a sentence.

Definitions

- *Sentence:* an independent unit of words that contains a subject and a verb and closes with a mark of punctuation. (For identification, subjects are underlined once and verbs twice.)

Writers write.

Sometimes the subject is understood.

Write! ["you" understood]

- *Subject:* that part of a sentence about which something is said or asked.

  <u>Students</u> write.

  The *simple subject* is the main noun or pronoun.

  <u>He</u> ran.

  The *complete subject* is the main noun or pronoun plus its associated modifiers.

  <u>The man in the red jumpsuit</u> ran.

  Subjects can be multiple.

  <u>The truck and the motorcycle</u> collided.

- *Verb:* the word or words in a sentence that show action, occurrence, or existence.

  Students <u>write</u>.
  Jones <u>is</u> in his office.

  Verbs can be multiple.

  College students <u>eat</u>, <u>sleep</u>, and <u>study</u>.

  Verbs can contain more than one word.

  Chemistry 101 <u>will be offered</u> at night.

  The *predicate* is the verb plus its modifiers and *complements.* The predicate is the part of the sentence that says or asks something about the subject. A complement completes the sentence meaning. Complements can be direct objects, indirect objects, subject complements, or object complements.

  The fire / <u>burned 42 acres of brush and trees.</u>
  predicate

Compound Sentences

Writers frequently combine related ideas into one sentence. If you want to join two or more ideas of equal importance, you may want to write a compound sentence. Look at the two sentences below:

> Mary graduated in June with a degree in geology.
> She now works as a computer programmer.

Suppose that you want to join those two ideas into one sentence, giving equal importance to each idea. You might say

> Mary graduated in June with a degree in geology; she now works as a computer programmer.

What are you implying by this combination? A simple chronology: Mary did that and now is doing this. Another choice might be this one:

> Mary graduated in June with a degree in geology, and she now works as a computer programmer.

But think about the *meaning* you want to communicate. Is that job as a programmer an expected outgrowth of a geology degree? Or do you want to imply a contrast by the way you combine those sentences? For example, you might say

> Mary graduated in June with a degree in geology, but she now works as a computer programmer.

or

> Mary graduated in June with a degree in geology; however, she now works as a computer programmer.

In each of the four sentences above, the two ideas are correctly combined into a compound sentence. However, the implications are different in the four sentences. That's what I mean by punctuating for meaning.

Definitions   Before you can learn the punctuation rule for compound sentences, you need to be sure of the meanings of five key terms:

- *Clause:* a group of related words that contains a subject and a verb. The subject may be understood. A short sentence is called a clause when it becomes part of a larger sentence.

  but he examined the precipitate

- *Independent clause:* a clause that can stand alone as a complete sentence.

  the report summarized data during a 6-month period

- *Coordinators:* the nine connecting words listed below. They are also called *coordinating conjunctions.* "Co" implies equal, as in the word "coworker."

and	either	yet
but	neither	so
or	nor	for

- *Adverbial connectives:* adverbs that can be used—along with punctuation—to join independent clauses into a compound sentence. They are also called *conjunctive adverbs.* They include the following:

accordingly	instead
also, too	meanwhile
anyhow	moreover
anyway	namely

besides	nevertheless
consequently	now
furthermore	otherwise
hence	still
however	then
indeed	therefore
	thus

Note: Phrases can also be used like adverbial connectives.

as a result	for example
in fact	in addition

- *Compound sentence:* a sentence containing two or more related independent clauses joined in some way.

Punctuation rule #1   I tell my students this is Rew's Rule #1 for punctuating a compound sentence:

A compound sentence can be punctuated in one of three ways:

1. independent clause   (, coordinator)   independent clause

The flywheel recovers in less than a second, and another power stroke begins.

2. independent clause   (;)   independent clause

The flywheel recovers in less than a second; another power stroke begins.

3. independent clause (; adverbial connective,) independent clause

The flywheel recovers in less than a second; then, another power stroke begins.

As the writer, you have the flexibility to influence your reader by the method you choose. What are the differences in meaning in the three sample sentences above? Be prepared to discuss them in class.

Note: The adverbial connective often starts the second clause, but it can also appear elsewhere in that clause.

The tests are not material performance tests; they are, instead, material quality tests.

EXERCISE F   Sentence Combining: Writing Compound Sentences

Directions: To review your understanding of key terms, answer each of the following questions. Try to answer without looking at the definitions. Then check your answers against the written definitions.

1. What is a clause?
2. What is an independent clause?

3. How would you define a compound sentence?

4. When should you write a compound sentence?

5. What are the three ways a compound sentence can be punctuated?

Directions: Following are some simple sentences. Join each pair into a compound sentence using *each* of the three ways of compounding. Be sure that you write a compound sentence, even though you could, perhaps, write a simple sentence. Mark the best one with a star and be prepared to defend your choice in class.

1. Normal range for the AM-FM radio is approximately 35 miles. On flat terrain and with powerful transmitters, the range could be considerably extended.

2. A home radio has constant voltage and is rarely subjected to wide variations in temperature. An FM car radio operates under more severe conditions.

3. The tandem master cylinder is mounted on the forward end of the booster. The rear end of the booster is connected to a frame-mounted bracket.

4. Two output lines are connected to the power section. One leads to the power steering gear and the other to the reservoir.

5. Brake adjustment is required after installation of new or relined brake shoes. Adjustment is also necessary whenever excessive travel of pedal is needed to start braking action.

6. Rear axles can be noisy if they are worn. They can be noisy if they lack proper lubrication.

Complex Sentences with Subordinators

Sometimes you want to combine two or more ideas in a sentence, and you want to make each idea equally important. Then you write a compound sentence. At other times, you want to combine ideas, but you want to make one idea dominant and the other subordinate. Then you write a complex sentence. Occasionally, you want to do both. Then you write a compound-complex sentence.

Look at the two sentences below.

J. Smith designed the SF 150 module.
He was promoted to manufacturing manager.

Which of these two ideas do you, as writer, want to be the more important?

Suppose you choose the idea of promotion as more important. That sentence will become your main, or *independent*, clause; the sentence about designing the module will become a subordinate, or *dependent*, clause. You can form a dependent clause by beginning it with a *subordinate conjunction*.

Because J. Smith designed the SF 150 module, he was promoted to manufacturing manager.

The independent clause can come at the beginning or the end of the sentence.

> J. Smith was promoted to manufacturing manager *because* he designed the SF 150 module.

If you chose the design idea as more important, you would begin the promotion clause with a subordinate conjunction.

> J. Smith designed the SF 150 module *because* he was promoted to manufacturing manager.

or

> *Because* he was promoted to manufacturing manager, J. Smith designed the SF 150 module.

Do you see how the meaning has changed? You changed it by the choice of the main clause. You could also change the meaning dramatically by choosing a different subordinate conjunction. Suppose you use *after:*

> J. Smith designed the SF 150 module *after* he was promoted to manufacturing manager.

or

> *After* he was promoted to manufacturing manager, J. Smith designed the SF 150 module.

Definitions

- *Clause:* a group of related words that contains a subject and a verb. The subject may be understood. A short sentence is called a clause when it becomes part of a larger sentence.

- *Independent clause:* a clause that can stand alone as a complete sentence.

- *Dependent clause:* a clause that cannot stand alone as a complete sentence. It is subordinate to an independent clause.

   if the product is announced

- *Complex sentence:* a sentence containing one independent clause and one or more dependent clauses.

- *Compound-complex sentence:* a sentence containing two or more independent clauses and one or more dependent clauses.

- *Subordinator:* a connecting word that begins a dependent clause and joins it to an independent clause to make a complex sentence. Subordinators are also called subordinate conjunctions. "Sub" implies "under." The following words can be subordinators:

after	unless
although, though	until, till
as, as . . . as	when
as if	whenever
as though	where
because	whereas
before	wherever
if, even if	while
in order that	why
once	
since	

Punctuation rule #2A I tell my students this is Rew's Rule #2A:

To punctuate a complex sentence when the dependent clause begins with a subordinate conjunction:

1. if the dependent clause stands first, it must be followed by a comma.

    Before high-speed maglevs will be practical, a breakthrough in electrical generation is necessary.

2. if the dependent clause stands after the independent clause, no punctuation is required unless the subordinator is "though," "although," or "whereas."

    A breakthrough in electrical generation is necessary before high-speed maglevs will be practical.

    The harbor entrance rapidly filled with sand, although it was dredged out every year.

EXERCISE G   Sentence Combining: Complex Sentences with Subordinators Showing Cause or Logic

Directions: To review your understanding of key terms, answer each of the following questions. Try to answer without looking at the definition. Then check your answers against the written definition.

1. What is a dependent clause?
2. What is a subordinate conjunction?
3. When should you subordinate a clause?
4. Write the punctuation rule for a complex sentence: Rule #2A.
5. Here are the common causal/logical subordinators:

because	since	unless
if	although	whereas
even if	though	

What relationships do they indicate?

Directions: Using a causal/logical subordinator, join the following independent clauses to make a complex sentence. Be sure to write a complex sentence, even though you could write a simple or compound sentence. Try to use a different conjunction for each one, and try the dependent clause in different positions. Punctuate correctly. Be prepared to defend your choices. Use standard copyediting marks.

Insert __A__   Delete __◝__   Make lowercase __E̸__
                                     Capitalize __e̲̲__

1. Darwin was a fledgling naturalist in 1832. His observations of plant and animal life were precise and detailed.
2. Darwin took extensive notes on his round-the-world journey in the *HMS Beagle.* He later published the *Journal of Researches* based on his journey.
3. The *Journal of Researches* is an invaluable guide. The reader wants to understand Darwin's *On the Origin of Species.*
4. Darwin's observations were limited to animals on friendly shores. Some obstreperous South American natives would not allow the crew of the *Beagle* to land.
5. Today scientists try to use only objective words in their technical descriptions. Darwin used terms like "hideous," "ugly," "repulsive," and "disgusting."
6. Students today do not appreciate the accomplishments of the 19th-century naturalists. They realize that Darwin and Huxley were breaking new ground in both concentration and technique.

EXERCISE H   Sentence Combining: Writing Complex Sentences with Subordinators Showing Time or Place

These are the common temporal/spatial subordinators:

where	after	until
wherever	before	till
when	since	while
whenever	once	

They are used in the same way as the causal/logical subordinators.

Directions: Combine each of the following sets of sentences into a single complex sentence by using one of the temporal/spatial subordinators. Be sure your dependent clause indicates time or place, and be sure to punctuate correctly. Use standard copyediting marks.

Insert __A__ Delete __◝__ Make lowercase __E̸__ Capitalize __e̲̲__

1. James Watson, an American molecular biologist, was in his early twenties. He went to Cambridge to study the structure of proteins.

2. He collaborated with English biologist Francis Crick. They discovered the chemical structure of DNA.

3. The discovery excited the scientific community. Watson and Crick won the Nobel Prize.

4. Watson became a professor of biology at Harvard. He published a popular account of their search in the book *The Double Helix*.

5. Watson and Crick were searching for the DNA model. Linus Pauling was also close to finding the answer.

6. Pauling sent a preliminary manuscript about his study from California to Cambridge. Pauling's son, Peter, was a biophysicist.

7. Watson looked at the Pauling manuscript. He realized that either Pauling had erred, or he was developing an unorthodox chemical solution to the problem.

Complex Sentences with Relative Pronouns

Dependent clauses used as adjectives Complex sentences consist of dependent and independent clauses. Some dependent clauses are formed by adding subordinating conjunctions to clauses, thus reducing their independence. Other dependent clauses are formed by adding relative pronouns to clauses to reduce their independence. These clauses may be used either as adjectives or as nouns. A clause used as an adjective stands after the word it modifies.

Common relative pronouns that can begin adjective clauses are

that	who	whose
which	whom	

For example:

The wallet *that I lost* contained all my credit cards.

The St. Claire Hotel, *which was rundown and dirty*, has been restored to its former splendor.

The student *who caught the shoplifter* received a substantial reward.

The candidate *whom we supported* was not elected.

The student *whose bike was stolen* had to commute by bus.

Each dependent clause above is used as an adjective; that is, it modifies or explains the noun that precedes it, answering the questions "What?" or "Who?"

Dependent clauses used as nouns Dependent clauses can also be used as nouns in the sentence. The clauses may look the same, but the punctuation rule is different. For example:

That I lost his confidence is painfully obvious. [subject]

I know in my heart *what I am doing.* [direct object]

He wrote the letter to *whom it might concern.* [object of preposition]

Punctuation rules Dependent clauses used as *nouns* are generally not separated from the rest of the sentence by commas. Dependent clauses used as *adjectives* sometimes must be set off by commas.

If the clause is necessary to establish the specific identity of the noun it follows, it is a *primary identifier:* it restricts or limits the meaning of the word it describes. Such a primary identifier is called "restrictive," and no commas surround it.

If the clause merely adds extra information to the noun it describes, it is a secondary identifier: it could be dropped from the sentence without materially affecting the meaning. Such a secondary identifier is called "nonrestrictive," and it should be surrounded by commas. For example:

The woman *whom you have just met* is in charge of the project. [restrictive]

Felicia Lopez, *whom you have just met,* is in charge of the project. [nonrestrictive]

The book *that I am now reading* is a collection of essays. [restrictive]

A Moveable Feast, which I am now reading, is a collection of essays. [nonrestrictive]

Note: Confused about when to use "that" and "which"?

"That" begins restrictive clauses (without commas).
"Which" begins nonrestrictive clauses (with commas).

Also note: Many writers use "that" and "which" interchangeably, so you can't rely on the words themselves to tell you if the clause is restrictive or not. If you follow the rule, however, you will help your reader.

Punctuation rule #2B I tell my students this is Rew's Rule #2B.

To punctuate a complex sentence when the dependent clause is introduced by a relative pronoun:

1. if the relative clause is a primary identifier (restrictive), no punctuation is required.

   A carburetor that will give better gas mileage is badly needed.

2. if the relative clause is a secondary identifier (nonrestrictive), it must be surrounded by commas.

   The Solex carburetor, which gives better gas mileage, is an ecological bonus.

EXERCISE I   Writing and Punctuating Complex Sentences with Relative Clauses

Directions: Write one sentence using each clause as a restrictive, or "primary," adjective modifier. Then write another sentence using the same clause as a nonrestrictive, or "secondary," adjective modifier. Punctuate correctly.

1. who had seen the entire incident
2. that   polluted the atmosphere
   which
3. whom he admired
4. whose term expired this year
5. that   stopped at ten o'clock
   which
6. who was elected in 1948
7. that   was sold yesterday
   which

EXERCISE J   Recognizing Relative Clauses and Relative Pronouns

Directions: Underline the relative clause in each sentence below. Circle the relative pronoun.

1. In addition to the contaminants that may be present in free air, the air compressor machinery may introduce contaminants.
2. Scuba tanks often are fitted with a rubber or plastic boot that should have holes to permit draining.
3. An operator who is unaware of the problems involved with compressed air systems can endanger the diver's life.
4. Portable high-pressure compressors, which are used for filling scuba tanks, should deliver a minimum of 2 standard cubic feet per minute of air at 2,000 psi.
5. Water-lubricated compressors that do not produce carbon monoxide or other contaminants are available on a limited basis.
6. A molecular sieve, which has an extremely large surface area to enhance its capacity for adsorption, removes harmful contaminants by causing them to adhere to its surface.
7. For field use, instruments are available that provide sufficiently accurate data to determine the safety of a gas as a breathing medium.
8. Oxygen is shipped in gas cylinders that are color-coded *green.*
9. Nitrogen, which is the most commonly used diluent, is limited because of its tendency to produce narcosis.
10. Helium has a high diffusivity that allows it to leak through penetrators and into equipment.

EXERCISE K Recognizing and Punctuating Restrictive and
Nonrestrictive Clauses

Directions: Underline the relative clause in each sentence and identify it
as restrictive or nonrestrictive. Punctuate it correctly. Correct the use of
"that" and "which" if necessary.

Insert comma ⟋   Insert word ⟋ⓐ
Delete comma ⟋   Delete word ⟋

1. Speech is one of those basic abilities that set us apart from animals.
2. We all use automatons like the dial telephone and automatic
   elevator which either get their instructions from us or report back
   to us on their operations.
3. The movements of the vocal organs generate a speech sound wave
   that travels through the air between speaker and listener.
4. Pressure changes at the ear activate the listener's hearing mechanism and produce nerve impulses that travel along the acoustic
   nerve to the listener's brain.
5. We can compare all the words of a language and find those sounds
   that differentiate one word from another.
6. There are some words, like "awe" and "a," which have only 1
   phoneme, and others that are made up of 10 or more phonemes.
7. The rules that outline the way sequences of words can be combined to form acceptable sentences are called the grammar of a
   language.
8. Peter Denes and Elliot Pinson who wrote *The Speech Chain* are
   both communication scientists.
9. *The Speech Chain* is part of a science study series by scientists
   who write for students and laymen.
10. The neuron has an expanded part, the cell body which contains
    the cell nucleus.
11. The figure shows a neuron similar to those neurons whose cell
    bodies are located in the spinal column.

1.3 CLASSY PUNCTUATION

What I call classy punctuation is those marks that, properly used, clarify
the meaning by adding subtle shades of emphasis or differentiation. This
section covers

- colons
- hyphens
- dashes

If you learn to use them effectively, your writing will take on professional polish.

Colons

College students seldom use colons, perhaps because they are unsure of how to use them correctly. Professional writers, however, use colons extensively, relying on them to convey subtle shades of meaning to the reader. Fortunately, using the colon is easy, and it can give your writing a touch of class.

You can use a colon in four situations:

- to introduce a formal list or series, especially one including the word "following."

   The following Pacific Coast bays are really drowned river valleys: San Francisco Bay, Coos Bay, Humboldt Bay, and Grays Harbor.

   Note: You should not use a colon to separate a verb from its object or complement or to separate a preposition from its object.

   NOT: The three most conspicuous plants in brackish water marshlands are: cattails, bulrushes, and sedges.
   BUT: The three most conspicuous plants in brackish water marshlands are cattails, bulrushes, and sedges.
   NOT: Collected animals should be put in: plastic pails, plastic bags, or sloshing water.
   BUT: Collected animals should be put in plastic pails, plastic bags, or sloshing water.

   However, you *can* use a colon if the objects or complements will be presented in a vertical list.

- to introduce a word group or clause that summarizes or explains the first word group. In this case, the colon can replace "that is" or "namely."

   There is only one way to preserve collected animals: keep the new surroundings as close as possible to the original habitat.

   The coloration of the upperside of the carapace varies extensively: Sometimes it is red or purplish, but in many specimens it is gray, brown, or olive.

   Note: If the clause following the colon is a complete sentence, it is often capitalized.

- to introduce a long or formal quotation.

   Kozloff's warning is plain: "Leaving a rock 'belly up' is an almost sure way to kill most of the animals that are living on its underside, and perhaps also the animals and plants on its upper side."

- to show typographical distinctions or divisions.

Dear Mr. Simpson:

4:30 P.M.

"The Pacific Northwest: An Ecological Study"

EXERCISE L   Editing for Colons

Directions: Using standard copyediting marks, insert colons in the following sentences if they are needed. Below the sentence, write the reason for inserting the colon. If no colons are needed, give the reason.

Insert colon ⟨:⟩

1. Of the DC and AC types, the AC variety is most prevalent It permits larger displays, although it's more difficult to build and requires more complex drive circuitry.
2. The main difference between AC and DC plasma panels is AC displays have dielectric layers separating the gas from the unit's activating electrodes.
3. The thin-film EL panels used in information displays usually fail in one of three ways package fracture from shock, dielectric breakdown from overvoltage, brightness decay from high temperatures.
4. The new development program could prove to be one of the most significant in the history of flat-panel displays the production of a full-color EL display.
5. Hix stated "The use of class 10 clean rooms allows production of panels with 100 percent functional pixels and no pin-hole defects."
6. VF panels basically employ the same principle as a triode vacuum tube the phosphor-coated anode emits light when struck by electrons from the cathode.
7. So far this technology has appeared in only two military applications a compass computer and a digital message device.
8. Applications for AC plasma displays have included the following Trident submarines, the Harpoon missile system, and an antenna-directing system in the E-4B emergency Presidential command aircraft.

Hyphens

Hyphen usage can be complicated, and neither dictionaries nor handbooks always agree on suggested rules. But technical writers and editors need solid information about hyphens because they are used so frequently in technical writing. Here are the most common uses of the hyphen. For further information see Karen Judd's *Copyediting: A Practical Guide* (Los Altos, Wm. Kaufmann, 1982) or the Government Printing Office *Style Manual.* You can use hyphens for these reasons.

To indicate syllable breaks This is the most common use of hyphens: to break a word at the end of a line. The only things you have to remember are these:

- Hyphenate between syllables only. If you don't know where the break is, look in the dictionary.
- Do not divide one-syllable words.
- Do not hyphenate names, contractions, abbreviations, acronyms, or numerals.
- Do not hyphenate the last word on a page or the last word in a paragraph.

In compound words Authorities and dictionaries often do not agree on whether a hyphen is required in a compound word. Usually what happens is that a compound word begins as two separate words, is then hyphenated, and eventually becomes a single word. You need to remember these things:

- Be consistent. Choose a single dictionary, handbook, or company style guide and follow it faithfully.
- Keep your reader in mind. Hyphens are supposed to make the reading easier. Sometimes you must use hyphens simply to avoid confusion. For example:

an un-ionized compound (not unionized)
re-cover it (make a new cover, not recover)
ball-like
cave-in

In compound words used as adjectives Use hyphens in the following compound words when they are used as adjectives *preceding* nouns:

- numbers from twenty-one to ninety-nine when they are spelled out

  Thirty-three engineers attended the meeting.

- a number plus a unit of measure used as an adjective

  9-minute interval

- spelled-out fractions used as adjectives

  one-third cup

- an adjective or a noun plus a past or present participle used as an adjective

  motion-activated alarm
  software-testing strategies

  Note: Do *not* use hyphens with compound adjectives that follow forms of the verb "to be."

  NOT: The alarm was motion-activated.
  BUT: The alarm was motion activated.

- an adjective plus a noun used as an adjective

  high-performance workstations
  dual-port design

- two colors used as an adjective

  yellow-green smoke

- words with prefixes if the root is a proper name or a number

  pre-1980
  post-Newtonian

In compound words used as nouns   These include:

- all compound nouns beginning with "self"

  a self-study

- a compound made of two nouns of equal value

  owner-operator
  motor-generator

- numbers from twenty-one to ninety-nine when written out

  The number seventy-three comes first.

Note: Do not use hyphens:

- with most prefixes and suffixes

  starlike
  pretest

- with names of chemicals used as adjectives

  sodium chloride deposits

- with an adverb ending in "ly" plus an adjective or participle

  a finely tuned engine

- if the individual words are modifying the noun separately

  a new digital analyzer

EXERCISE M   Editing for Hyphens

Directions: Insert or delete hyphens in the following sentences if necessary and indicate the reason for your action.

Insert hyphen ___=___   Delete hyphen ___↗___

1. The chemist performed the complex analysis by using his balance and finely-graduated weights.

2. During the decision making process, one needs to think through alternative actions.
3. The procedure has been used for the last twentyfive years.
4. That refinery has a long term contract with Exxon Corp.
5. The powerplant uses 16 cylinder diesel electric marine modules.
6. The power modules combine on a single skid GE's custom 8000 selfventilated generator.
7. The car buying public has responded well to front wheel drive.
8. That business requires highly skilled and technically trained management.
9. A 5.7-liter engine is used in that model.
10. Fuel mileage is rated at 16 mpg, which is a 3 mpg improvement.

Dashes

What we usually call the dash is more specifically known as the "em dash." The name comes from the width of the em dash—about the same width as the capital M. In typing, the em dash is shown by two hyphens, and no spaces are used before or after the em dash.

Dashes are often used as substitutes for other marks of punctuation, but they are not as precise. As a writer, challenge yourself to use the exact mark; don't compromise with an all-purpose dash. At the same time, when you edit someone else's writing, respect the writer's choice; don't change the dash without good reason. The dash tends to be informal, especially if it is used frequently, but it is most effective when it is used sparingly.

Use dashes for the following reasons.

To set off parenthetical material Commas and parentheses can also set off parenthetical material, but for the strongest emphasis on the parenthetical material, use the dash.

> Certain cells of the retina—the familiar rods and cones—undergo photochemical changes that enable you to see in various light levels.

For equal emphasis between the sentence and the parenthetical material, use the comma.

> Certain cells of the retina, the familiar rods and cones, undergo photochemical changes that enable you to see in various light levels.

To reduce the importance of the parenthetical material, use parentheses.

> Certain cells of the retina (the familiar rods and cones) undergo photochemical changes that enable you to see in various light levels.

Which version do you think the author used for this sentence? If you guessed the first example, you are right.

To show a break in thought   A dash helps to set off an emphatic idea or to summarize an idea.

> Modern arithmetic was the result of a long search—"research"—for a tool that would handle complicated quantitative problems.

> The human eye is a dual organ—two eyes working together to transmit information to the brain.

For clarity   When commas are already used within parenthetical material, use dashes to set off the parenthesis.

> Complex sounds—such as musical notes, spoken words, or an engine's roar—all produce a more or less definite pitch.

EXERCISE N   Editing for Dashes

Directions: If necessary, insert or delete dashes in the following sentences or change inappropriate dashes to other punctuation marks. Be prepared to explain why you took the action you did.

Insert dash ⸺   Insert semi-   Insert colon ⊙
Insert comma ⌃   colon ⌃

1. If the computer is a mini- or super-mini class, it should have the ability to use micros as emulators—this will prevent it from being limited to system application software.
2. This makes available to the user a large library of less expensive, already available microcomputer software—the 9816 is capable of running CPM software in addition to H-P software.
3. Unlike those types of instruments whose sound source is produced by plucking, striking, or blowing, electronic instruments, the synthesizer for example, produce sound through an electrical circuit that generates an oscillating electrical current.
4. The synthesizers in this report are called hybrids, a combination of analog and digital technology.
5. Each letter of the ADSR generator represents a separate stage of its four major functions—Attack, Decay, Sustain, and Release.
6. Each envelope generator has four different time durations that can be programmed to modify a sound—these time durations control the speed and volume level of the programmed event and are expressed in rates and levels, seconds, and milliseconds.
7. When one or more of the control sources, the keyboard, LFO, or envelope generator, is triggered, the audio chain is enacted, resulting in audio output.
8. The upper midsection of the board of an ARP Odyssey II is devoted to a panel of slide switches that control many of the gut components—the low-frequency oscillator (LFO), the digitally con-

trolled oscillator (DCO), high-pass filter (HPF), voltage-controlled filter (VCF), voltage-controlled amplifier (VCA), and envelope generator (EG).

9. There is apparently (I'm not sure) no way to enter hard spaces or characters.
10. This "desktop mainframe" is a 32-bit super-mini computer, which currently is the only super-mini to use a "super-chip," a 32-bit processor on a single chip.

2.0 *Numbers*

As a technical writer, you must use numbers of all kinds: counting numbers, ordering numbers, fractions, measurements, and percentages. You need to establish some simple rules so that you know when to use the numerical symbols (1, 7, 360) as opposed to word equivalents (one, seven, three hundred sixty) and always treat numbers consistently.

No single standard has been accepted in industry or publishing; therefore, you must use common sense in setting up guidelines for number use. When you must write about numbers:

1. Determine your company's accepted style and follow it.
2. If your company has no accepted style or you are working on your own, follow these guidelines, which have been adapted from the Government Printing Office *Style Manual*. The assumption throughout this section is that while words may make smoother sentences, numbers (numerical symbols) are easier for readers to understand.

Use Numbers

Use numerical symbols in the following situations:

- in counting 10 items or more

  14 agricultural workers

- in ordering items from the 10th on

  the 14th floor

- in sentences containing 2 or more numbers when 1 number is 10 or more

  Yesterday we shipped 15 orders, but today we only managed 8.

- in all units of time, measurement, or money

  His shift ends at 4:00 P.M.

  The diameter is 250 mm.

  Total cost of the repair was $1452.37.

  They forecast a 20 percent rise in sales over the next 5 years.

  If the diameter is 2.469 inches or less, replace the shaft and both rotors.

  Note: If a sentence contains both numbers under nine and units of time, measurement, or money, follow the general rule of spelling out up to nine for the numbers and the rule of using numerals for the units.

  Each of the six volunteers was tested for 45 minutes.

- for very large round numbers. In this case, use a number followed by a word.

  2 billion years
  $4 million

- when two numerals appear together. In this case, spell out the lower number and use numerals for the higher number.

  14 three-foot sections
  six 100-lb. bags of flour

  Note: Use a hyphen between a number and a measurement when the two are used together to modify a noun. Also see the Punctuation section on hyphens.

Use Words

Always use words:

- in counting up to nine items

  three maps
  seven undergraduates

- in ordering items up to the ninth

  the third building on the left
  Fifth Avenue

- to begin a sentence

  Forty-three geographical areas have been identified.
  Six examples are included in the interim report.

  Note: If the number is more than two words, it's better to reorder the sentence.

  NOT: Six hundred thirty-three patents were processed.
  BUT: The patent office processed 633 patents.

- to indicate general approximations

  during the sixties

  but not when the approximation is used with a modifier

  almost 70 errors
  approximately 300 experiments

- if two numerals occur together. In this case, spell out the lower number and use numerals for the higher number.

  three 12-inch rods
  144 nine-ounce containers

- to indicate fractions, unless the fraction is used as a modifier

  three-fourths of an inch
  BUT: 3/4-inch pipe

  Note: When possible, change the fraction to a decimal.

  NOT: 2 1/2 million
  BUT: 2.5 million

EXERCISE O Numbers—Revising for Clarity and Consistency

Directions: Evaluate each of the following sentences to see if it conforms to the rules for number use. Make necessary changes, using standard copy-editing marks.

Insert ___A___   Make lowercase ___E___
Delete ___✲___   Spell out ___(lb.)___

1. Florida has the nation's highest concentration of millionaires: nineteen for every 1,000 people.
2. Most millionaires are ordinary people in their early sixties; they have worked hard for 30 years, six days a week.
3. Jupiter, the largest of the planets, has a diameter only 1/10 of the sun.
4. Venus, though similar to Earth in size and mass, has a dense atmosphere and a surface temperature of over 400 degrees C.

5. The Earth makes a complete revolution around the sun every 365 days, five hours, forty-eight minutes, and forty-six seconds.
6. Two and one-half million dollars were earned by the Spacki Corp. during fiscal 1988.
7. Some employees now work 4 10-hour days instead of 5 8-hour days.
8. These employees frequently begin work as early as 6:00 A.M.; however, they may begin work at eight in the morning and leave as late as seven in the evening.
9. Araxco's offices are on the ninth floor at 153 W. Tenth Street, Topeka, Kansas.
10. Lake Texoma, created by the damming of the Red River, covers more than ninety-three thousand acres and borders on the two states of Texas and Oklahoma.
11. The vehicle must have a fully charged twelve-volt battery (minimum specific gravity 1.220, temperature corrected) for the tester to accurately analyze the ignition system.

3.0 Capitalization and Spelling

3.1 CAPITALIZATION

Like good punctuation, good capitalization should be invisible; that is, it should help the reader understand subtle distinctions in meaning without drawing attention to itself.

The first rule in capitalization is to follow your company style sheet; however, if no style sheet exists, follow these guidelines to set up your own style guide.

General Rules

Capitalize the following.

- Names of:
  people
  organizations
  structures
  vehicles
  nationalities
  languages
  religions
  months
  days of the week
  holidays (but not seasons like winter)
  geographic areas (but not directions like north)
  courses or studies (if they are official titles)
- Titles of:
  people
  books
  articles
  chapters

Specific Rules

- In lists that complete a sentence, do *not* capitalize the list items.

  Preflight procedures include:

  1. visual inspection of the exterior
  2. draining of fuel sumps
  3. checking of oil
  4. visual inspection of the engine

- In titles and headings, capitalize all major words. Capitalize a preposition, article, or conjunction if it contains five or more letters or is the first or last word of the title or heading.

  Roll Around a Point
  "A Manual of Style"

  Note: Special rules apply to listing titles in reference lists. See Chapter 13.

- Capitalize trade names, even those that have taken on common usage.

  Kleenex   Teflon
  Plexiglas   Formica
  Scotch tape   Ping-Pong

- Capitalize references to figures, tables, and chapters when they are followed by a number or letter. Do not capitalize them when you make a general reference.

See Table 1.
All the tables that follow are from the U.S. Census Bureau.

3.2 SPELLING

For a technical writer, one of the advantages of a word processor may be its spelling checker. Spelling checkers generally have a large vocabulary of common words and will flag misspellings and typos so they can be corrected. However, as the writer you still have to make final judgments, and you can't always count on having access to word processing. Therefore, it will benefit you to pay attention to your spelling.

No magic formula for good spelling exists, but here are some guidelines to help you become a better speller.

- Buy a good dictionary, like *Webster's Third International*, and look up every word you're not sure of. Even better, get your company to buy you one as an investment in your writing. Use it all the time.
- Keep a list of your own problem words, either on 3 × 5 cards, on the wall by your desk, or on your computer terminal. Use it as a quick reference. Few people have more than 20 problem spelling words, and a list will save you time.

Here are some of my problem words. Do you recognize any of yours?

accessible	comparative	definitely
competitive	separate	judgment
accommodate	receive	occurrence
consistent	descendant	liaison

- When you write a document, proofread for spelling by reading it backward, one sentence at a time: read the whole last sentence; then the next-to-last sentence, and so on. This will force you to pay attention to individual words.
- Ask someone whose judgment you trust to read your documents to check for spelling. In return, offer to read their work.

EXERCISE P   **Editing for Capitalization and Spelling**

Directions: Edit the following sentences for capitalization and spelling errors. Consult a dictionary. Use standard copyediting marks to make any necessary changes.

Make lowercase ___Ɇ___   Insert ___A___
Capitalize ___e___   Delete ___ℓ___

1. After flight 854 to Dallas was cancelled, vice-president R. A. Jones and the sales manager booked a flight to Houston.
2. With each occurence of failure, the clock stops.
3. The court's judgement is that school district 109 already complys with the desegregation order.
4. Plant no. 10 in Oshkosh replaced all its fluoresent bulbs with sodium vapor bulbs during the Summer.
5. Sales Manager Shockman serves as liason with the university on the cooperative education program.
6. When we receive acknowledgment of the benefit increase, we will send you a xerox copy of the original.
7. Professor Haloran diagramed the process on the board, but the students in chemistry 101 still didn't understand why naptha was used as the solvent.
8. If you mispell any words on your resumé, you will not be contacted by our personel office.
9. Mortality From Principle Types of Accidents (heading)
10. "Comunicating Ideas through Writen Language" (title)

USEFUL BOOKS

Bell, Paula. 1985. *Hightech writing: How to write for the electronics industry.* New York: John Wiley. Sound information for the writer in the electronics field.

Brady, John. 1977. *The craft of interviewing.* New York: Vintage. Easy-to-read advice about interviewing and writing up interviews. Primarily designed for journalists, but gives helpful advice about dealing with people and the written word.

Browning, Christine. 1984. *Guide to effective software technical writing.* Englewood Cliffs: Prentice-Hall. Specific advice for writing software manuals.

Brusaw, Charles T., Gerald J. Alred, and Walter E. Oliu. 1987. *Handbook of technical writing.* 3rd ed. New York: St. Martin's. Alphabetical entries make it easy to find information on everything from resumés to demonstrative adjectives.

The Chicago manual of style. 1982. 13th ed. Chicago: University of Chicago Press. A widely used and comprehensive style guide covering information on manuscript preparation, documentation, and other matters of style. Also see Chapter 13 of this text for a listing of other style guides.

Day, Robert A. 1983. *How to write and publish a scientific paper.* Philadelphia: ISI Press. Specific information for writers of research reports for scientific journals.

Fielden, John S., Jean D. Fielden, and Ronald E. Dulek. 1984. *The business writing style book.* Englewood Cliffs: Prentice-Hall. Defines style, tone, and organization. Tells how to choose style to match goals and includes an analysis of types of readers.

Gastel, Barbara. 1983. *Presenting science to the public.* Philadelphia: ISI Press. Advice for those in scientific fields who want to communicate technical information to generalists. Deals both with media interviews and direct writing of books and articles.

Government Printing Office. 1984. *Style manual.* Washington: Government Printing Office. A comprehensive guide to writing documents for government agencies.

Judd, Karen. 1982. *Copyediting: A practical guide.* Los Altos, CA: Kaufmann. Advice about editing and revising documents.

Michaelson, Herbert B. 1986. *How to write and publish engineering papers and reports.* Philadelphia: ISI Press. A short, clearly written book specifically for writing in the engineering field.

Price, Jonathan. 1984. *How to write a computer manual: A handbook of software documentation.* Menlo Park, CA: Benjamin/Cummings. An informal guide for computer manual writing.

Sabin, William A. 1979. *The Gregg reference manual.* 5th ed. New York: McGraw-Hill. A comprehensive guide to details of spacing and layout in documents. Punctuation, grammar, and accepted format.

Williams, Joseph M. 1985. *Style: Ten lessons in clarity and grace.* 2nd ed. Glenview, IL: Scott, Foresman. Advanced lessons in grammar, style, and punctuation.

GLOSSARY

abstract—a summary of a longer technical document, written primarily for a specialist or technician. It can be either descriptive or informational.

abstract word—a word naming a general condition, quality, concept, or act. Abstract words are difficult to visualize.

active voice—sentence construction in which the subject of a sentence *performs* the action of the verb.

adverbial connectives—adverbs that can be used—along with punctuation—to join independent clauses to make a compound sentence; also called conjunctive adverbs.

analytical resumé—a resumé organized to emphasize your qualifications and skills as they relate to your stated job or professional objective.

appositive—a second noun that amplifies the meaning of the first noun.

back matter—peripheral information placed at the end of a formal proposal, report, or manual. It can include a reference list, glossary, and appendixes such as technical data, examples, and supporting material.

boilerplate—standard segments of a report, proposal, or manual that can be used over and over again with only minor changes.

chronological resumé—a resumé organized in reverse chronological order to show what you have done in the past.

clause—a group of related words that contains a subject and a verb.

cliché—an overused and worn-out expression that has lost much of the force of its meaning.

comparative report—a document that examines two or more products or options and matches one against the other or against a set of criteria.

complement—a word, phrase, or clause found in the predicate of a sentence that completes the sentence meaning. Complements can be direct objects, indirect objects, subject complements, or object complements.

complete subject—the main noun or pronoun about which something is said or asked, plus its associated modifiers.

complex sentence—a sentence containing one independent clause and one or more dependent clauses.

compound-complex sentence—a sentence containing two or more independent clauses and one or more dependent clauses.

compound sentence—a sentence containing two or more related independent clauses joined in some way.

concrete word—a word naming a specific place, object, action, or person. Concrete words call up visual images.

conjunctive adverbs—see *adverbial connectives.*

connotative meanings—associated meanings of a word that are indirect, subjective, and often emotionally loaded.

coordinators—the connecting words used to join words, phrases, or (in a compound sentence) clauses; also called coordinating conjunctions.

criteria—standards for judgment or decision-making.

demonstration—a manual or portion of a manual that is written for an experienced user; sometimes called the "cookbook method."

denotative meaning—the direct, objective, and neutral meaning of a word.

dependent clause—a clause that cannot stand alone as a complete sentence. It is subordinate to an independent clause.

descriptive abstract—a very short summary of a document that tells what the contents are in general terms without giving the key points of the information itself.

diagram—a graphic presentation of data that is effective for showing the relationship of parts to one another and to the whole. Types include pie, organization, hierarchical, flow, block, schematic, and wiring.

diction—the choice of words and the force and accuracy with which they are used; word choice.

direct object—a noun or noun equivalent that completes the meaning of a transitive verb by answering the question "What?" or "Whom?"

documentation—the process of identifying sources of information; also called citation.

executive summary—a condensation of a technical document intended for managers. It may give more definitions and more on the project's purpose, scope, methods, and relationship to management concerns than a comparable abstract.

expletives—introductory words "It" and "There" that appear in the usual subject position in a sentence but provide no content in the sentence.

extemporaneous speech—a planned spoken communication in which the speaker uses an outline or notes rather than a written transcript.

feasibility report—a document that looks at options, ideas, or products to see if they can be done or used and which one is preferable.

field report—an informal report presenting data collected from some location.

figure—any graphic presentation of information other than a table. The main types of figures are graphs, diagrams, photographs, drawings, maps, and printouts.

form—the way a piece of writing looks based on arrangement and structure.

formal report—a document written in an objective, impersonal style and following an established form, which usually includes front and back matter as well as the main body of content.

formal style—writing that is highly structured, impersonal, and deliberate.

format—the way a piece of writing looks based on type face, type size, and page design.

front matter—peripheral information placed at the beginning of a formal proposal, report, or manual. It can include a transmittal document, abstract, executive summary, title page, table of contents, list of figures, and list of tables.

gobbledygook—meaningless writing characterized by excess words and inappropriate jargon.

grammatical parallelism—use of the same structure in sentences or parts of sentences for points that are alike, similar, or direct opposites.

graph—a visual presentation of numerical data. Types of graphs include line graphs, bar graphs, segmented bar graphs, and pictographs.

graphics—any visual form for presenting information—pictures or arranged numbers as opposed to sentences; also called visuals or illustrations.

heading—in a business letter, the writer's street address, city, state, ZIP code, and the date.

IFB (information for bid)—a document that supplies data needed to bid on a project. IFBs are usually for physical objects.

impromptu—unplanned speech.

independent clause—a clause that can stand alone as a complete sentence.

informal report—a written document of fewer than 10 pages, which is often directed to a decision-maker. If intended for internal use, it may follow memo form or be attached to a transmittal memo. If directed outside the organization, it may be in the form of a letter or attached to a transmittal letter.

informal style—writing that is casual and conversational, like the everyday language of speakers.

informational abstract—a summary of a document that includes major facts, conclusions, and recommendations.

inside address—in a business letter, the intended reader's name and position, company name, street address, city, state, and ZIP code.

instructions—writing that tells the reader how to do something; the purpose is to teach performance of a specific activity.

interview—a meeting between two or more people for the purpose of obtaining information.

jargon—(1) the highly specialized vocabulary of a particular group, trade, or profession; (2) meaningless or pretentious terminology.

job application letter—a short business letter directed to a specific person at a company that (1) tells of your interest in a specific job, (2) highlights applicable information from your accompanying resumé, and (3) requests an interview. The purpose of the letter is to obtain an interview.

key words—the major items of information pertaining to a topic that are used in an abstract or in information gathering. Often three to six key words are listed after an abstract to help in indexing.

laboratory report—an informal report explaining the procedure by which an operation was carried out and the results of that procedure.

letter—a short document addressed personally to a reader or readers who are usually outside the writer's organization.

manual—a document that gives both information and instructions. Often it combines definitions, descriptions, analyses, and explanations with instructions and procedures.

memorandum—a short document directed to a reader within the same organization as the writer. The plural form is either memorandums or memoranda, and the word is often shortened to memo.

narrative report—an informal report, or segment of a report, that describes events in the order in which they occurred in the past.

objects—nouns or noun equivalents used in a sentence to answer the questions "What?" "Whom?" "To what?" or "To whom?" Objects can be direct objects, indirect objects, or objects of a preposition.

oral presentation—a speech planned and practiced before presentation.

paragraph—a group of related sentences, complete in itself but also usually part of a larger whole.

parallellism—use of the same structure in sentences or parts of sentences for points that are alike, similar, or direct opposites.

passive voice—sentence construction in which the subject of the sentence *receives* the action of the verb.

phrase—a group of related words.

popular article—a report that simplifies and explains a scientific or technical subject so that it can be understood by generalists. Popular articles are published in newspapers and general magazines.

predicate—in a sentence, the verb plus its modifiers and complements. The predicate is the part of the sentence that says or asks something about the subject.

procedures—writing that explains how something happens or is done; the purpose is to give information about ongoing or repetitive activities. Sometimes procedures are called process descriptions.

progress report—a written account of your accomplishments at stated intervals in a project, such as the end of a test phase. If internal, it often follows memo form.

proposal—a written offer to solve a problem or provide a service by following a specified procedure, using identified people, and adhering to an announced timetable and budget. Proposals can be solicited or unsolicited, external or internal, formal or informal.

purpose—in writing, the goal: the intended or desired result, the reason for doing the writing.

reader—the primary target of a piece of writing; also called the audience.

recommendation report—a document that studies and sorts products or solutions in order to choose one.

redundancy—unnecessarily saying the same thing in more than one way.

report—a document that explains, describes, or recommends action based on research or analysis by the writer.

research report—a document that presents original research results and reviews existing research, often for publication in a professional journal; also called a journal article or a scientific, investigative, or technical paper.

resumé—a short and highly organized summary of your education, activities, and experience for the purpose of telling a potential employer of your skills and abilities; also called a data sheet.

RFP (request for proposal)—a document that solicits proposals to solve problems for companies or for branches of the government. An RFP is usually for services or studies.

schedule—a detailed work plan for a writing project that extends from the beginning date to the completion date and shows intermediate deadlines.

scope—what will and what will not be covered in a piece of writing; the limits.

semiformal style—writing that is carefully structured but uses the first or second person (*I* and *you*) and short, uncomplicated sentences in the active voice.

sentence—an independent unit of words that contains a subject and a verb and closes with a mark of punctuation.

short-term memory—the mind's temporary storage, with a capacity of seven plus or minus two items.

simple subject—in a sentence, the main noun or pronoun about which something is said or asked.

speech—a planned spoken communication to listeners for a given purpose; also called an oral presentation.

stance—the position of the writer in relation to the reader.

status report—a written account of your accomplishments on the job during a designated period of time. Usually status reports are written weekly, biweekly, or monthly, and they often follow memo form.

stipulated meaning—a specified definition of a word for a specific context.

style—the manner in which a document is written. Style can be formal, informal, or semiformal.

subject—that part of a sentence about which something is said or asked.

subordinator—a connecting word that begins a dependent clause and joins it to an independent clause to make a complex sentence; also called a subordinate conjunction.

summary—the section of a document that reviews the facts and adds up the main points already covered.

syntax—the pattern, structure, and arrangement of words in a sentence; word order.

table—a graphic presentation of data (often numbers) in columns for easy understanding or comparison.

technical article—a report that describes applied research or development, often giving information about new techniques or products and published in the semitechnical journals of professional societies or the trade journals supported by advertising. Also called a trade article.

thesis—the main idea you want the reader to get. It provides the answer to the problem or question being investigated. It can be either a statement of content or a proposition to be defended.

time frame—the amount of time you want to cover in a report and how far into the past you want to go.

tone—the attitude you project to your readers through content and writing style.

topic—what is to be treated in a document or part of a document.

topic sentence—a sentence that contains the main point of a paragraph.

trade article—see *technical article.*

transition—a word, phrase, sentence, or paragraph that moves the reader from one idea to another.

transmittal letter—a formal business letter directed to the person who will receive (or approve) a report or proposal; sometimes called a cover letter. Within a company, this document can also be written in memo form.

tutorial—a manual or portion of a manual that is written for a first-time user. It often takes the reader through a sample or typical operation to illustrate the procedure.

verb—the word or words in a sentence that show action, occurrence, or existence.

verbal—a word made from a verb but not acting as a verb. Infinitives and participles are verbals.

verbalize the action—to put the action of a sentence into the verb itself.

vogue words—words that are currently fashionable and overused.

voice—the verb form that shows whether the subject is acting or acted upon.

word processor—a computer (the hardware) into which one can type words that will be displayed on a screen and that can be saved for later reproduction in printed form.

INDEX